T0314295

Electron Beam-Specimen Interactions and Simulation Methods in Microscopy

Current and future titles in the Royal Microscopical
Society—John Wiley Series

Published

*Principles and Practice of Variable Pressure/Environmental Scanning
Electron Microscopy (VP-ESEM)*
Debbie Stokes

Aberration-Corrected Analytical Electron Microscopy
Edited by Rik Brydson

*Diagnostic Electron Microscopy—A Practical Guide to Interpretation
and Technique*
Edited by John W. Stirling, Alan Curry & Brian Eyden

Low Voltage Electron Microscopy—Principles and Applications
Edited by David C. Bell & Natasha Erdman

Standard and Super-Resolution Bioimaging Data Analysis: A Primer
Edited by Ann Wheeler and Ricardo Henriques

*Electron Beam-Specimen Interactions and Simulation Methods in
Microscopy*
Budhika G. Mendis

Forthcoming

Understanding Practical Light Microscopy
Jeremy Sanderson

Atlas of Images and Spectra for Electron Microscopists
Edited by Ursel Bangert

Focused Ion Beam Instrumentation: Techniques and Applications
Dudley Finch & Alexander Buxbaum

Electron Beam-Specimen Interactions and Simulation Methods in Microscopy

Budhika G. Mendis

Department of Physics
Durham University, UK

Published in association with the Royal
Microscopical Society

Series Editor: Susan Brooks

WILEY

Registered Office(s)
John Wiley & Sons, Inc., 111 River Street, Hoboken, NJ 07030, USA
John Wiley & Sons Ltd, The Atrium, Southern Gate, Chichester, West Sussex, PO19 8SQ, UK

Editorial Office
The Atrium, Southern Gate, Chichester, West Sussex, PO19 8SQ, UK

For details of our global editorial offices, customer services, and more information about Wiley products visit us at www.wiley.com.

Wiley also publishes its books in a variety of electronic formats and by print-on-demand. Some content that appears in standard print versions of this book may not be available in other formats.

Library of Congress Cataloging-in-Publication Data applied for
Hardback : 9781118456095

Cover design: Wiley
Cover image: **g** = bright field image of overlapping stacking faults in a Cu-Si alloy. The shear (**R**) at the stacking fault plane causes a phase shift $2\Pi \mathbf{g} \cdot \mathbf{R}$ of the diffracted beam, resulting in the black-white interference contrast. If sufficient number of faults overlap such that the net **g·R** is an integer, the contrast is suppressed. From Mendis, B.G, Jones, I.P. and Smallman, R.E. (2004), Journal of Electron Microscopy 53, 311. Reproduced with permission of Oxford University Press.

Set in 10.5/13pt SabonLTStd by SPi Global, Chennai, India
Printed and bound in Singapore by Markono Print Media Pte Ltd

10 9 8 7 6 5 4 3 2 1

Contents

Preface

In writing this book, I have attempted to introduce electron beam scattering in the context of simulation methods, since the practicing microscopist is usually concerned with the latter. Despite the availability of user-friendly software, some understanding of the underlying physics is essential, both to 'optimise' the simulation and recognise its limitations and to correctly interpret and/or generalise the results. In highlighting applications, I have selected examples that have been made possible by the tremendous advances in instrumentation, such as aberration correction and monochromation. The examples are limited to those of general interest, that is, mainly imaging and spectroscopy. The field is, however, progressing rapidly and it is only a matter of time before electron beam simulation methods proliferate into new areas, such as time-resolved microscopy, in situ microscopy in gaseous and liquid environments and unconventional probes geometries in the form of vortex beams.

The book is intended to be as self-contained as possible, with derivations provided for the main results. The level of mathematics and physics assumed is largely limited to graduate level calculus and quantum mechanics. Certain topics, such as Maxwell's equations, may need refreshing, and in such cases I have referenced some of the many excellent textbooks that are widely available. It is also assumed that the reader has some familiarity with electron microscopes and basic techniques, such as HREM, HAADF imaging and EDX, EELS spectroscopy.

The interactions I have had over the years with colleagues and students have helped shape my own understanding of the subject. I am also grateful to Ian Jones, Mervyn Shannon, Rik Brydson, Alan Craven and Tom Lancaster for providing helpful comments on individual chapters.

Finally, I would like to dedicate this effort to my parents, who have unconditionally supported me in my formative years.

Budhika G. Mendis
Durham

1

Introduction

Electron beam scattering has had a long and distinguished history. Some of the essential physics was investigated even before the first electron microscope was built. The unsuspecting reader may find it surprising to come across familiar names such as Bethe, Bohr, Rutherford, Fermi and Mott in this book. While electron beam scattering is a mature theory its widespread use in electron microscopy measurements is arguably a more recent phenomenon. This is primarily due to two reasons. The first is the processing speed of modern computers; even a standard desktop computer can now produce useful results within a reasonable time and thankfully there are many software packages that take advantage of this. The second reason is the emergence of a new generation of electron microscopes that can resolve atom columns that are less than an angstrom apart, that have ~10 meV energy resolution or less for measuring vibronic modes and that can record events separated in time by femtoseconds. With such a wealth of new information there is a strong emphasis on extracting quantitative information about the sample. Electron beam scattering calculations are often indispensable for correct data interpretation.

Two examples help illustrate the advantages of combining experimental results with simulation. The first is using high angle annular dark field (HAADF) imaging in a scanning transmission electron microscope (STEM) to characterise the interface in an AlAs–GaAs superlattice (Robb and Craven, 2008; Robb *et al.*, 2012). Figure 1.1a shows the HAADF image of a [110]-oriented, epitaxial AlAs–GaAs superlattice acquired using an aberration corrected STEM. In this orientation the Group III–V elements are distributed as closely spaced (i.e. ~1.4 Å) atom column pairs or 'dumbbells'. The HAADF signal increases monotonically with

---

Electron Beam-Specimen Interactions and Simulation Methods in Microscopy,
First Edition. Budhika G. Mendis.
© 2018 John Wiley & Sons Ltd. Published 2018 by John Wiley & Sons Ltd.

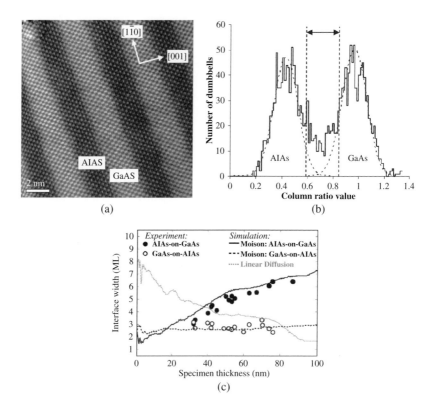

Figure 1.1   (a) HAADF STEM image of a [110]-oriented, AlAs–GaAs superlattice. The background subtracted column intensity ratio values for all dumbbells in (a) are shown in (b) as a histogram. (c) plots experimental and multislice simulated values for the AlAs–GaAs interface width, as defined from the column intensity ratio profile, as a function of specimen thickness. Supercells for the multislice simulation are constructed assuming a linear diffusion model and a more accurate diffusion model (Moison *et al.*, 1989) valid for the AlAs–GaAs system. Results are shown for superlattices grown on GaAs (labelled 'AlAs-on-GaAs') and AlAs (labelled 'GaAs-on-AlAs') substrates respectively. (a) and (b) From Robb and Craven (2008). Reproduced with permission; copyright Elsevier. (c) From Robb *et al.* (2012). Reproduced with permission; copyright Elsevier.

the atomic number of the scattering element, so that using AlAs dumbbells as an example, the intensity of an As column is larger than that of Al. Figure 1.1b is a histogram of the background subtracted column intensity ratio values for all dumbbells in Figure 1.1a. The two prominent peaks are due to dumbbells in 'bulk' AlAs and 'bulk' GaAs respectively. However, there are also intermediate values for the column intensity ratio (arrowed region in Figure 1.1b) and further analysis reveals these to be due to dumbbells located at the AlAs–GaAs interface region (Robb and Craven, 2008). An interface 'width' can be defined based on the 5–95% variation in column intensity ratio across the interface. The interface width is found to be independent of the specimen thickness for super-lattices grown on an AlAs substrate, but not on GaAs substrate.

It is not clear if the interface width is due to chemical inter-diffusion, electron beam spreading within the sample or interfacial roughness. This can, however, be tested by constructing supercells representing the different scenarios and performing multislice simulations (Chapter 3). Figure 1.1c shows the simulated results for chemical diffusion. The interface width, as deduced from the column intensity ratio values, is plotted as a function of specimen thickness for a linear composition profile and a more realistic diffusion model valid for the AlAs–GaAs system (Moison *et al.*, 1989). The latter accurately reproduces the experimental results, suggesting diffusion as a likely candidate. In fact, simulations for a saw tooth-shaped and smooth interface did not agree with experiment, so that interfacial roughness and beam spreading have only a secondary effect on the measurement (Robb *et al.*, 2012).

The second example is the use of electron energy loss spectroscopy (EELS) to extract the local electronic density of states for a silicon dopant atom in graphene (Ramasse *et al.*, 2013). As illustrated in Figure 1.2a, the silicon atom can be incorporated either through direct substitution (i.e. threefold coordination) or as a fourfold coordinated atom in defect regions of the graphene sheet. The dopant atom can be readily identified using HAADF imaging in an aberration corrected STEM, taking advantage of the higher atomic number of silicon compared to carbon. The solid lines in Figure 1.2b are the Si $L_{2,3}$-EELS edges measured from the two different dopant atom configurations. Owing to the nature of inelastic scattering (Chapter 5) the shape of the EELS spectrum is governed by the angular momentum resolved unoccupied density of electronic states. The filled spectra in Figure 1.2b are the results obtained from density functional theory simulation. There is excellent agreement between theory and experiment for the fourfold coordinated atom. For the threefold coordinated atom, however, accurate results are only obtained if it is assumed that the silicon dopant atom is displaced out of the graphene sheet (Figure 1.2a). This can be justified by the slightly longer Si–C bond length compared to graphene (note that the structure was relaxed to its lowest energy configuration prior to EELS simulation; Ramasse *et al.*, 2013). The out-of-plane displacement of the silicon atom is not evident in the HAADF image and was only revealed through a careful quantitative analysis of the EELS result with the aid of simulation.

1.1 ORGANISATION AND SCOPE OF THE BOOK

There are many ways to simulate electron beam scattering. Although the fundamental physics is unchanged, there are differences in the manner

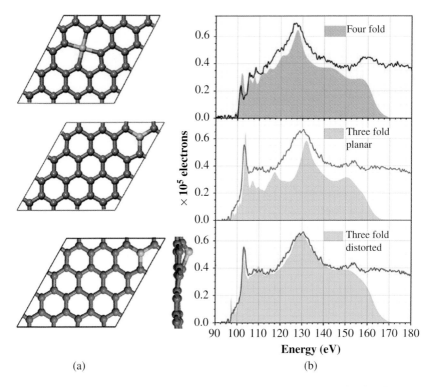

(a)   (b)

Figure 1.2 (a) Supercells used for simulating the EELS edge shape for fourfold and threefold coordinated silicon dopant atoms in graphene. For the latter, a planar structure and a distorted structure, where the silicon atom is displaced out of the graphene sheet, are assumed. (b) shows the corresponding EELS spectra, with the experimental measurement represented as a solid line and the simulated result superimposed as a filled spectrum. From Ramasse *et al.* (2013). Reproduced with permission; copyright American Chemical Society.

in which it is implemented and consequently the information that can be extracted. For example, if the interest is in images formed from high energy electrons passing through a thin foil, such as in transmission electron microscopy (TEM), then the strongest signal will be due to elastically scattered electrons. The less probable inelastic scattering events can be treated phenomenologically or in certain cases (e.g. a single graphene sheet) ignored altogether. This approach considerably simplifies and speeds up the calculation while still providing the required information.

Four different simulation methods are discussed in this book, namely Monte Carlo (Chapter 2), multislice (Chapter 3), Bloch waves (Chapter 4) and electrodynamic theory (Chapter 6). Chapter 5 deals with inelastic scattering of core atomic electrons and extends the multislice,

Bloch wave methods to include simulation of inelastic images in the form of chemical maps. Together these form a core body of techniques for analysing a large range of electron microscopy data. Electronic structure calculations, based on either density functional theory or multiple scattering, are also widely used for simulating the fine structure of EELS spectra, but are not discussed here in any great detail. This is a vast area separate from the main topic of this book and the interested reader should consult textbooks such as Martin (2004) for further details. Table 1.1 lists some of the advantages and disadvantages of each of the simulation techniques. It should give some indication of which technique to use for a given problem and which techniques to avoid.

Finally, it should be noted that it is impossible to give an exhaustive treatment of electron beam scattering in a book of this size. Instead, the emphasis is on describing the essential physics, so that the reader is able to comprehend a large part of the literature and develop an understanding of simulation packages beyond a mere 'black box'. As for the vast literature on the subject, a word of caution is appropriate: unfortunately, there is no universally accepted notation in scattering theory. In this book, I have tried to be as consistent as possible, following the procedure outlined below:

(i) The relativistic mass of the high energy electron is distinguished from its rest mass by using m for the former and m_o for the latter. There are, however, several examples in the text where the kinetic energy is expressed as $\frac{1}{2}mv^2$, rather than the correct relativistic formula $(\gamma - 1)m_o c^2$, where v is the speed of the electron, c the speed of light and $\gamma = [1 - (v/c)^2]^{-\frac{1}{2}}$. This is standard practice in most of the literature, although for a 300 keV electron beam the fractional error is as large as 18%. The magnitude of the momentum $p = mv$ is, however, relativistically correct, and consequently there is no error in the de Broglie wavelength.

(ii) An electron plane wave is represented as $\exp(2\pi i \mathbf{k} \cdot \mathbf{r})$, with $\mathbf{k}$ being the wave vector and $\mathbf{r}$ the position vector; some texts may use the form $\exp(i\mathbf{k} \cdot \mathbf{r})$. With the notation adopted in this book the integrand of the Fourier transform of a function $f(\mathbf{r})$ is $f(\mathbf{r})\exp(-2\pi i \mathbf{q} \cdot \mathbf{r})$, where $\mathbf{q}$ is the reciprocal space variable.

(iii) The potential energy of an electron in a potential field $V(\mathbf{r})$ is $-eV(\mathbf{r})$, where $-e$ is the charge of the electron. The Schrödinger equation then becomes $[\hbar^2 \nabla^2 / 2m + eV(\mathbf{r}) + E]\psi(\mathbf{r}) = 0$, where $\hbar$ is the reduced Planck's constant, E the energy and $\psi(\mathbf{r})$ the electron wavefunction. In some texts, the potential energy is denoted by $V(\mathbf{r})$ and the

Table 1.1 The simulation methods discussed in this book and some of their advantages and disadvantages

Simulation method	Advantages	Disadvantages
Monte Carlo (probabilistic scattering of particles, e.g. incident electrons)	• Both elastic and inelastic scattering are readily incorporated • Applicable for a wide range of specimen geometries (SEM[a] and TEM) • Large range of signals can be simulated (e.g. images, electron beam induced current, X-ray generation, etc.)	• Cannot reproduce channelling or diffraction in a crystal • Inelastic scattering is often modelled as a continuous energy loss (i.e. stopping power). Hence 'straggling' is not observed
Multislice (physical optics approach based on transmission and propagation of the incident electron wave)	• Used in both TEM and STEM image simulations • Channelling and dynamic diffraction in a crystal are reproduced • Supercells can be constructed, such as defects, amorphous materials, etc.	• Most software packages only simulate elastic scattering, with thermal diffuse scattering modelled as a pseudo-elastic scattering event (i.e. frozen phonon). Core level inelastic scattering can be included, but at the expense of computing time • Computing time increases with thickness of the specimen • Can be difficult to interpret the underlying scattering mechanisms

Bloch wave (based on Schrödinger's equation for an electron in a periodic potential)	• Used in both TEM and STEM image simulations • Channelling and dynamic diffraction in a crystal are reproduced • Intuitive description of dynamic diffraction • Computing time does not increase with specimen thickness (unless information at different depths is required)	• Most software packages only simulate elastic scattering, with thermal diffuse scattering modelled phenomenologically via an 'optical potential'. Core level inelastic scattering can be included, but at the expense of computing time • Only useful as a computational technique for periodic crystals with small unit cells. Column approximation and/or perturbation methods can nevertheless be used to obtain useful information about defect crystals
Electrodynamic theory (based on Maxwell's equations)	• Vastly simplifies inelastic scattering events involving many electrons, for example, plasmons • Only the specimen dielectric function is required to calculate the energy loss • Analytical solutions are available for relatively simple specimen geometries (e.g. thin films, spheres, etc.) • Radiative phenomena, such as Cerenkov and transition radiation, can also be analysed	• Only valid when the energy loss is a negligible fraction of the incident electron energy • Simulation methods exist for arbitrary specimen geometries, but can be computationally intensive

[a]Scanning electron microscope.

Schrödinger equation modified accordingly. The term 'potential' is also frequently used to mean 'potential energy', although they are not the same, though closely related. The notation here makes the distinction more transparent.

(iv) The magnitude of the elastic scattering vector is $q = 2\sin\theta/\lambda$, where θ is the scattering semi-angle and λ the electron wavelength. The Debye–Waller factor (B) in the thermal smearing term 'exp($-Bq^2$)' is then equal to $2\pi^2\overline{x^2}$, where $\overline{x^2}$ is the mean square atom displacement (Chapter 3). The definition $q = \sin\theta/\lambda$ is also sometimes used, in which case, $B = 8\pi^2\overline{x^2}$.

(v) The inelastic scattering vector is $\mathbf{q} = k_m\mathbf{n}_m - k_0\mathbf{n}_0$, where k_0, k_m are wave numbers (see also [ii] for wave vector definition) for the incident and inelastically scattered electron along the unit vector directions $\mathbf{n}_0$ and $\mathbf{n}_m$ respectively. In many texts, $\mathbf{q} = 2\pi[k_m\mathbf{n}_m - k_0\mathbf{n}_0]$. The motivation for using the alternative notation here is that it is consistent with the definition of the elastic scattering vector, but more importantly because the results from quantum mechanics (Chapter 5) can be directly compared with electrodynamics (Chapter 6). The latter frequently uses Fourier transforms, which follow the notation outlined in (ii).

(vi) Maxwell's equations are expressed in SI units. The equivalent expressions in Gaussian units can be found in Jackson (1998).

REFERENCES

Jackson, J.D. (1998) *Classical Electrodynamics*, John Wiley & Sons, New York.

Martin, R.M. (2004) *Electronic Structure- Basic Theory and Practical Methods*, Cambridge University Press, UK.

Moison, J.M., Guille, C., Houzay, F., Barthe, F. and Van Rompay, M. (1989) *Phys. Rev. B* **40**, 6149.

Ramasse, Q.M., Seabourne, C.R., Kepaptsoglou, D.M., Zan, R., Bangert, U. and Scott, A.J. (2013) *Nano Lett.* **13**, 4989.

Robb, P.D. and Craven, A.J. (2008) *Ultramicroscopy* **109**, 61.

Robb, P.D., Finnie, M. and Craven, A.J. (2012) *Ultramicroscopy* **118**, 53.

2

The Monte Carlo Method

The Monte Carlo method is a convenient starting point for discussing electron beam–specimen interactions. It is unique in that conceptually it treats electron scattering as a 'collision' between two particles, the other particle being either the nucleus or electrons of the atoms in the solid. The fundamentals of scattering can then be described within the framework of classical physics. The wave nature of the electron is overlooked, although results from quantum theory can be incorporated into the method in order to improve its accuracy. Despite these limitations Monte Carlo simulations are capable of simultaneously accounting for all electron interactions, elastic as well as inelastic, within a solid in a computationally tractable manner and produce results that are consistent with experiment. They are most widely used in SEM simulations. Interestingly, the Monte Carlo method has its origins in the somewhat similar problem of neutron penetration through matter. It was developed by Stanislaw Ulam, John von Neumann and Nicholas Metropolis as part of the Manhattan Project. Being classified, 'Monte Carlo' was chosen as a code name based on the well-known casino in Monte Carlo, Monaco.

The Monte Carlo method is a statistical theory of electron beam–specimen interactions. Consider for example a game of billiards – we are interested in the deflection and speed of the struck ball, which in classical mechanics can be calculated exactly if we knew the trajectory of the cue ball. If the target ball is placed at exactly the same position and several attempts were made at striking it with the cue ball the results for each individual event are likely to differ from one

Electron Beam-Specimen Interactions and Simulation Methods in Microscopy,
First Edition. Budhika G. Mendis.
© 2018 John Wiley & Sons Ltd. Published 2018 by John Wiley & Sons Ltd.

another. Given the deterministic *and* random elements in the process the only meaningful information that can be extracted is statistical in nature, such as the mean angle of deflection of the target ball and its standard deviation. This information is nevertheless useful as it can, for example, differentiate players according to their skill provided the sampling is large enough. A similar argument applies to electron beams in a solid. In an SEM where the probe current is, say, 1 nA there are a large number of electrons (10^9–10^{10} per second) impinging on the sample and the resulting signals (e.g. X-rays, backscattered electrons, etc.) from the elastic/inelastic interactions for all these electrons are collected within the acquisition time of the experiment. Furthermore, in analogy with the billiards example, a random element is also present in SEM measurements. Since electrons have charge they interact via a Coulomb force with the positively charged nuclei and atomic electrons in the solid. The Coulomb interaction has an inverse square radial dependence and hence incident electrons passing a nucleus or atomic electron at different distances will undergo varying degrees of deflection and energy loss. In a Monte Carlo simulation the deflection angles are calculated with the aid of random numbers generated in a computer; the random number effectively takes into account the random distribution in the distance of separation between the incident electron and the target particle (nucleus or atomic electron). However, the calculated deflection angles are such that preference is given to the most probable event, which in turn is governed by the underlying physics of the interaction. This is similar to recording the outcomes of a biased die. If all faces of the die were identical then all events have equal probability and the process is purely random in nature. However, with a biased die certain events are more likely to occur than others, although all physically allowed events have some probability. An example of an unlikely, although not improbable, event is backscattering, where the deflection angle is so large the incident electron reverses its direction of motion. Monte Carlo simulations give preference to forward scattering events with only a few electrons undergoing backscattering, the statistical ratio of forward to backscattering for a large number of incident electrons being determined by exact physical laws.

In Section 2.1 the physics of elastic and inelastic scattering are introduced and their implementation in a Monte Carlo model described. Some applications of Monte Carlo simulations in electron microscopy are described in Section 2.2, while in Section 2.3 further advanced topics, such as the quantum mechanical description of scattering, are outlined.

2.1 PHYSICAL BACKGROUND AND IMPLEMENTATION

An incident electron within a solid can interact with a heavy, positively charged nucleus and/or an atomic electron with identical charge and mass (interaction with many electrons, for example, plasmons, are ignored for the moment). In both cases, the interaction is electrostatic in nature, although the outcomes are different on account of the charge and mass of the target particle. They are both two-body problems and are more conveniently analysed using centre of mass (or C-) coordinates, where the origin is fixed to the centre of mass so that the total linear momentum of the system is always zero. A lucid account of the two-body problem in C-coordinates can be found in Evans (1955), although in the following sections scattering is analysed without much recourse to the theory.

2.1.1 Elastic Scattering By an Atomic Nucleus

Consider the case of scattering of the incident electron by an atomic nucleus. The theory is due to Rutherford (1911) and was originally derived to explain the scattering of α-particles through a thin gold foil as observed in the Geiger–Marsden experiment. It can be reasonably assumed that the centre of mass coincides with the atomic nucleus, since it is considerably heavier than the incident electron. By the same reasoning the energy transferred to the heavy atomic nucleus is negligible and the electron kinetic energy is therefore unchanged before and after the collision. However, in the vicinity of the nucleus the kinetic energy of the incident electron momentarily increases as the potential energy due to the Coulomb interaction decreases, but the total energy of the electron is conserved throughout. Since no net energy is transferred the scattering is elastic.

Figure 2.1 shows the trajectory of the incident electron of speed v as it passes the atomic nucleus of charge $+Ze$ initially at a distance b (also known as the impact parameter). The trajectory is most easily analysed using polar coordinates since the Coulomb force is a function of the distance of separation r between the nucleus and the electron. The origin is placed at the nucleus which is also the centre of mass. The acceleration of the electron along the radial direction (a_r) is given by

$$a_r = \frac{d^2 r}{dt^2} - r\left(\frac{d\varphi}{dt}\right)^2 = -\frac{1}{m}\left(\frac{Ze^2}{4\pi\varepsilon_o r^2}\right) \tag{2.1}$$

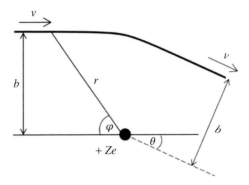

Figure 2.1 Rutherford elastic scattering of an electron of speed v by a nucleus of charge $+Ze$. The impact parameter is b and the electron trajectory is described using (r, φ) polar coordinates. θ is the scattering angle.

where t is time, φ the polar angle, ε_o the permittivity of free space and m, e are the mass and charge of the electron respectively. The first term is the linear acceleration along r, while the second term is the centripetal acceleration, with $\omega = d\varphi/dt$ being the angular speed. The negative sign for the Coulomb force represents the fact that the electron is attracted towards the nucleus (decreasing r). Since there is no force normal to r the magnitude of the angular momentum[1] $L = [m(r\omega)]r = mr^2(d\varphi/dt)$ is a constant. Furthermore, at large distances from the nucleus the angular momentum is mvb (Figure 2.1) so that

$$L = mr^2 \left(\frac{d\varphi}{dt} \right) = mvb \qquad (2.2)$$

Equation (2.1) can be solved for r with the help of Eq. (2.2) to obtain the trajectory of the electron. The final result is (the interested reader can refer to Rutherford (1911), Evans (1955) or Eisberg and Resnick (1985) for the derivation; note however that these derivations are for an α-particle where the Coulomb interaction is a repulsion whereas in our case it is an attractive force)

$$\frac{1}{r} = \frac{1}{b} \sin \varphi + \frac{D}{2b^2}(1 - \cos \varphi) \qquad (2.3)$$

$$D = \frac{Ze^2}{2\pi\varepsilon_o mv^2} \qquad (2.4)$$

---

[1] The angular momentum vector (**L**) is defined as the cross product between the position vector **r** and linear momentum ($m\mathbf{v}$), that is $\mathbf{L} = \mathbf{r} \times (m\mathbf{v})$, where **v** is the velocity.

The parameter D is the distance of closest approach for a head-on collision between two charged particles having the same sign (positive or negative), where the kinetic energy is completely converted to potential energy, that is $\frac{1}{2}mv^2 = Ze^2/(4\pi\varepsilon_o D)$. Equation (2.3) is a hyperbola in polar coordinates and is illustrated in Figure 2.1. The angle of scattering θ is evaluated by solving for φ as $r \to \infty$ in Eq. (2.3) and setting $\theta = (\varphi - \pi)$. This gives

$$\tan\left(\frac{\theta}{2}\right) = \frac{D}{2b} \tag{2.5}$$

Equation (2.5) states that the scattering angle θ depends on the impact parameter b such that electrons passing close to the nucleus undergo large deflections. For a large deflection such as backscattering, where $\theta > \pi/2$, the impact parameter must be smaller than $(D/2)$, which from Eq. (2.4) is only 0.01 Å for 20 kV electrons in a Cu target. Since the random variable b can in principle take any value from zero to infinity (in practice an upper limit for b would be half the inter-atomic spacing in a crystalline solid) this means that the majority of electrons are scattered in the forward direction. A useful physical parameter is the fraction of scattering events within a narrow range of deflection angles (i.e. between θ and $\theta + d\theta$) for the case of uniform illumination over unit area of the target. This can be expressed as

$$\frac{dI}{I_o} = \left(\frac{N_A \rho t}{A}\right) \cdot d\sigma(\theta) \tag{2.6}$$

where dI is the intensity scattered from a total incident intensity of I_o, N_A is Avogadro's constant, ρ the density of the target material with atomic/molecular weight A and t is the thickness of the target of unit area. The term $(N_A \rho t/A)$ is the number of scattering atoms/molecules within the target. $d\sigma(\theta)$ is a differential scattering 'cross-section' and can be understood with the aid of Figure 2.2, which shows electrons uniformly illuminating a target of unit area and thickness t. For scattering between the angles θ and $\theta + d\theta$ the electrons must have an impact parameter within a thin annular ring centred about a given atomic nucleus, the radius and width of the ring being determined by Eq. (2.5). $d\sigma(\theta)$ is the area of the annular ring and is therefore

$$d\sigma(\theta) = 2\pi b(db) = \frac{\pi D^2}{4} \frac{\cos(\theta/2)}{\sin^3(\theta/2)} d\theta \tag{2.7}$$

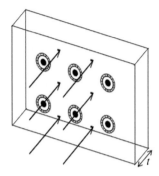

Figure 2.2 Schematic illustrating the differential scattering cross-section. Electrons uniformly illuminate a target of unit area and thickness t. Solid circles represent individual atoms in the target. An electron incident within the thin annular ring centred about a given atom is scattered between the angles θ and $(\theta + d\theta)$. The differential scattering cross-section $d\sigma(\theta)$ is the area of an annular ring and is proportional to the fraction of electron intensity scattered within the angular range θ to $(\theta + d\theta)$.

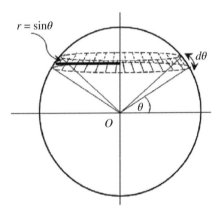

Figure 2.3 The solid angle Ω is defined as the area subtended on a sphere of unit radius by a given angle. Consequently, $d\Omega$ is the shaded area in the figure, that is, the difference in subtended areas for the angles θ and $(\theta + d\theta)$. The shaded area can be unfolded into a rectangular strip with length $2\pi \sin \theta$ and width $d\theta$ to give $d\Omega = (2\pi \sin \theta)d\theta$.

Note that the scattering cross-section has dimensions of area. Frequently, scattering is expressed in terms of solid angles and in such cases the angular range θ to $\theta + d\theta$ corresponds to the solid angle $d\Omega$, which from Figure 2.3 is equal to $(2\pi \sin \theta)d\theta$, that is, the area subtended on a sphere of unit radius. Equation (2.7) can therefore be expressed in the alternative form:

$$\frac{d\sigma}{d\Omega} = \frac{D^2}{16} \frac{1}{\sin^4(\theta/2)}$$

$$= \left(\frac{Ze^2}{8\pi\varepsilon_o mv^2}\right)^2 \frac{1}{\sin^4(\theta/2)} \qquad (2.8)$$

Equation (2.8) is the differential cross-section for Rutherford elastic scattering of an electron by an atomic nucleus of charge Ze. The result is based purely on classical mechanics by assuming the incident electron has predominantly particle-like properties. The cross-section decreases with increasing energy of the electron beam due to the v^4 term in the denominator and, as discussed previously, forward scattering is more probable than large deflections such as backscattering. However, as $\theta \to 0$ the cross-section becomes infinite. This is because the Rutherford model does not take into account screening of the nuclear charge by the atomic electron cloud for incident electrons with impact parameter greater than the first Bohr orbit radius. Without screening, electrons incident a large distance away from the atomic nucleus can still undergo scattering to small angles and the cumulative effect over the full range of impact parameters leads to cross-sections that are unreasonably large (note that since the scattering angle has an 'inverse' dependence with respect to b in Eq. (2.5) both the radius and thickness db of the annular ring in Figure 2.2 increase continuously as θ decreases). The effects of screening are modelled by including an exponential decay term to the standard Coulomb potential, so that the new potential $V(r)$ is

$$V(r) = \frac{Ze}{4\pi\varepsilon_o r} \exp\left(-\frac{r}{r_o}\right) \qquad (2.9)$$

where r_o is a decay constant that varies as $Z^{(-1/3)}$ (Cosslett and Thomas, 1964a,b). The differential scattering cross-section for $V(r)$ is derived by treating the electron as a wave and applying the Born approximation, which is applicable for weak scattering (see Appendix A for a concise description of the Born approximation; note that we have now moved away from a particle-based classical mechanics approach). Through this method it can be shown that the screened Rutherford scattering cross-section is given by

$$\frac{d\sigma}{d\Omega} = \left(\frac{Ze^2}{8\pi\varepsilon_o mv^2} \right)^2 \left[\frac{1}{\sin^2(\theta/2) + \alpha} \right]^2$$

$$\alpha = \left(\frac{1}{4\pi k r_o} \right)^2 \tag{2.10}$$

where $k = 1/\lambda = mv/h$ is the wave number of the electron with wavelength λ and h is the Planck's constant. For no screening (i.e. $r_o \to \infty$) Eq. (2.10) becomes equal to Eq. (2.8) as required. Furthermore, the additional term α in the denominator prevents the cross-section from diverging at small scattering angles. It should also be pointed out that both the screened and unscreened Rutherford models do not take into account the finite size of the nucleus, so that the true potential deviates from that of a point charged particle in the vicinity of the nucleus. However, this discrepancy does not lead to an unphysical behaviour in the cross-section for large scattering angles (i.e. Eqs. (2.8) and (2.10) are finite even for $\theta = \pi$). The scattering is nevertheless overestimated due to the actual Coulomb force being smaller than the standard inverse square (i.e. $1/r^2$) law. A rigorous calculation is significantly more complicated owing to the need to include electron spin effects (see, however, Mott scattering in Section 2.3.2, which does include spin). Equation (2.10) is therefore a good compromise between accuracy and complexity and is hence widely used.

Having analysed scattering by the nucleus of an atom it is worth considering scattering by the atomic electrons for a more complete description of the incident electron trajectory. The difference in charge and mass of an atomic electron compared to a nucleus means that scattering is different for the two cases. In fact, atomic electron scattering is inelastic and this aspect will be discussed in more detail in the next section, but here the main focus is on the scattering angle. The arguments used previously can be applied to the atomic electron as well, with the simplification that the scattering is due to an unscreened Coulomb potential acting between the incident and atomic electrons. The atomic electron is assumed to be unbound from its nucleus, a criterion that is approximately valid for the outer lying valence electrons. For scattering by an atomic nucleus, laboratory (L-) and centre of mass (C-) coordinates can be made identical by fixing the origin and centre of mass at the stationary nucleus. For scattering by atomic electrons, however, such a simplification is not possible and the problem is easiest to analyse in C-coordinates. The interested reader is referred to Evans (1955) for a full description. The important results are

$$\tan \theta = \frac{D'}{2b} \tag{2.11}$$

$$D' = \frac{e^2}{\pi \varepsilon_o m v^2} \tag{2.12}$$

$$\left(\frac{d\sigma}{d\Omega}\right)_{elec} = \left(\frac{e^2}{2\pi \varepsilon_o m v^2}\right)^2 \frac{\cos \theta}{\sin^4 \theta} \tag{2.13}$$

These must be compared with Eqs. (2.5), (2.4) and (2.8) respectively. For non-relativistic scattering of identical particles (e.g. electrons) there is no backscattering as this would violate conservation of energy, so that the maximum scattering angle in Eqs. (2.11) and (2.13) is $\theta = \pi/2$. In a neutral target atom Eq. (2.13) must also be multiplied by Z, so that scattering by the Z number of atomic electrons is accounted for. The electron to nuclear differential scattering cross-section ratio in a single target atom is therefore

$$\frac{Z(d\sigma/d\Omega)_{elec}}{(d\sigma/d\Omega)_{nucl}} = \frac{1}{Z} \left[\frac{\cos \theta}{\cos^4(\theta/2)} \right] \tag{2.14}$$

where Eq. (2.8) was substituted for $(d\sigma/d\Omega)_{nucl}$. The term within the square brackets decays monotonically from unity to zero as θ is increased to the maximum scattering angle of $\pi/2$. Furthermore, the inverse dependence on Z means that for most target elements the scattering by atomic electrons is only a small fraction of that due to the nucleus for all values of θ. When calculating the trajectory of electrons through matter it is therefore frequently assumed that the deflections are due to scattering by the atomic nuclei alone.

Before concluding our discussion on elastic scattering by an atomic nucleus let us re-examine Figure 2.1 in more detail. It was assumed that the kinetic energy of the incident electron at large values of r is unchanged before and after scattering owing to the significantly larger mass of the nucleus. However, we have overlooked the fact that the electron is a charged particle so that the change in momentum (or equivalently acceleration) caused by deflection close to the nucleus will result in the emission of electromagnetic radiation and therefore a reduction in the kinetic energy. The emitted radiation is known as 'Bremsstrahlung' (German for braking radiation) and forms a continuum with maximum energy equal to the incident electron kinetic energy. In classical physics Bremsstrahlung is emitted for every scattering event (Griffiths, 1999; Jackson, 1998), while in the quantum mechanical theory (Bethe and Heitler, 1934) a rather 'surprising' result is obtained. The majority of the scattering events do not emit radiation at all and are purely elastic, while those events that do emit constitute only a small fraction. The total radiative energy is equal in both the classical and quantum mechanical theories, although their spectral shape varies. In the former there are a large number of low energy radiative events, while in the latter the number of events is considerably smaller, although the average radiative energy is larger. Experimental observations of the Bremsstrahlung spectrum are consistent with the quantum mechanical theory but not classical physics. It is therefore reasonable to describe the electron–nucleus scattering as being largely elastic.

2.1.2 Inelastic Scattering by Atomic Electrons

Although atomic electrons produce small deflections compared to nuclei they are the main source of *inelastic* collisions. The differential cross-section for elastic nuclear scattering is a function of electron energy (through the v^{-4} term in Eqs. (2.8) and (2.10)), so that the energy loss due to inelastic collisions must be taken into account when simulating the electron trajectory through matter. This is especially true for low atomic number specimens (e.g. biological materials and polymers) since from Eq. (2.14) the relative cross-section for inelastic electron–electron interactions is increased with respect to elastic electron–nuclear interactions. Classical physics will again be used as a starting point to analyse the problem that is illustrated in Figure 2.4. The incident electron is scattered at an angle θ with speed v_1 while the atomic electron, assumed to be unbound to the nucleus, is scattered at

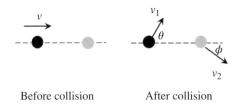

Before collision   After collision

Figure 2.4   Collision parameters for inelastic electron–electron scattering. The incident electron is denoted by the black circle and the atomic electron is the grey circle.

an angle ϕ with speed v_2. The energy transferred during the collision is therefore $\tfrac{1}{2}mv_2{}^2$. Conservation of energy and linear momentum leads to the following relations:

Conservation of energy : $v^2 = v_1^2 + v_2^2$ $\qquad$ (2.15)

Conservation of momentum in horizontal direction :

$$v = v_1 \cos \theta + v_2 \cos \phi \qquad (2.16)$$

Conservation of momentum in vertical direction : $v_1 \sin \theta = v_2 \sin \phi$
$$\qquad (2.17)$$

Comparing expressions for v^2 from Eqs. (2.15) and (2.16) gives

$$v_1^2 \sin^2 \theta + v_2^2 \sin^2 \phi = 2 v_1 v_2 \cos \theta \cos \phi \qquad (2.18)$$

Substitution of Eq. (2.17) then results in

$$\frac{v_1}{v_2} = \frac{\cos \theta \cos \phi}{\sin^2 \theta} = \frac{\sin \phi}{\sin \theta} \qquad (2.19)$$

where the second equality is due to rearrangement of Eq. (2.17). Equation (2.19) indicates that $\cot \theta = \tan \phi$ and hence $\theta = (\pi/2) - \phi$, that is, the scattered incident and atomic electron trajectories are normal to one another. Using Eqs. (2.16) and (2.17) to calculate v_1^2

$$v_1^2 = v_1^2 \sin^2 \theta + v_1^2 \cos^2 \theta = v_2^2 \sin^2 \phi + (v - v_2 \cos \phi)^2$$
$$= v^2 + v_2^2 - 2 v v_2 \cos \phi \qquad (2.20)$$

which on substituting the expression for v^2 from Eq. (2.15) yields $v_2 = v \cos \phi = v \sin \theta$. The energy transfer Q is therefore

$$Q = \frac{1}{2} m v_2^2 = \frac{1}{2} m v^2 \sin^2 \theta \qquad (2.21)$$

It is now possible to calculate the differential scattering cross-section $d\sigma$ in terms of Q rather than scattering angle θ. From Eq. (2.13) and $d\Omega = (2\pi \sin \theta)d\theta$,

$$d\sigma = 2\pi \left(\frac{e^2}{2\pi \varepsilon_o m v^2} \right)^2 \frac{\cos \theta \, d\theta}{\sin^3 \theta} \qquad (2.22)$$

Substituting Eq. (2.21) and its differential $dQ = (mv^2 \sin \theta)\cos \theta \, d\theta$ in the above gives

$$\frac{d\sigma}{dQ} = \left(\frac{e^4}{8\pi\varepsilon_o^2 mv^2}\right)\frac{1}{Q^2} \qquad (2.23)$$

The cross-section for inelastic scattering diverges as $Q \to 0$. Since a small Q corresponds to small scattering angles θ (Eq. (2.21)) and hence large impact parameters b (Eq. (2.11)) the discrepancy is somewhat similar to the divergence of the elastic nuclear scattering cross-section, as expressed in the form of Eq. (2.8), although the physical origins are quite different. An important parameter in the study of particle trajectories through matter is the rate at which the particle loses energy – otherwise known as the 'stopping power', which is equal to (dE/ds), where dE is the energy lost in traversing an infinitesimal distance ds through the solid. The stopping power is given by

$$\frac{dE}{ds} = -\left(\frac{N_A\rho}{A}\right)Z\int_{Q_{min}}^{Q_{max}} Q\left(\frac{d\sigma}{dQ}\right)dQ$$

$$= -Z\left(\frac{N_A\rho}{A}\right)\left(\frac{e^4}{8\pi\varepsilon_o mv^2}\right)\ln\left(\frac{Q_{max}}{Q_{min}}\right) \qquad (2.24)$$

where $(N_A\rho/A)$ is the number of atoms per unit volume (for a definition of terms, see Eq. (2.6)). The Z term takes into account the Z electrons per neutral atom and the negative sign represents the fact that energy is lost during propagation. The maximum energy transfer Q_{max} is, by Eq. (2.21), equal to $\frac{1}{2}mv^2$ and corresponds to the case of a head-on collision where the incident electron comes to a complete rest by transferring all its kinetic energy to the atomic electron. Estimation of the minimum energy transfer Q_{min} is less straightforward. Clearly, $Q_{min} > 0$ for Eq. (2.24) to converge. Bohr first realised that Q_{min} is determined by the binding energy of the atomic electron to its nucleus, an interaction that is not taken into account in deriving Eq. (2.23). If the atomic electron is displaced from its equilibrium position, say by the Coulomb repulsion of the incident electron, then the binding force of the nucleus results in simple harmonic motion of the atomic electron with frequency f. If the collision time ($\sim b/v$) is much smaller than the natural vibration period ($1/f$) the atomic electron is essentially 'free' and energy is transferred.[2] The

---

[2]The collision time follows from the duration of the transverse electric field 'pulse' due to a charged particle moving at constant velocity (McDaniel, 1989; see also Chapter 6).

maximum impact parameter (b_{max}) for energy transfer is therefore of the order of (v/f). In a simple harmonic oscillator the resonance frequency f increases as the square root of the spring constant and is therefore greater for electrons with larger binding energies. If $f = 10^{15}$ Hz is substituted for loosely bound electrons, determined from refractive index data of solids (Hecht, 2002), a b_{max} value of ~60 nm is obtained for a 20 keV electron beam in the SEM, indicating that energy transfer can take place over large distances for small energy losses (see however Section 6.1.3 for more accurate estimates). The energy loss Q can be expressed in terms of the impact parameter b by combining Eqs. (2.11) and (2.21):

$$Q = \frac{1}{2}mv^2 \left[\frac{1}{1 + (2b/D')^2} \right] \qquad (2.25)$$

so that

$$Q_{min} = \frac{1}{2}mv^2 \left[\frac{1}{1 + (2b_{max}/D')^2} \right] = \frac{1}{2}mv^2 \left[\frac{1}{1 + (2v/fD')^2} \right] \qquad (2.26)$$

It should be noted that a more exact expression for b_{max} is available (Bohr, 1913; Evans, 1955) but only differs from (v/f) by a constant multiplication factor. Substituting Eq. (2.26) in Eq. (2.24) with $Q_{max} = \frac{1}{2}mv^2$ and reasonably assuming that $b_{max} \gg D'$

$$\frac{dE}{ds} = -Z \left(\frac{N_A\rho}{A} \right) \left(\frac{e^4}{4\pi\varepsilon_0^2 mv^2} \right) \ln \left(\frac{2\pi\varepsilon_o mv^3}{e^2 f} \right) \qquad (2.27)$$

If each of the Z electrons has its own unique vibration frequency f_i due to differences in binding energy then the above equation must be modified to

$$\frac{dE}{ds} = - \left(\frac{N_A\rho}{A} \right) \left(\frac{e^4}{4\pi\varepsilon_0^2 mv^2} \right) \sum_{i=1}^{Z} \ln \left(\frac{2\pi\varepsilon_o mv^3}{e^2 f_i} \right)$$

$$= -Z \left(\frac{N_A\rho}{A} \right) \left(\frac{e^4}{4\pi\varepsilon_0^2 mv^2} \right) \ln \left(\frac{2\pi\varepsilon_o mv^3}{e^2 \bar{f}} \right) \qquad (2.28)$$

where $\bar{f}$ is the geometric mean vibration frequency, that is,

$$Z \ln \bar{f} = \sum_{i=1}^{Z} \ln f_i \qquad (2.29)$$

Equation (2.28) represents the stopping power due to electron–electron inelastic collisions determined within the classical regime by assuming particle properties for the incident electron. A key feature of the classical model is the ability to define an impact parameter. However, in the quantum mechanical regime the wave properties of the electron are dominant so that the concept of an impact parameter is meaningless. In Section 2.3.1, the conditions for classical versus quantum mechanical behaviour are discussed in more detail. It will be shown that for most electron microscopy measurements (TEM as well as SEM) the wave properties of the electron are dominant and hence, strictly speaking, scattering should be analysed within the quantum mechanical framework. The stopping power for bound electrons has been calculated by Bethe (1930) within the Born approximation, assuming a weak perturbation model for the Hamiltonian, and in non-relativistic form is given by

$$\frac{dE}{ds} = -Z\left(\frac{N_A\rho}{A}\right)\left(\frac{e^4}{4\pi\varepsilon_o^2 mv^2}\right)\ln\left(\frac{0.583mv^2}{J}\right) \qquad (2.30)$$

Equation (2.30) must be compared with Eq. (2.28). J is the geometric mean excitation and ionisation potential and is a statistical parameter representing the average energy loss per scattering event. J is mathematically defined as

$$Z\ln J = \sum_{n,l} f_{nl} \ln A_{nl} \qquad (2.31)$$

where f_{nl} is a transition probability (i.e. oscillator strength, Inokuti, 1971; see also Chapter 5) for electrons in the n, l atomic shell and A_{nl} is the transition energy. J has a similar form to $\bar{f}$ in Eq. (2.29). An analysis by Bloch (1933) has shown J to be proportional to Z. The variation of the Bethe stopping power as a function of incident electron energy is illustrated in Figure 2.5 for the case of copper. The decrease in stopping power at higher energies explains the lower rates of radiolysis beam damage (i.e. chemical bond breaking) in polymer materials, as the accelerating voltage is increased in the TEM.

It is also of interest to consider other sources of energy loss apart from inelastic scattering by atomic electrons. The most intense feature in an energy loss spectrum, as measured in the TEM, is the plasmon, which is a collective oscillation of unbound electrons in the solid, and therefore not modelled by Eq. (2.30). Phonon losses, due to thermal vibration of atoms in the solid, can also occur, but the energy loss is comparatively

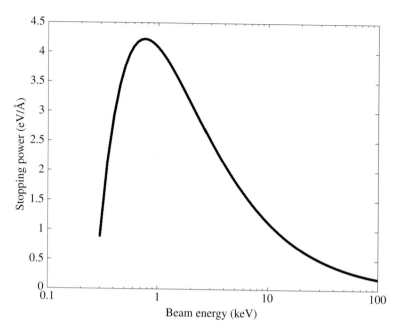

Figure 2.5   Bethe continuous energy loss stopping power for electrons in a copper target, calculated using Eqs. (2.30) and (2.43), plotted as a function of incident electron energy.

small (<100 meV) owing to the low energy of a phonon. Furthermore, Bremsstrahlung radiation due to scattering of the incident electron by the atomic nucleus is also not included in the stopping power. Above a particular (reasonably small) incident electron energy the stopping power due to atomic electron collisions, as given by Eqs. (2.28) and (2.30), decreases monotonically with energy (Figure 2.5). However, the stopping power due to Bremsstrahlung radiation shows the opposite trend and increases with electron energy. The incident electron energy at which the two stopping powers are equal is extremely large, for example, ~10 MeV for Pb (Evans, 1955). Hence, for SEM and TEM the stopping power can be calculated purely based on inelastic scattering by atomic electrons alone, while neglecting the relatively small contribution from Bremsstrahlung radiation. Section 2.3.3 also describes a method to include solid state effects, such as plasmons, in the stopping power.

2.1.3 Implementation of the Monte Carlo Algorithm

The background physics covered in Sections 2.1.1 and 2.1.2 will now be used to develop the Monte Carlo method for simulating electron beam–specimen interactions. For simplicity, it is often assumed that the incident electron deflection is governed by scattering due to the atomic nuclei only, while the energy loss is due to scattering by the atomic electrons. The validity of these assumptions has already been demonstrated; however, it must also be noted that these assumptions are by no means restrictive. For example, deviations in the electron trajectory due to scattering by atomic electrons, as well as Bremsstrahlung energy losses, can be easily incorporated in a Monte Carlo simulation, although this will only increase its complexity without much gain in accuracy. Practical aspects of designing and implementing the algorithm are covered in Joy (1995), Heinrich et $al.$ (1976) and Hovington et $al.$ (1997).

For Monte Carlo simulations we need to define a further parameter λ_{el}, the elastic mean free path, which is the average distance the electron travels through the solid before undergoing elastic scattering by an atomic nucleus. Consider uniformly illuminating a target of unit area and thickness t so that the intensity scattered within the angular range θ to $\theta + d\theta$ is given by Eq. (2.6). If the thickness of the target is λ_{el} then all of the incident intensity is scattered to some angle between 0 and π, that is,

$$1 = \left(\frac{N_A \rho}{A}\right) \lambda_{el} \int_0^{\pi} \left(\frac{d\sigma}{d\Omega}\right)(2\pi \sin \theta)d\theta \qquad (2.32)$$

where the substitution $d\Omega = (2\pi \sin \theta)d\theta$ has been made. The above equation can be simplified as

$$\lambda_{el} = \frac{A}{N_A \rho \sigma_E} \qquad (2.33)$$

where the total scattering cross-section (σ_E) can be calculated from the screened Rutherford differential scattering cross-section, Eq. (2.10), and is given by

$$\sigma_E = \left(\frac{Ze^2}{8\pi\varepsilon_o mv^2}\right)^2 \cdot \int_0^{\pi} \frac{2\pi \sin \theta}{[\sin^2(\theta/2) + \alpha]^2}d\theta = \left(\frac{Ze^2}{8\pi\varepsilon_o mv^2}\right)^2 \frac{4\pi}{\alpha(\alpha + 1)} \qquad (2.34)$$

The screening parameter α can be evaluated using an analytical approximation such as (Joy, 1995)

$$\alpha = 0.0034\frac{Z^{2/3}}{E_o} \qquad (2.35)$$

where E_o is the incident electron energy in kiloelectron volts. Scattering can now be analysed via the schematic shown in Figure 2.6. Following

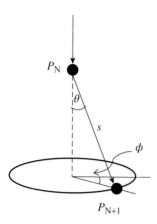

Figure 2.6 Schematic illustrating the collision sequence assumed in a Monte Carlo simulation. The incident electron is first scattered at P_N with scattering angle θ and azimuthal angle ϕ, the angular deviations being defined with respect to the electron incident direction (downward vertical arrow). The scattered electron travels a distance s before being re-scattered at the point P_{N+1}.

a scattering event by an atomic nucleus at P_N the incident electron travels a distance s before being re-scattered by a different nucleus at P_{N+1}. Scattering at P_N results in a deflection angle of θ and an azimuthal angle of ϕ. In a Monte Carlo simulation the objective is to *estimate* values for s, θ and ϕ, such that for a large number of scattering events the statistical distributions of these parameters are consistent with the physical models. For example, consider the scattering distance s. The distribution of s values follows a Poisson distribution, with average equal to λ_{el} (Eqs. (2.33)–(2.35)). The probability $p(s)\cdot ds$ that the scattering distance is between s and $(s + ds)$ is therefore

$$p(s) \cdot ds = \frac{1}{\lambda_{el}} \exp\left(-\frac{s}{\lambda_{el}}\right) ds \qquad (2.36)$$

It is now required to generate a series of scattering distances s for the individual scattering events based on the above probability distribution. In a computer, this is most easily done by generating random numbers that are uniformly distributed within the range $[0, 1]$. Consider the ratio RND defined as

$$\text{RND} = \int_0^s \frac{1}{\lambda_{el}} \exp\left(-\frac{s}{\lambda_{el}}\right) ds \bigg/ \int_0^\infty \frac{1}{\lambda_{el}} \exp\left(-\frac{s}{\lambda_{el}}\right) ds \qquad (2.37)$$

If a series of random numbers are generated then the fraction whose value is less than or equal to RND represents the probability that the scattering distance will be less than or equal to s. Rearranging Eq. (2.37),

$$s = -\lambda_{el} \ln(1 - RND) = -\lambda_{el} \ln(RND1) \tag{2.38}$$

Since Eq. (2.38) is derived from Eq. (2.37) a series of random numbers RND1 can be generated in the computer and the resulting scattering distance s, as calculated from Eq. (2.38), is guaranteed to have the required Poisson distribution in Eq. (2.36). Similarly, a different random number (RND2) can be used to generate the scattering angles θ with the correct distribution, that is,

$$RND2 = \int_0^\theta \left(\frac{d\sigma}{d\Omega} \right) (2\pi \sin \theta) d\theta \bigg/ \int_0^\pi \left(\frac{d\sigma}{d\Omega} \right) (2\pi \sin \theta) d\theta \tag{2.39}$$

Using the screened Rutherford differential cross-section in Eq. (2.10) to evaluate the above expression,

$$\sin^2 \left(\frac{\theta}{2} \right) = \frac{\alpha(RND2)}{(1 + \alpha - RND2)} \tag{2.40}$$

Since the scattering angle is restricted to the range $[0, \pi]$, Eq. (2.40) is usually expressed in the alternative form:

$$\cos \theta = 1 - \frac{2\alpha(RND2)}{(1 + \alpha - RND2)} \tag{2.41}$$

The azimuthal angle ϕ can take any value within the range $[0, 2\pi]$ uniformly and is therefore easily generated using a third random number RND3 according to

$$\phi = 2\pi(RND3) \tag{2.42}$$

Equations (2.38), (2.41) and (2.42) determine the trajectory from P_N to P_{N+1} and are applied iteratively to generate subsequent trajectories (e.g. from P_{N+1} to the next scattering event at P_{N+2}). The parameters s, θ and ϕ are defined with respect to the incident electron direction prior to scattering, so that a fixed frame of reference is required to keep track of the electron coordinates, as it (potentially) undergoes multiple scattering events within the solid. This is done via a standard transformation of coordinates, the procedure for which can be found in Joy (1995). Furthermore, it is also necessary to calculate the energy loss along the path length s in Figure 2.6, since the elastic scattering at P_{N+1} depends on the incident electron energy *at that position*. Since the scattering distance is usually small (e.g. λ_{el} is 2 nm for a 20 keV beam in Cu) the energy loss can be calculated by simply multiplying the Bethe stopping power (dE/ds) in Eq. (2.30) by the scattering distance s. Note, however, that

(dE/ds) is only an expectation value for the energy loss. In reality, core electron ionisation and Bremsstrahlung radiation can account for a significant fraction of the incident electron energy in one scattering event, so that large deviations from the statistical average are possible, though less likely. This effect is called 'straggling' and is not directly taken into account when using the Bethe stopping power. Experimental measurements have been used to derive an empirical relationship for J (expressed in kiloelectron volt; Berger and Seltzer, 1964) in Eq. (2.30):

$$J = \left(9.76Z + \frac{58.5}{Z^{0.19}}\right) \times 10^{-3} \qquad (2.43)$$

J varies from a few tens of electron volts for low Z material to several hundred electron volts for high Z values (e.g. for Cu it is $\sim$320 eV). The Bethe stopping power is valid for energies that are large with respect to J, but breaks down close to J (for small electron energies many of the electronic transitions defining J in Eq. (2.31) are not feasible due to conservation of energy). For example, when $0.583mv^2 < J$ the logarithm term in Eq. (2.30) becomes negative and the electron is incorrectly predicted to gain energy with distance. Joy and Luo (1989) have proposed a modified stopping power based on experimental results, which avoids any mathematical inconsistencies:

$$\frac{dE}{ds} = -Z\left(\frac{N_A\rho}{A}\right)\left(\frac{e^4}{4\pi\varepsilon_o^2 mv^2}\right)\ln\left(k' + \frac{0.583mv^2}{J}\right) \qquad (2.44)$$

where the value of k' is close to unity. In practice, the incident electron can either be backscattered out of the specimen or 'absorbed' within the solid as an electron in the conduction band. For the latter, the calculation is terminated at a small, but non-zero, electron energy (of the order of the mean energy loss J), since for small energies λ_{el} (Eq. (2.33)) is greatly reduced by large values of σ_E (Eq. (2.34)) and no significant net displacement is added to the electron trajectory. The trajectory of many incident electrons (typically 10^5 or larger) must be simulated in order to obtain statistically meaningful results for electron beam–specimen interactions.

2.2 SOME APPLICATIONS OF THE MONTE CARLO METHOD

2.2.1 Spatial Resolution and Backscattered Imaging

One of the most basic and important applications of the Monte Carlo method is to determine how electron beam–specimen interactions affect the spatial resolution. For secondary and Auger electrons, where the measured signal is generated close to the free surface owing to the short mean free path of the ejected electron, the spatial resolution is effectively the lateral size of the interaction volume near the free surface. For more 'bulk' techniques, such as backscattered electrons and X-rays, the size of the interaction volume along the depth direction must also be taken into account. The transmission of high energy electrons through a thin film, such as a TEM foil, is relatively easy to analyse, since any energy loss of the incident electron can be neglected. The Monte Carlo simulated electron trajectories through a 100 nm thin film of Si and Au are shown in Figure 2.7 for electron beam energies of 100 and 400 keV respectively. The interaction volume increases with increasing atomic number of the target and decreasing beam energy. This behaviour is consistent with the elastic scattering cross-section for atomic nuclei, Eqs. (2.8) and (2.10), that governs the angular deviation of the incident electrons. However, even in a thin foil the electron can be scattered multiple times before exiting the specimen at an angle θ' with respect to the incident beam direction. The mean square angle $\langle \theta^2 \rangle$ for single scattering is given by

$$\langle \theta^2 \rangle = \int_0^\pi \theta^2 \frac{d\sigma}{d\Omega} d\Omega \bigg/ \int_0^\pi \frac{d\sigma}{d\Omega} d\Omega \qquad (2.45)$$

Elastic scattering events are independent of one another, so that the angular distribution of the scattered electrons has a Gaussian profile with the peak centred about the incident beam direction and mean square angle $\langle \Theta^2 \rangle = (t/\lambda_{el})\langle \theta^2 \rangle$, where (t/λ_{el}) is the mean number of elastic scattering events in a film of thickness t (strictly speaking, Poisson statistics must be used if the mean number of elastic scattering events is small). $\langle \Theta^2 \rangle$ is measured relative to the incident beam direction. The normalised electron distribution, $P(\theta')d\theta'$, is therefore given by

$$P(\theta')d\theta' = \frac{1}{\sqrt{\pi \langle \Theta^2 \rangle}} \exp\left[-\frac{(\theta')^2}{< \Theta^2 >} \right] d\theta' \qquad (2.46)$$

Detailed discussion of the angular distribution of 5–30 keV electrons transmitted through thin-film targets can be found in Cosslett and Thomas (1964a and b).

Bulk solids (e.g. SEM specimens) are far more complex since here energy loss of the incident electrons is important. The size of the interaction volume is largely determined by those electrons that lose energy and come to 'rest' within the solid. Hence, the SEM interaction volume is

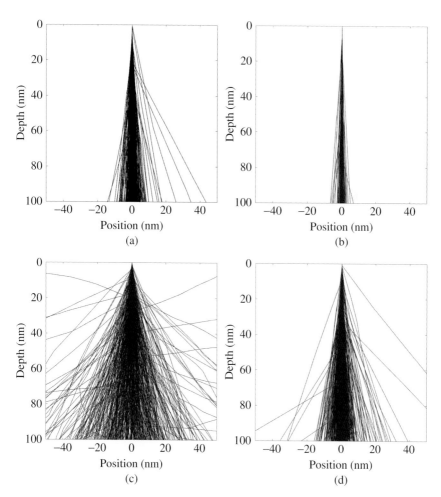

Figure 2.7 Monte Carlo simulated trajectories for (a) 100 keV, (b) 400 keV electrons in Si and (c) 100 keV, (d) 400 keV electrons in Au. The target is a thin film of 100 nm thickness. For visual clarity only 500 electron trajectories are shown.

larger for higher electron beam energies and low atomic number targets (note that this is the opposite trend to TEM). This is illustrated in the Monte Carlo simulated electron trajectories for bulk Si and Au at beam energies of 5 and 20 keV (Figure 2.8). Examination of the electron trajectories reveals that some of the electrons are scattered out of the sample entrance surface; these electrons constitute the backscattered electron signal. Monte Carlo methods are particularly suited for studying the properties of backscattered electrons. Dapor (1992), for example, has compared Monte Carlo predictions on backscattering with experimental

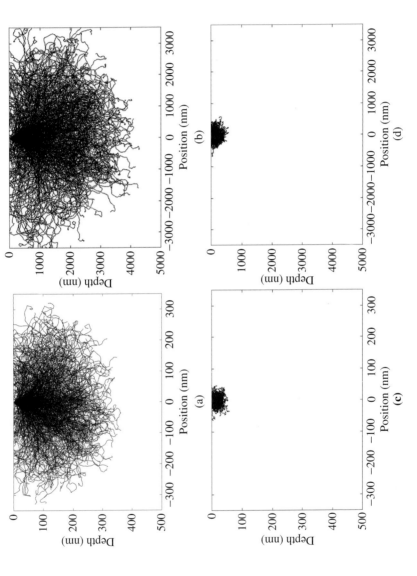

Figure 2.8 Monte Carlo simulated trajectories for (a) 5 keV, (b) 20 keV electrons in Si and (c) 5 keV, (d) 20 keV electrons in Au. The target is a bulk solid. For visual clarity, only 500 electron trajectories are shown. Note the change in scale between figures (a), (c) and

measurements on a large range of atomic number targets for electron beam energies in the range 5–30 keV. The agreement between theory and experiment is excellent except when the beam energy is too low (e.g. 5 keV) and the target atomic number large ($Z > 50$). This is, however, due to inaccuracy of the screened Rutherford cross-section used in the Monte Carlo simulations and can be improved by adopting the more accurate Mott cross-sections as described in Section 2.3.2.

The backscattering coefficient η is the ratio of the number of electrons backscattered out of the sample to the number of incident electrons. Figure 2.9a plots the variation of η as a function of target atomic number for electrons that are incident normal to a flat bulk specimen. The monotonic increase in η with atomic number means that the grey level in the backscattered image can be used to identify individual phases with different mean atomic number. The 'chemical' contrast is easily

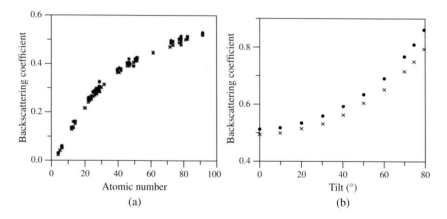

Figure 2.9 (a) Variation of the backscattering coefficient with target atomic number for 20 keV electrons at normal incidence and (b) backscattering coefficient as a function of beam tilt angle for 20 keV electrons incident on Au ('×' symbols are Monte Carlo calculated values, while the other symbols represent experimental measurements). From Dapor (1992). Reproduced with permission; copyright John Wiley and Sons. (c) Monte Carlo simulated trajectories for a 20 keV electron beam incident at 70° with respect to the surface normal of a Si target. The arrow represents the electron beam direction. Scattering is predominantly in the 'forward' direction (i.e. from left to right) and close to the surface. For visual clarity, only 200 electron trajectories are shown. (d) and (e) are the experimentally measured probability densities for backscattering from a 30 keV electron beam at normal incidence to Au and Si targets respectively. The backscattered electron energy E is normalised with respect to the incident energy E_o. The number (in degrees) indicated in each graph is the emission angle of the backscattered electrons (0° is anti-parallel to the electron beam). From Gérard et al. (1995). Reproduced with permission; copyright John Wiley & Sons.

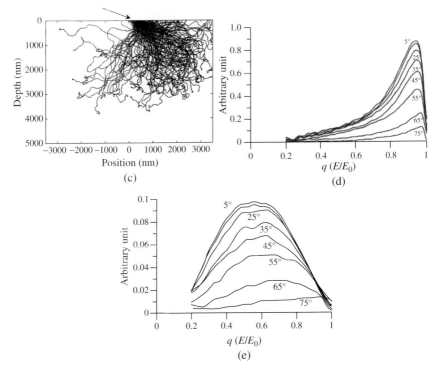

Figure 2.9 (*Continued*)

understood by noting that backscattering is due to large deflections of the incident electrons that pass close to the atomic nucleus; hence the larger the atomic number of the target the greater the cross-section for backscattering. However, for correct interpretation it is important that the specimen surface is flat, since as shown in Figure 2.9b, η increases with specimen tilt angle (i.e. the angle between the incident beam direction and specimen normal), so that rough regions of the specimen produce more backscattered electrons than flat regions. The reason for this increase in backscattering is clear from a Monte Carlo simulation of electron trajectories for an inclined beam (Figure 2.9c). Since elastic scattering predominantly occurs in the forward direction, the interaction volume is concentrated just below the specimen surface along the beam direction, so that incident electrons need only undergo relatively small deflections in order to escape out of the solid (compare this with normal incidence where much larger deflection angles are required for backscattering). At normal incidence, the spatial distribution of backscattered electrons escaping the solid is radially symmetric about the incident beam

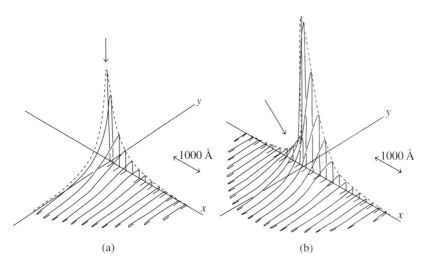

(a)   (b)

Figure 2.10   Spatial variation of emitted backscattered electron intensity for 20 keV
electrons incident on Au. In (a) the electron beam is normal to the target while in
(b) the beam tilt angle is 45°. Figure (a) is symmetrical with respect to each quad-
rant, while figure (b) is symmetrical only about the xz-plane. From Murata (1974).
Reproduced with permission; copyright AIP Publishing LLC.

direction, such that the number of backscattered electrons decreases
monotonically away from the beam impact point (Figure 2.10a; Murata,
1974). The tail of the distribution is due to electrons that have pene-
trated deep into the solid before being backscattered. The extent of the
tail increases with decreasing target atomic number and increasing beam
energy, consistent with the trends observed for the interaction volumes
in Figure 2.8. The tail of the distribution not only determines the lateral
spatial resolution for backscattered imaging, but is also an important
source for secondary and Auger electron generation, especially in
high atomic number solids where the backscattering coefficient is high
(Goldstein *et al.*, 2003). In contrast, the backscattered electron spatial
distribution for an inclined beam is asymmetrical, that is, backscatter
emission is peaked in the electron beam direction, but is otherwise
symmetrical about the plane containing the incident beam and surface
normal as shown in Figure 2.10b (cf. Figure 2.9c). This has important
consequences for detection of the backscattered signal, since only 'line of
sight' backscattered electrons are measured (Goldstein *et al.*, 2003). For
example, a backscattered detector placed at the level of the objective lens
will be less sensitive to the sloped regions of a rough surface, compared
to an Everhart–Thornley detector that is directly facing the slope.

It is also interesting to examine the energy distribution of backscattered electrons, both experimentally and through Monte Carlo simulations, since this is directly related to the depth of penetration of the incident electrons prior to backscattering. Experimentally measured energy distributions for bulk Au and Si targets are shown in Figures 2.9d and e respectively (Gérard *et al.*, 1995). The energy distribution for high atomic number targets, such as Au, is sharply peaked at a value close to the incident beam energy, whereas for lower atomic numbers (e.g. Si) the distribution is broader and is shifted further away from the incident beam energy. Monte Carlo simulations can be used to analyse the depth generation profile of backscattered electrons; as expected, the profiles show a similar qualitative behaviour to the energy distributions (i.e. for high atomic numbers generation is sharply peaked close to the surface, while for lower atomic numbers the distribution is broader and extends deeper into the solid, Murata, 1974). The elastic scattering cross-section (Eq. (2.10)) indicates that scattering to large angles is more probable for higher atomic numbers and smaller electron beam energies. Furthermore, the electron stopping power increases with atomic number (Eq. (2.30)). A high energy electron incident on a low atomic number target will initially be predominantly forward scattered. As it travels through the solid and loses energy the probability of being scattered through large angles increases, thereby giving rise to an increasing number of backscattered electrons. This cannot, however, continue indefinitely, since electrons backscattered deep within the solid do not have sufficient energy to reach the free surface and are either absorbed or are backscattered again into the solid. Alternatively, in a high atomic number target where the probability for large angle scattering and stopping power is greater, the depth generation profile will be narrower and closer to the surface. Furthermore, large angle scattering increases as Z^2 (Eq. (2.10)) whereas the rate of energy loss increases as Z (Eq. (2.30)), so that the net effect is a shifting of the backscattered energy distribution towards the incident beam energy in higher atomic number targets (Figures 2.9d and e).

2.2.2 Characteristic X-Ray Generation

The ability to extract structural and chemical information from the same region of interest at high spatial resolution is one of the most important attributes of electron microscopy. Monte Carlo simulations can also be adapted to analyse X-ray generation (characteristic as well as Bremsstrahlung) due to electron beam–specimen interactions. In order to do so, the cross-section for X-ray emission must first be determined.

For characteristic X-rays the quantum mechanical cross-section (σ_{nl}) for ionisation of an atomic electron is due to Bethe (1930):

$$\sigma_{nl} = \frac{\pi e^4 n_{nl}}{(4\pi\varepsilon_o)^2 E_o E_{nl}} b_{nl} \ln\left(\frac{c_{nl} E_o}{E_{nl}}\right) \tag{2.47}$$

where E_o is the incident electron energy, E_{nl} the critical ionisation energy of an electron with quantum numbers (n, l), n_{nl} the number of electrons in the shell/sub-shell (e.g. 2 for K-shell electrons) and b_{nl}, c_{nl} are constants depending on the electronic transition (for more details on the b_{nl}, c_{nl} constants, see Jones, 1992). The term $U_{nl} = E_o/E_{nl}$ is called the overpotential. The above formula is only valid for intermediate values of the overpotential, since at small overpotentials the Born approximation used for its derivation breaks down and at large overpotentials additional spin and exchange effects need to be taken into account. Nevertheless, Bethe's formula is widely used in Monte Carlo simulations and quantitative chemical analysis. Figure 2.11a plots the variation in cross-section for Cu K X-rays as a function of the overpotential. The curve passes through a maximum at $U_{nl} = \exp(c_{nl})/c_{nl}$, which is approximately 2.7 given that c_{nl} is typically of the order of unity (Jones, 1992).

In Monte Carlo simulations the number of characteristic X-rays ($I_{X\text{-ray}}$) generated along a step length s, such as, for example, the scattering distance between points P_N and P_{N+1} in Figure 2.6, is given by Joy (1995)

$$I_{X\text{-ray}} = s\sigma_{nl}\left(\frac{N_A \rho}{A}\right)\omega \tag{2.48}$$

where ω is the so-called 'fluorescence yield' and represents the probability that the ionised atom will return to the ground state by emitting a characteristic X-ray rather than an Auger electron. An approximate relationship for the fluorescence yield is

$$\omega = \frac{Z^4}{a + Z^4} \tag{2.49}$$

where 'a' is a constant. Hence X-ray emission is preferred for high atomic number (Z) elements, while Auger electron emission becomes increasingly important as the atomic number decreases.

Characteristic X-ray emission can only occur provided the electron energy is not below the critical energy E_{nl} (i.e. $U_{nl} \geq 1$). The volume for X-ray generation, and hence the X-ray spatial resolution, will therefore

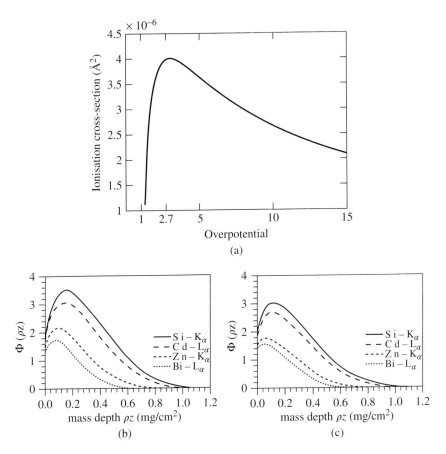

Figure 2.11 (a) Ionisation cross-section for Cu K X-rays as a function of overpotential, calculated using Eq. (2.47). (b) and (c) are respectively the Monte Carlo simulated and experimental $\phi(\rho z)$ curves for characteristic X-rays from tracer elements in a Ag matrix. The electron beam energy is 20 keV. From Ze-jun and Ziqin (1993). Reproduced with permission; copyright Institute of Physics.

invariably be smaller than the total electron beam–specimen interaction volume. Even for a given element, the spatial resolution may vary depending on the characteristic X-ray line. For example, high energy Cu K X-rays have higher spatial resolution than low energy Cu L X-rays, although it must be noted that Cu L X-rays generated from deep within the sample may not contribute appreciably to the measured signal due to absorption within the solid. Monte Carlo simulations can be used to generate the so-called $\phi(\rho z)$ curves, which plot the depth generation profile of X-rays within a solid (ρ is the density and z is the depth within the solid). $\phi(\rho z)$ curves are used for ZAF corrections in quantitative chemical analysis of bulk samples, albeit in a simplified analytical form in order to improve the computational efficiency (Bishop, 1974; Jones, 1992). Examples of Monte Carlo simulated and experimental $\phi(\rho z)$ curves for several characteristic X-ray lines from tracer elements in a Ag matrix are shown in Figures 2.11b and c respectively (Ze-jun and Ziqin, 1993). Note the greater generation depth of 1.7 keV Si K X-rays compared to 10.8 keV Bi L X-rays. The general shape of a $\phi(\rho z)$ curve is determined by two factors: the rate of energy loss of electrons through the solid (Eq. (2.30)) and the dependence of the ionisation cross-section on the overpotential (Eq. (2.47)). The maximum of a $\phi(\rho z)$ curve is located approximately at the depth where the electron energy corresponds to the maximum in the ionisation cross-section (Figure 2.11a; Goldstein et al., 2003).

2.2.3 Cathodoluminescence and Electron Beam Induced Current Microscopy

In cathodoluminescence (CL) and electron beam induced current (EBIC) microscopy the important physical process is the generation of electron–hole pairs by the electron beam in a material containing a band gap. The incident electrons impart a *fraction* of their energy in promoting a valence band electron in the solid into the conduction band (compare this to photoabsorption where the entire energy of the photon is expended). In CL the radiative decay of electron–hole pairs is detected (Yacobi and Holt, 1990), while in EBIC the minority carriers that diffuse to a space charge region (either a built in p–n junction or a Schottky barrier contact) is measured as an electric current (Leamy, 1982). CL and EBIC can be used to measure the electrical properties of defects (see, e.g. Yacobi and Holt, 1990; Donolato, 1983; Mendis et al., 2010), while Monte Carlo simulations play an important role in understanding the defect contrast (Holt and Napchan, 1994). The spatial resolution in CL and EBIC is dependent on two factors: the generation of electron–hole pairs within the incident beam interaction volume and the subsequent diffusion/drift of the carriers within the solid. The acquisition time per pixel of a typical CL or EBIC image is of the order of microseconds, which is significantly longer than the lifetime (~nanoseconds) of most semiconductor materials. The distribution of carriers will therefore reach a quasi-steady state, with the electron beam effectively treated as a time invariant excitation source (Mendis and Bowen, 2011). The quasi-steady state carrier distribution defines the spatial resolution for CL and EBIC. However, in phosphor materials the lifetime can be significantly longer than the acquisition time per pixel, so that the spatial resolution is better than that given by the quasi-steady state, and similarly for time-resolved CL experiments where the pulsed electron beam can be as short as a few picoseconds (Merano *et al.*, 2005).

The electron–hole pair generation rate $G(\mathbf{r})$ at position vector $\mathbf{r}$ due to the electron beam is given by Yacobi and Holt (1990)

$$G(\mathbf{r}) = \langle g(\mathbf{r}) \rangle \frac{I_b \gamma}{e} \qquad (2.50)$$

where $\langle g(\mathbf{r}) \rangle$ is the average distribution of electron–hole pairs created by a single incident electron, I_b is the incident beam current, γ is a correction term that takes into account the energy lost due to backscattering (=1 − (total energy of backscattered electrons/total energy of incident electrons)) and e is the electronic charge. γ and $\langle g(\mathbf{r}) \rangle$ are calculated using Monte Carlo simulations. For the latter, the total energy loss due to the stopping power (Eq. (2.30)) along a trajectory segment (e.g. along the scattering distance from P_N to P_{N+1} in Figure 2.6) is divided by the average energy it takes to create an electron–hole pair. The total number of electron–hole pairs thus calculated is distributed uniformly over the trajectory segment. This process is repeated for all incident electrons in the Monte Carlo simulation and the result normalised to obtain the average electron–hole pair distribution for a single electron (i.e. $\langle g(\mathbf{r}) \rangle$). The average energy required for electron–hole pair creation is typically ~3 times the material bandgap (see Klein (1968) for a discussion of the underlying semiconductor physics). Once $G(\mathbf{r})$ is known, the distribution of carriers at any given time (t) is determined by solving the continuity equation (Sze, 1985):

$$\frac{\partial n(\mathbf{r}, t)}{\partial t} = D\nabla^2 n(\mathbf{r}, t) + G(\mathbf{r}) - \frac{n(\mathbf{r}, t)}{\tau} \pm \mu \nabla(\varepsilon . n(\mathbf{r}, t)) \qquad (2.51)$$

where $n(\mathbf{r},\ t)$ is the minority carrier concentration with diffusion coefficient D, mobility μ and lifetime τ. The minority carrier drift due to the electric field ε is positive for electrons and negative for holes. Equation (2.51) is 'integrated', via finite difference methods, over the pixel acquisition time to give the CL/EBIC spatial resolution. If quasi-steady state is reached (i.e. $\partial n / \partial t = 0$), then it is computationally simpler to integrate Eq. (2.51) over a finite number of carrier lifetimes τ;

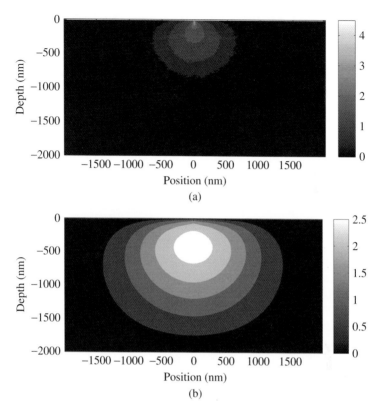

Figure 2.12 (a) Monte Carlo simulated electron–hole pair generation volume for a 15 keV electron beam in CdTe. The electron beam current is 1.6 nA and the carrier density is plotted on a logarithmic scale. (b) is the 'quasi-steady state' carrier distribution after numerically integrating for three lifetimes. The surface recombination velocity is 5×10^5 cm/s and the minority carrier diffusion length is 500 nm.

the accuracy can be estimated from the $\exp(-t/\tau)$ decay of carrier concentration when the excitation source is 'switched off'. Appropriate boundary conditions, such as the recombination velocity at the beam entrance surface, must also be included in order to determine the true carrier distribution. A Monte Carlo simulated carrier generation profile for a 1.6 nA current, 15 keV energy, electron beam incident on bulk CdTe is shown in Figure 2.12a. Figure 2.12b shows the carrier distribution at quasi-steady state (strictly speaking three carrier lifetimes). Note the large decrease in spatial resolution following carrier diffusion (the diffusion length is 500 nm).

2.3 FURTHER TOPICS IN MONTE CARLO SIMULATIONS

2.3.1 Classical or Quantum Physics?

In the previous sections, electron scattering was first treated classically, although the more accurate screened Rutherford elastic cross-section (Eq. (2.10)) and Bethe stopping power (Eq. (2.30)) were ultimately derived from quantum mechanics. In classical physics the impact parameter plays a crucial role in the scattering of the electron *particle*, but is not applicable to electrons with predominantly wave-like properties. It is therefore important to define the conditions for classical versus quantum behaviour. Consider a thought experiment where an opaque screen with a narrow slit is used to collimate a beam of electrons with speed v (Figure 2.13). The electrons passing through the narrow slit approach a single scattering centre with atomic number Z at an impact parameter b. A Coulomb force $\sim Ze^2/(4\pi\varepsilon_o b^2)$ acts on the electron for the duration of the collision ($\sim b/v$). The change in momentum is therefore $\sim Ze^2/(4\pi\varepsilon_o bv)$. If this is taken as an estimate of the uncertainty in the electron momentum the uncertainty in its position Δb is, by Heisenberg's principle, $\sim(2\pi\varepsilon_o hbv)/Ze^2$, where h is the Planck's constant. For the concept of an impact parameter to be valid Δb must be small in comparison to b, that is,

$$\left[\frac{\Delta b}{b} = \left(\frac{2\pi\varepsilon_o hc}{e^2}\right)\frac{\beta}{Z} \approx \frac{137\beta}{Z}\right] \ll 1 \tag{2.52}$$

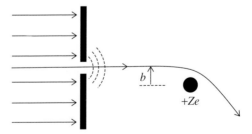

Figure 2.13 Schematic of a thought experiment used to define classical versus quantum mechanical scattering regimes. An opaque screen with a narrow slit is used to collimate a beam of electrons. If the electron is strictly a particle only a single 'ray' will be selected with a well-defined impact parameter of b. However, owing to the wave nature of the electron diffraction from the narrow slit can take place. The spreading of the Hüygen wavelets gives rise to an uncertainty Δb in the impact parameter.

where the speed has been rewritten as $v = \beta c$, with c being the speed of light. For a 20 keV electron and Cu target $(\Delta b/b)$ is greater than unity and hence it is clear that classical physics does not apply even at typical SEM beam energies. Physically, the uncertainty Δb can be interpreted as being due to diffraction of the electron wave from the narrow slit in Figure 2.13.

In the quantum mechanical description of scattering, the electron wavefunction consists of two parts: a plane wave of the form $e^{2\pi i k z}$ representing the unscattered wave propagating along the z-axis direction with wave number k (unit amplitude is assumed for simplicity) and an outgoing (distorted[3]) spherical wave due to scattering. The latter has the form $f(\theta)e^{2\pi i k r}/r$ at large distances r from the atomic scattering centre that is placed at the origin. $f(\theta)$ is the so-called atom scattering factor. The scattered electron intensity within the infinitesimal volume $d\tau$ at position coordinates (r, θ, ϕ) is given by $|f(\theta)e^{2\pi i k r}/r|^2 d\tau$ (for an unpolarised beam of electrons and radially symmetric potential the scattering is independent of the azimuthal angle ϕ). The flux of electrons scattered between angle θ and $(\theta + d\theta)$ is $v|f(\theta)e^{2\pi i k r}/r|^2 \cdot (r^2 d\Omega)$ or $v|f(\theta)|^2 d\Omega$ (note that for elastic scattering the speed v of the incident electrons is unchanged). Similarly the flux of incident electrons passing through unit area is $v|e^{2\pi i k z}|^2$ or simply v. The differential scattering cross-section $d\sigma$ can then be defined with the aid of Eq. (2.6) to obtain

$$d\sigma = \frac{v|f(\theta)|^2 d\Omega}{v} \quad \text{or} \quad \frac{d\sigma}{d\Omega} = |f(\theta)|^2 \qquad (2.53)$$

Note that since only a single scattering centre is being considered, the $(N_A \rho t/A)$ term in Eq. (2.6) is redundant. Equation (2.53) indicates that the differential scattering cross-section can be calculated if the atom scattering factor is known. Details of the analysis can be found in McDaniel (1989) or Mott and Massey (1965) but only the salient points are presented here.

The total electron wavefunction (ψ) must satisfy Schrödinger's equation $\nabla^2\psi + [4\pi^2k^2 + U(r)]\psi = 0$, where $U(r)$ is related to the scattering potential $V(r)$ through $U(r) = (8\pi^2 me/h^2)V(r)$. In spherical polar coordinates the Laplacian is expressed as

$$\nabla^2 = \frac{1}{r^2}\frac{\partial}{\partial r}\left(r^2\frac{\partial}{\partial r}\right) + \frac{1}{r^2\sin\theta}\frac{\partial}{\partial\theta}\left(\sin\theta\frac{\partial}{\partial\theta}\right) + \frac{1}{r^2\sin^2\theta}\frac{\partial^2}{\partial\phi^2} \qquad (2.54)$$

---

[3]Each electron in the atom scatters the incident electron as an outgoing spherical wave, but interference between all such spherical waves gives rise to an overall wavefunction of the form $f(\theta)e^{2\pi i k r}/r$.

The last term can be neglected since scattering is independent of ϕ. Adopting a separation of variables method, so that $\psi(r, \theta) = L(r)Y(\theta)$, and substituting in Schrödinger's equation gives

$$\frac{1}{L}\frac{d}{dr}\left(r^2\frac{dL}{dr}\right) + r^2[4\pi^2 k^2 + U(r)] = -\frac{1}{Y}\left[\frac{1}{\sin\theta}\frac{d}{d\theta}\left(\sin\theta\frac{dY}{d\theta}\right)\right] \quad (2.55)$$

Equation (2.55) can only be valid if the left-hand and right-hand sides equal a constant, which for simplicity is denoted as $l(l+1)$. The differential equation for the right-hand side is equivalent to Legendre's differential equation and produces finite solutions only if l is zero or any positive integer. In this case, the solution for $Y_l(\theta)$ is given by the Legendre polynomial $P_l(\cos\theta)$, which is a polynomial of order l in $\cos\theta$. In comparison, the differential equation for the left-hand side yields only an approximate solution for $L_l(r)$, which is valid at large r, and is of the form

$$L_l(r) \approx \frac{1}{2\pi kr}\sin\left(2\pi kr - \frac{l\pi}{2} + \eta_l\right) \quad (2.56)$$

Strictly speaking, the asymptotic form for $L_l(r)$ is valid for potentials $U(r)$ that converge faster than $1/r$ as $r \to \infty$ (this includes the screened Coulomb potential in Eq. (2.9)). The phase shift η_l is dependent on the wave number k and scattering potential $V(r)$. Its physical significance will be discussed later on. The total wavefunction is therefore

$$\psi(r,\theta) = \sum_{l=0}^{\infty} A_l P_l(\cos\theta)L_l(r) \quad (2.57)$$

where the constant A_l represents the contribution of the lth partial wave. As mentioned previously, the total wavefunction is the sum of an unscattered plane wave and a scattered spherical wave. Weak scattering is assumed, so that the unscattered wave has approximately unit amplitude, similar to the incident wave. The unscattered wave $e^{2\pi ikz}$ can also be expressed as

$$e^{2\pi ikz} = e^{2\pi ikr\cos\theta} = \sum_{l=0}^{\infty}(2l+1)i^l P_l(\cos\theta)j_l(2\pi kr) \quad (2.58)$$

where $j_l(2\pi kr)$ are the spherical Bessel functions that have peak intensity around $2\pi kr \approx 1.5l$, followed by weaker subsidiary maxima at larger values of $2\pi kr$. Equation (2.58) can be subtracted from Eq. (2.57) to give the scattered wavefunction ψ_{scatt} at large r. The constants A_l are determined from the condition that the solution for ψ_{scatt} must only contain outgoing waves of the form $e^{2\pi ikr}$ and no incoming waves of the form $e^{-2\pi ikr}$. The atom scattering factor can then be extracted by equating ψ_{scatt} to $f(\theta)e^{2\pi ikr}/r$ at large r, that is,

$$f(\theta) = \frac{1}{4\pi ik} \sum_{l=0}^{\infty} (2l + 1)(e^{2i\eta_l} - 1)P_l(\cos\theta) \qquad (2.59)$$

The only physical parameters in the above equation are the wave number k and phase shift η_l. From Eq. (2.58) it is clear that the individual partial waves constituting a plane wave incident on a scattering centre will have intensity concentrated at a characteristic distance, which increases monotonically with the order l of the partial wave. Thus, although an impact parameter cannot be defined in the classical sense the partial waves do have unique expectation values for the distance. The partial waves therefore sample different regions of the scattering potential field $V(r)$. When the incident electron penetrates the potential field of the atomic nucleus its kinetic energy increases at the expense of a decreasing potential energy, that is, there is a shortening of the electron de Broglie wavelength within the potential field compared to free space. The shorter wavelength means that the scattered electron is phase shifted by η_l with respect to the incident plane wave. Clearly, the phase shift is dependent on the region of the potential field being sampled and hence will vary with the order l of the partial wave. Only those partial waves that have undergone a phase shift will contribute to $f(\theta)$ and hence the differential scattering cross-section (from Eq. (2.59) if $\eta_l = 0$ for a given partial wave its contribution to $f(\theta)$ is zero as required). As an example if $V(r)$ is highly localised then to a good approximation the $l = 0$ partial wave will be the dominant term in the scattering.

2.3.2 Spin–Orbit Coupling and the Mott Cross-Section

The partial wave analysis of scattering in the previous section did not take into account relativistic effects such as spin of the incident electrons. This requires solving the Dirac equation (rather than the Schrödinger equation) as was first done by Mott (1929). As in the previous section only the relevant background physics is discussed; mathematical derivation of the Mott cross-section can be found in Mott (1929) and the book by Mott and Massey (1965).

Spin affects electron scattering through the phenomenon of spin–orbit coupling (Eisberg and Resnick, 1985). When viewed from the frame of reference of the incident electron the atomic nucleus of charge $+Ze$ moves with speed v in a direction opposite to that of electron motion. By Biot–Savart's law a moving charge, in this case the nucleus, generates a magnetic field $\mathbf{B}$. The direction of $\mathbf{B}$ is parallel to the angular momentum vector $\mathbf{L}$ of the electron about the nucleus (see footnote 1 for a definition of $\mathbf{L}$). The electron also has an intrinsic magnetic dipole moment μ_s due to its spin angular momentum $\mathbf{S}$ (μ_s and $\mathbf{S}$ are anti-parallel). Consequently, there is a potential energy ΔE arising from the interaction of the spin dipole moment μ_s with the electron angular momentum induced magnetic field $\mathbf{B}$ given by

$$\Delta E = -\boldsymbol{\mu}_s \cdot \mathbf{B}$$

$$= \left[-\frac{e}{2mc^2 r} \frac{dV(r)}{dr} \right] \mathbf{S} \cdot \mathbf{L} \qquad (2.60)$$

Equation (2.60) is the potential energy due to spin–orbit coupling. Scattering is modified since the incident electron experiences a potential energy ΔE in addition to the potential energy $-eV(r)$ of the scattering atom. This is illustrated schematically in Figure 2.14, which shows the potential energy around an atom nucleus placed at the origin with and without spin–orbit coupling. The incident electron direction is into the plane of the paper. The angular momentum $\mathbf{L}$ points upwards for the electron incident on the right-hand side and vice versa for the electron incident on the left-hand side. Consider the former electron. Since the bracketed term in Eq. (2.60) is positive, if the electron spin $\mathbf{S}$ is parallel to $\mathbf{L}$ (spin up) the scattering potential will be larger compared to the case of anti-parallel $\mathbf{S}$ (spin down). The force F acting on the electron is directed towards the atom and has magnitude equal to the gradient of the potential energy curve, which by inspection of Figure 2.14 is larger for the spin down case. Hence, spin down electrons incident on the right-hand side of the atomic nucleus will be more strongly deflected to the left-hand side compared to spin up electrons. Similarly, spin up electrons incident to the left of the nucleus are more strongly scattered to the right-hand side than spin down electrons. In an unpolarised beam, which is equally populated by spin up and spin down electrons, the asymmetry in the left–right scattering will be averaged out (i.e. there is no azimuthal dependence to the scattering), although the scattered electrons will be polarised (i.e. higher fraction of spin down/spin up electrons scattered to the left/right respectively). However, for electrons incident along the y-axis (x-coordinate of zero) the direction of $\mathbf{L}$ is normal to $\mathbf{S}$, so that the lack of any spin–orbit coupling results in an unpolarised scattered beam. Finally, if the incident

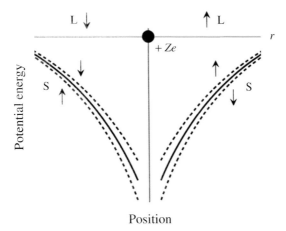

Figure 2.14 Effect of spin–orbit coupling on electron scattering by an atomic nucleus of charge +Ze. The electron is incident into the plane of the paper at a distance r from the atom nucleus. The direction of the angular momentum L for an electron passing the nucleus on the left- and right-hand sides is indicated. The potential energy due to spin–orbit coupling (dashed lines) either adds to or subtracts from the intrinsic Coulomb potential energy (solid line) depending on the relative orientation of the electron spin S with respect to L. This in turn modifies scattering. See text for further details.

electron beam is polarised (it is not in conventional electron microscopes) the scattered intensity will be a function of the azimuthal angle.

The Mott elastic differential scattering cross-section for an unpolarised electron beam is given by Mott (1929) and Mott and Massey (1965):

$$\frac{d\sigma}{d\theta} = |f(\theta)|^2 + |g(\theta)|^2$$

$$f(\theta) = \frac{1}{4\pi i k} \sum_{l=0}^{\infty} [(l+1)\left(e^{2i\eta_l\uparrow} - 1\right) + l\left(e^{2i\eta_l\downarrow} - 1\right)] P_l(\cos\theta)$$

$$g(\theta) = \frac{1}{4\pi i k} \sum_{l=0}^{\infty} \left[e^{2i\eta_l\downarrow} - e^{2i\eta_l\uparrow}\right] P_l^1(\cos\theta) \qquad (2.61)$$

where $\eta_l\uparrow$ and $\eta_l\downarrow$ are the lth partial wave phase shifts for spin up and spin down electrons and $P_l^1(\cos\theta)$ is the associated Legendre polynomial of order one $(= d[P_l(\cos\theta)]/d\theta)$. Note that if the spin up and spin down phase shifts are equal, the $g(\theta)$ term vanishes and the non-relativistic differential cross-section (Eqs. (2.53) and (2.59)) is obtained as required.

In practice, this condition is satisfied when the spin–orbit coupling contribution is relatively small, that is, for light atoms and high energy electrons, particularly at small scattering angles when the electron wave is on average travelling further away from the atomic nucleus. The criterion of high energy electrons appears counter-intuitive, as we would expect relativistic effects to be dominant at high velocities, but in fact the extra deflection due to the relatively small spin–orbit coupling energy term is less effective for swift electrons (Ding and Shimizu, 1996). A comparison of Rutherford and Mott scattering cross-sections for different elements can be found in McKinley and Feshbach (1948).

In order to calculate the Mott differential scattering cross-section the spin up and spin down phase shifts for each partial wave must be determined numerically by solving the radial part of the Dirac equation. Thomas–Fermi–Dirac, Hartree–Fock and relativistic Hartree–Fock–Slater type potentials are typically assumed for the scattering atom (Drouin *et al.*, 1997). The phase shift is also dependent on the electron energy. The large computational cost associated with employing Mott cross-sections in Monte Carlo simulations means that it is often more convenient to use the Rutherford scattering formula instead, especially if some accuracy can be sacrificed. For Monte Carlo simulations involving Mott cross-sections the standard procedure is to tabulate its value for a range of elements and electron beam energies and use the tables to interpolate to the parameters of interest. For example, the total cross-section is required to calculate the elastic mean free path (Eq. (2.33)). Furthermore the random number for generating the scattering angle (Eq. (2.39)) does not have an analytical solution for Mott cross-sections and hence a tabulation method is again required (Drouin *et al.*, 1997).

2.3.3 Dielectric Model of Stopping Power and Secondary Electron Emission

Secondary electrons are generated when the incident primary electron imparts sufficient energy to atomic (bound as well as valence) electrons such that they escape the sample surface. Prior to escaping, the ejected atomic electrons may themselves generate further secondary electrons in a cascade process. Decay of bulk and surface plasmons, excited by the incident electron, by interband transitions is another source of secondary electrons and is in fact the dominant generation mechanism for nearly free electron metals such as Al (Chung and Everhart, 1977). Experimentally the secondary electron signal is collected from those electrons that escape the sample surface with energies less than $\sim$50–100 eV, since it is these electrons that are easily deflected towards a positively biased Everhart–Thornley detector. For metals the secondary electron yield as a function of energy has a peak at $\sim\phi_w/3$ (see Cazaux, 2010, which also contains expressions for semiconductors and insulators) and the intensity of this peak increases monotonically with the work function ϕ_w (Baroody, 1950). The fact that secondary electrons have low energy and that collective plasmon oscillations are involved in their generation means that the Bethe stopping power is inadequate for this aspect of Monte Carlo modelling.

An alternative approach is to derive a stopping power based on the energy lost by a moving charged particle in a dielectric medium (Ritchie, 1957; more details can be found in Section 6.1.1). It can be shown that (see Eq. (6.18))

$$\frac{\partial^2(1/\lambda_{inel})}{\partial(\hbar\omega)\partial q} = \frac{me^2}{\varepsilon_o h^2 E_o}\frac{1}{q}Im\left\{\frac{-1}{\varepsilon(q,\omega)}\right\} \qquad (2.62)$$

where $\hbar\omega$ and q are the energy loss and magnitude of momentum change of the incident electron of energy E_o. $\varepsilon(q,\omega)$ is the dielectric function of the solid (a complex number) and 'Im' refers to the imaginary part of $(-1/\varepsilon)$. For simplicity, it is assumed here that the dielectric function is isotropic with respect to the momentum vector $\mathbf{q}$. λ_{inel} is the inelastic mean free path, so that from a relationship similar to Eq. (2.33), $1/\lambda_{inel}$ is proportional to the total inelastic cross-section. The stopping power is then (Ding and Shimizu, 1996)[4]

$$\frac{dE}{ds} = -\int_0^{E_o}\hbar\omega\frac{d(1/\lambda_{inel})}{d(\hbar\omega)}d(\hbar\omega)$$

$$= \frac{-me^2}{\varepsilon_o h^2 E_o}\int_0^{E_o}\hbar\omega\left[\int_{q_{min}}^{q_{max}}\frac{1}{q}Im\left\{\frac{-1}{\varepsilon(q,\omega)}\right\}dq\right]d(\hbar\omega) \qquad (2.63)$$

The minimum and maximum momentum changes (q_{min}, q_{max}) occur when the wave vectors of the incident electron before and after the scattering event are parallel and anti-parallel respectively (see Section 5.1.1 for more details). The dielectric function $\varepsilon(q,\omega)$ is typically only available in the optical limit (i.e. $q \to 0$) and suitable models must be adopted

---

[4]Strictly speaking, Eq. (2.62) is only valid for small energy loss, so that it is not correct to integrate up to E_o if the incident electron energy is high. Nevertheless, low energy loss scattering events (e.g. plasmons) are the most intense features in an energy loss spectrum, so that the error in Eq. (2.63) is expected to be small.

to extrapolate to non-zero values of q (e.g. the plasmon energy varies quadratically with q, while for interband transitions the energy is independent of q; Egerton, 1996). A method due to Penn (1987) is used by Ding and Shimizu (1996), while the optical data model of Ashley (1988) is used by Kuhr and Fitting (1999). In a Monte Carlo secondary electron simulation it is assumed that the energy lost by an incident electron within a scattering distance s, as calculated from Eq. (2.63), is transferred to a single atomic electron. If the energy transfer ΔE is sufficiently large core electrons can be ejected with kinetic energy equal to ΔE minus the binding energy, while for energy transfers smaller than the binding energy the outermost conduction electrons gain kinetic energy. It should be noted that in the Monte Carlo model of Ding and Shimizu (1996) the continuous energy loss stopping power of Eq. (2.63) was not used. Instead, a parameter RND4 was first defined as

$$\text{RND4} = \int_0^{\Delta E} \frac{d(1/\lambda_{\text{inel}})}{d(\hbar\omega)} d(\hbar\omega) \bigg/ \int_0^{E_o} \frac{d(1/\lambda_{\text{inel}})}{d(\hbar\omega)} d(\hbar\omega) \qquad (2.64)$$

RND4 can then be used as a random number input to generate a probable energy loss ΔE similar to the method outlined in Section 2.1.3. This method is able to model straggling, unlike a continuous energy loss equation. Besides the energy loss, random numbers were also used to model two further scenarios: (i) determine if a particular scattering event is elastic or inelastic[5] and (ii) the deflection of the incident electron during inelastic scattering using a modified version of Eq. (2.62), rather than assuming no deflection. For inelastic scattering, the trajectory of the ejected atomic electron can be assumed to be perpendicular to that of the scattered incident electron with azimuthal angle chosen at random (as predicted by classical mechanics; Section 2.1.2) or isotropic; the latter is based on the observation that secondary electron emission from the sample surface has no preferred direction. The ejected atomic electrons can themselves generate further secondary electrons as part of a cascade process and hence their paths within the solid must be tracked using the standard Monte Carlo approach. In fact, Monte Carlo simulations show that the cascade process is largely responsible for the very low energy secondary electrons emitted from the sample (Ding et al., 2001). The cascade electrons are tracked until they come to 'rest' within the solid or are emitted from the sample surface. For the latter, the kinetic energy of

---

[5]The overall scattering mean free path (λ_m) is given by $\lambda_m^{-1} = \lambda_{\text{el}}^{-1} + \lambda_{\text{inel}}^{-1}$. Therefore, if a random number is less than ($\lambda_{\text{el}}^{-1}/\lambda_m^{-1}$) the scattering is elastic; otherwise, it is inelastic.

the electrons due to the velocity component normal to the surface must be larger than the surface potential energy barrier U_o. Under these conditions classical mechanics predicts that the electron will always escape into the vacuum, but in quantum physics a certain fraction of the electrons can be reflected back into the solid, especially if the electron energy is only slightly larger than the barrier height. The quantum mechanical transmission probability $T(E, \beta)$ for an electron travelling at an angle β with respect to the surface normal with energy E is (Eisberg and Resnick, 1985; Ding and Shimizu, 1996)

$$T(E, \beta) = \frac{4\sqrt{1 - U_o/E\cos^2\beta}}{[1 + \sqrt{1 - U_o/E\cos^2\beta}]^2} \tag{2.65}$$

The above equation is valid for $E\cos^2\beta > U_o$; if $E\cos^2\beta < U_o$ then $T(E, \beta) = 0$. A random number is generated and if its value is less than $T(E, \beta)$ the electron is emitted or else it is reflected back into the solid. The escaped electron undergoes refraction such that its angle β_o with respect to the surface normal in the vacuum is given by a modified version of Snell's law:

$$\sqrt{E - U_o}\sin\beta = \sqrt{E}\sin\beta_o \tag{2.66}$$

Monte Carlo simulation results for various elements, especially the energy distribution of emitted secondary electrons, can be found in Ding *et al.* (2001) and Kuhr and Fitting (1999).

2.4 SUMMARY

The Monte Carlo method uses computer-generated random numbers to simulate probable outcomes in electron beam–specimen interactions. The random number simulates the effect of a variable parameter, such as impact parameter if the two colliding species are treated as particles. The trajectories of many incident electrons are simulated in order to obtain a statistically averaged and physically meaningful result. The advantage of the Monte Carlo method lies in the fact that both elastic and inelastic scattering are readily incorporated and specimens of any shape or size can be simulated. Typically, elastic scattering is simulated using the screened Rutherford cross-section, while the Bethe continuous energy loss stopping power is used for inelastic scattering. The relativistic Mott cross-section, which includes electron spin effects, is used for more accurate elastic scattering calculations, especially for large atomic number elements and small electron energies. Similarly, a stopping power based on the dielectric loss function can be used to model inelastic scattering in the low energy regime, where the Bethe expression is not valid. Monte Carlo methods can be applied to a variety of problems, such as backscattered, secondary electron and X-ray generation as well as CL, EBIC microscopies.

REFERENCES

Ashley, J.C. (1988) *J. Electr. Spectr. Rel. Phenom.* **46**, 199.

Baroody, E.M. (1950) *Phys. Rev.* **78**, 780.

Berger, M.J. and Seltzer, S.M. (1964) *Studies in Penetration of Charged Particles in Matter*, Nuclear Science Series Report #39, NAS-NRC Publication 113 (Natl. Acad. Sci.: Washington D.C.), p. 205.

Bethe, H.A. (1930) *Ann. Phys.* **5**, 325.

Bethe, H.A. and Heitler, W. (1934) *Proc. Royal Soc.* **A146**, 83.

Bishop, H.E. (1974) *J. Phys. D: Appl. Phys.* **7**, 2009.

Bloch, F. (1933) *Z. Physik.* **81**, 363.

Bohr, N. (1913) *Phil. Mag.* **25**, 10.

Cazaux, J. (2010) *Ultramicroscopy* **110**, 242.

Chung, M.S. and Everhart, T.E. (1977) *Phys Rev B* **15**, 4699.

Cosslett, V.E. and Thomas, R.N. (1964a) *Brit. J. Appl. Phys.* **15**, 883.

Cosslett, V.E. and Thomas, R.N. (1964b) *Brit. J. Appl. Phys.* **15**, 235.

Dapor, M. (1992) *Phys. Rev. B* **46**, 618.

Ding, Z.J. and Shimizu, R. (1996) *Scanning* **18**, 92.

Ding, Z.J., Tang, X.D. and Shimizu, R. (2001) *J. Appl. Phys.* **89**, 718.

Donolato, C. (1983) *J. Appl. Phys.* **54**, 1314.

Drouin, D., Hovington, P. and Gauvin, R. (1997) *Scanning* **19**, 20.

Egerton, R.F. (1996) *Electron Energy-Loss Spectroscopy in the Electron Microscope*, 2nd edition, Plenum Press, New York.

Eisberg, R. and Resnick, R. (1985) *Quantum Physics of Atoms, Molecules, Solids, Nuclei and Particles*, 2nd edition, John Wiley & Sons, USA.

Evans, R.D. (1955) *The Atomic Nucleus*, McGraw-Hill, New York.

Gérard, P., Balladore, J.L., Martinez, J.P. and Ouabbou, A. (1995) *Scanning* **17**, 377.

Goldstein, J.I., Newbury, D.E., Joy, D.C., Lyman, C.E., Echlin, P., Lifshin, E., Sawyer, L. and Michael, J.R. (2003) *Scanning Electron Microscopy and X-ray Microanalysis*, 3rd edition, Springer, USA.

Griffiths, D.J. (1999) *Introduction to Electrodynamics*, Prentice-Hall, New Jersey.

Hecht, E. (2002) *Optics*, 4th edition, Addison Wesley, California.

Heinrich, K.F.J, Newbury, D.E. and Yakowitz, H. eds. (1976) *Use of Monte Carlo calculations in Electron Probe Microanalysis and Scanning Electron Microscopy*, National Bureau of Standards Special Publication #460, p. 5.

Holt, D.B. and Napchan, E., (1994) *Scanning* **16**, 78.

Hovington, P., Drouin, D. and Gauvin, R. (1997) *Scanning* **19**, 1.

Inokuti, M. (1971) *Rev. Mod. Phys.* **43**, 297.

Jackson, J.D. (1998) *Classical Electrodynamics*, John Wiley & Sons, New York.

Jones, I.P. (1992) *Chemical Microanalysis using Electron Beams*, Institute of Materials, London.

Joy, D.C. (1995) *Monte Carlo Modelling for Electron Microscopy and Microanalysis*, Oxford University Press, Oxford.

Joy, D.C. and Luo, S. (1989) *Scanning* **11**, 176.

Klein, C.A. (1968) *J. Appl. Phys.* **39**, 2029.

Kuhr, J.-Ch. and Fitting, H.-J. (1999) *J. Electr. Spectr. Rel. Phenom.* **105**, 257.

Leamy, H.J. (1982) *J. Appl. Phys.* **53**, R51.

McDaniel, E.W. (1989) *Atomic Collisions: Electron and Photon Projectiles*, John Wiley & Sons, USA.

McKinley, W.A. and Feshbach, H. (1948) *Phys. Rev.* **74**, 1759.

Mendis, B.G. and Bowen, L. (2011) *J. Phys.: Conf. Series* **326**, 012017.

Mendis, B.G., Bowen, L. and Jiang, Q.Z. (2010) *Appl. Phys. Lett.* **97**, 092112.

Merano, M., Sonderegger, S., Crottini, A., Collin, S., Renucci, P., Pelucchi, E., Malko, A., Baier, M.H., Kapon, E., Deveaud, B. and Ganiere, J.D. (2005) *Nature* **438** 479.

Mott, N.F. (1929) *Proc. Royal Soc.* **A124**, 425.

Mott, N.F. and Massey, H.S.W. (1965) *Theory of Atomic Collisions*, 3rd edition, Oxford University Press, Oxford.

Murata, K. (1974) *J. Appl. Phys.* **45**, 4110.

Penn, D.R. (1987) *Phys. Rev. B* **35**, 482.

Ritchie, R.H. (1957) *Phys. Rev.* **106**, 874.

Rutherford, E. (1911) *Phil. Mag.* **21**, 669.

Sze, S.M. (1985) *Semiconductor Devices: Physics and Technology*, John Wiley & Sons, USA.

Yacobi, B.G. and Holt, D.B. (1990) *Cathodoluminescence Microscopy of Inorganic Solids*, Plenum Press, New York.

Ze-jun, D. and Ziqin, W. (1993) *J. Phys. D: Appl. Phys.* **26**, 507.

3

Multislice Method

The Monte Carlo method, discussed previously, is based on scattering by a single atom, rather than the crystal as a whole (i.e. Bragg diffraction). It therefore works well for crystals where the planes are oriented away from the Bragg condition. While this might be a reasonable approximation for predicting electron beam–specimen interactions averaged over many crystal orientations (average backscattering coefficient, average size of interaction volume, etc.) it breaks down for measurements that rely directly on Bragg diffraction, such as electron backscattered diffraction (EBSD) in the SEM or high resolution electron microscopy (HREM) in the transmission electron microscope (TEM). Furthermore, since the incident electrons are charged the Coulomb interaction with matter is particularly strong, so that even in a thin TEM foil oriented along a zone-axis there can be multiple scattering, where a Bragg diffracted beam is scattered back into the unscattered beam or else a different Bragg reflection. A new theory is therefore required that not only takes into account scattering by the entire crystal but also dynamic diffraction (i.e. multiple scattering). There are two theories that are frequently used: the multislice method based on physical optics and the Bloch wave method discussed in Chapter 4. These two theories can be shown to be equivalent under certain conditions, but the numerical procedure is different, and this offers certain advantages, as well as disadvantages, for a given theory over the other. The first theory, the multislice method, is the subject of this chapter.

The principle of the multislice method is illustrated in Figure 3.1, where the crystal is oriented along a zone-axis and consists of alternate rows of atoms displaced by half the atomic spacing. Two such atomic rows constitute the periodic repeat pattern along the stacking direction.

Electron Beam-Specimen Interactions and Simulation Methods in Microscopy,
First Edition. Budhika G. Mendis.
© 2018 John Wiley & Sons Ltd. Published 2018 by John Wiley & Sons Ltd.

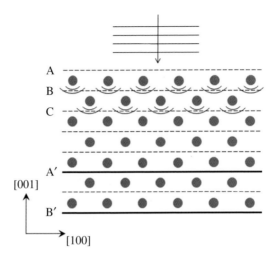

Figure 3.1 Schematic illustration of the multislice technique for an incident plane wave. The specimen is divided into thin slices, such as, slices AB, BC, etc., which correspond to a single atom row. The electron beam is scattered by the atoms in a given slice before propagating through free space to the next slice. The electron exit wavefunction is calculated by repeated application of scattering and propagation for all slices within the specimen.

The electrons are incident as a plane wave (i.e. parallel beam), but could alternatively be a focused probe depending on the experimental set-up. The crystal is divided into thin slices, such as, for example, slices AB, BC, etc.; in this scheme the slice thicknesses are all equal, although this need not be the case. When the electron beam is incident on the first slice AB scattering by the atoms within that slice 'perturb' the electron wavefunction. The perturbation is due to the incident electrons being accelerated by the positive potential of the atomic nuclei, so that there is a shortening of the de Broglie wavelength in the vicinity of the atoms. This is similar to the refractive index effect in light optics, except that here the origin is electrostatic in nature. For high energy electrons, such as in a TEM, atomic scattering is primarily in the forward direction (Section 2.1.1) and hence backscattering can be neglected, although this will not be the case for low energy electrons or highly tilted electron beams (for the latter, the angular deflection required for the incident electrons to exit the specimen is relatively small; cf. Section 2.2.1). Backscattering is therefore important in surface diffraction techniques, such as low energy electron diffraction (LEED) and potentially reflection high energy electron diffraction (RHEED). The perturbed electron wavefunction in slice AB must propagate through free space towards the next set of atom scattering centres in the adjoining slice BC. Propagation is modelled using Hüygen's principle, where each point in the electron wavefunction acts as a source for secondary wavelets (in Figure 3.1 only Hüygen wavelets emitted by the scattering atoms are depicted for clarity). The process of atom scattering and propagation of the electron wavefunction is repeatedly carried out for successive slices until the electron beam finally exits the specimen.

In the above description it is clear that atom scattering can take place in all directions, notwithstanding the approximations typically made for backscattering. However, in a crystal periodic arrangement of the atoms means that in the far field (i.e. diffraction plane) only those reciprocal lattice reflections that are intersected by the Ewald sphere can have non-zero intensity (here, broadening of the reciprocal lattice points by the finite size of the sample applies). Certain Bragg reflections are extinguished by lattice centring as well as any screw axes and/or glide planes. Consider forbidden reflections due to lattice centring, which have zero intensity even under dynamic diffraction conditions. As an example for a body-centred crystal the 100 reflection is forbidden. Referring to Figure 3.1 this arises due to scattering in the 100 beam direction by two displaced, neighbouring atomic layers being out of phase by π radians, so that destructive interference takes place in the far field. Dividing the crystal into slices AB, BC, etc. in a multislice simulation captures the essential features of this process. However, alternative slicing could also be used, such as, for example, a slice that corresponds to the periodic repeat unit along the stacking direction (e.g. slice A'B' in Figure 3.1). In the multislice method propagation through free space is modelled only between neighbouring slices, not between atoms within a given slice. This effectively means that the incident electrons are scattered by the *projected* potential of all atoms within the slice. For the A'B' slicing the incident electrons do not experience a change in scattering potential between successive slices and hence the intensity of forbidden reflections will not be correctly reproduced. A similar argument also applies to higher order Laue zone (HOLZ) reflections, where the variation in specimen potential along the zone-axis direction is important. The accuracy of the multislice method therefore depends on the slicing scheme, with thinner slices producing more accurate results. Finally, it should also be noted that the specimen need not be a periodic crystal; the method is equally applicable to amorphous specimens as well as crystals containing defects. This flexibility is one of the main advantages of the multislice method over Bloch wave calculations.

In Section 3.1, atom scattering and free space propagation are analysed mathematically. Practical implementation of the multislice algorithm is also discussed. Section 3.2 deals with applications, including extension of the technique to focused electron probes, such as, convergent beam electron diffraction (CBED) and scanning transmission electron microscopy (STEM). More detailed topics, such as accuracy of the multislice method, are discussed in Section 3.3. In this chapter, the emphasis is on multislice simulations of high energy electrons in TEM, and any energy lost by the incident electron to the thin TEM foil is assumed to be negligible. Calculations are therefore based on elastic scattering, although inelastic thermal diffuse scattering can also be included in order to improve the accuracy; in certain cases, such as CBED and STEM, it is in fact indispensable.

3.1 MATHEMATICAL TREATMENT OF THE MULTISLICE METHOD

A mathematical framework (Cowley and Moodie, 1957; Goodman and Moodie, 1974) underpinning the multislice method will now be developed. Let $\psi_n(\mathbf{R})$ denote the electron wavefunction incident on the $(n + 1)$th slice, where $\mathbf{R}$ is the two-dimensional position vector normal to the slice thickness. Scattering by all atoms within the $(n + 1)$th slice modifies the wavefunction to $\psi_n(\mathbf{R})Q_{n+1}(\mathbf{R})$, where Q is the so-called *transmission function* or *phase grating function* of the specimen. It should be noted that Q is determined by the projected potential of atoms within a given slice and is therefore a function of $\mathbf{R}$ only (i.e. the depth of the individual atoms within the slice is not taken into consideration). Without any loss of generality, it can be assumed that atom scattering takes place at the entrance surface of a given slice. The modified wavefunction propagates through free space before exiting the $(n + 1)$th slice. Modelling propagation as the interference of neighbouring Hüygen wavelets gives

$$\psi_{n+1}(\mathbf{R}) = \left[\psi_n(\mathbf{R}) \cdot Q_{n+1}(\mathbf{R})\right] \otimes P_{n+1}(\mathbf{R}) \qquad (3.1)$$

where the $\otimes$ symbol denotes the convolution operation and P is the Fresnel free space propagator function that represents spreading of a Hüygen wavelet.[1] The exit wavefunction for the $(n + 1)$th slice (ψ_{n+1})

---

[1] Note that in some texts Eq. (3.1) is written as $\psi_{n+1}(\mathbf{R}) = [\psi_n(\mathbf{R}) \otimes P_{n+1}(\mathbf{R})] \cdot Q_{n+1}(\mathbf{R})$, that is, atomic scattering is assumed to take place at the bottom of the slice following propagation. The two equations produce similar results, within the error of the multislice method (Section 3.3.1).

can therefore be calculated given the wavefunction at its entrance surface (ψ_n). Since the incident wavefunction (ψ_0) is known from the experimental set-up the electron wavefunction can be calculated at any depth within the specimen by applying Eq. (3.1) iteratively. Fourier transforming the above equation,

$$\tilde{\psi}_{n+1}(\mathbf{u}) = \left[\tilde{\psi}_n(\mathbf{u}) \otimes \tilde{Q}_{n+1}(\mathbf{u})\right] \cdot \tilde{P}_{n+1}(\mathbf{u}) \qquad (3.2)$$

where the tilde indicates a Fourier transform and $\mathbf{u}$ is a 2D vector in Fourier space. The specimen transmission function is determined by the atomic potentials, so that in a crystalline solid, where the atomic arrangement is periodic, $\tilde{Q}(\mathbf{u})$ has non-zero values only at the reciprocal lattice vectors ($\mathbf{g}$). For parallel beam illumination of a perfect crystal this considerably simplifies the convolution term in Eq. (3.2), since $\left[\tilde{\psi}_n(\mathbf{u}) \otimes \tilde{Q}_{n+1}(\mathbf{u})\right]$ is non-zero only at the reciprocal lattice points, that is,

$$\left[\tilde{\psi}_n(\mathbf{u}) \otimes \tilde{Q}_{n+1}(\mathbf{u})\right]_{\mathbf{u}=\mathbf{g}} = \sum_{\mathbf{g}'} \tilde{\psi}_n(\mathbf{g}')\tilde{Q}_{n+1}(\mathbf{g} - \mathbf{g}') \qquad (3.3)$$

where the summation is carried out over all reciprocal lattice vectors $\mathbf{g}'$, including those that satisfy extinction rules for Bragg diffraction. Equations (3.2) and (3.3) indicate that for a given slice all diffracted beams $\mathbf{g}'$ of the incident wavefunction $\tilde{\psi}_n$ can contribute to a given diffracted beam $\mathbf{g}$ of the slice exit wave $\tilde{\psi}_{n+1}$ by undergoing scattering via the vector $(\mathbf{g} - \mathbf{g}')$. This is illustrated in Figure 3.2 for the case of a plane wave propagating through the first two slices of a specimen. For normal incidence the plane wave $\exp(2\pi i\mathbf{k}\cdot\mathbf{r})$ can be represented as $\psi_0(\mathbf{R}) = 1$, or in Fourier space, $\tilde{\psi}_0(\mathbf{u}) = \delta(\mathbf{u})$, where δ is the Dirac delta function,[2] that is, in the far field only the unscattered beam is present at $|\mathbf{u}| = 0$ while diffracted beams have zero intensity. To simplify the discussion it is assumed that only three beams are involved, that is, the unscattered beam and two $\pm\mathbf{g}$ diffracted beams. Furthermore, for a centrosymmetric crystal $\tilde{Q}(\mathbf{g}) = \tilde{Q}(-\mathbf{g})$, provided the centre of inversion is at the origin, while $\tilde{P}(\mathbf{g}) = \tilde{P}(-\mathbf{g})$ for a spherical Hüygen wavelet and an electron beam incident normal to the specimen. Equations (3.2) and (3.3) then give $\tilde{Q}(\mathbf{g})\tilde{P}(\mathbf{g})$ for the $\pm\mathbf{g}$ diffracted beams and $\tilde{Q}(\mathbf{0})\tilde{P}(\mathbf{0})$ for the unscattered beam after the first slice.

---

[2]Since only phase differences are important in interference phenomena the constant phase of the plane wave across all values of $\mathbf{R}$ can be ignored. Furthermore the intensity of the plane wave has been normalised to unity.

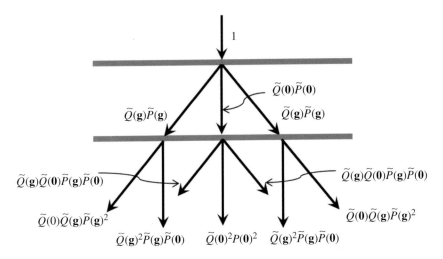

Figure 3.2 Unscattered (0) and ±g diffracted beam wavefunctions predicted by the multislice method. The incident wavefunction is a plane wave of unit amplitude and only the first two slices are shown. See text for details.

Figure 3.2 indicates the individual wave contributions after the incident electrons have propagated through the second slice. The +g diffracted beam has partly re-scattered (i.e. $\tilde{Q}(g)^2\tilde{P}(g)\tilde{P}(0)$ wave contribution) back into the unscattered beam. This double diffracted intensity is weak on account of the small intensity of the diffracted beam after the first slice. On the other hand, there is a much larger scattering contribution of $\tilde{Q}(g)\tilde{Q}(0)\tilde{P}(g)\tilde{P}(0)$ from the unscattered beam into the +g beam. Hence, there is a net increase in diffracted beam intensity at the expense of the unscattered beam following propagation through the second slice. However, this trend cannot continue indefinitely, since at a certain depth the diffracted beam would be sufficiently intense to make double diffraction to the unscattered beam the dominant process. There is then net transfer of intensity from the diffracted beams to the unscattered beam. This process is cyclic and leads to the well-known pendellösung phenomenon in electron diffraction (Hirsch *et al.*, 1965).

Numerical implementation of the multislice algorithm requires expressions for the specimen transmission function and Fresnel free space propagator. These are derived in the following sections. Practical aspects of multislice programming can be found in Kirkland (2010) and Self and O'Keefe (1988).

3.1.1 Specimen Transmission Function

The relativistic wavelength (λ) of an electron accelerated through a potential V_{vac} in vacuum can be derived from the standard relationships:

$$E = eV_{vac} + m_o c^2 \tag{3.4}$$

$$E^2 = (pc)^2 + (m_o c^2)^2 \tag{3.5}$$

$$\lambda = \frac{h}{p} = \frac{h}{\sqrt{2m_o e V_{vac}\left(1 + \frac{eV_{vac}}{2m_o c^2}\right)}} \tag{3.6}$$

where E is the total energy, m, m_o are the relativistic and rest mass of the electron of charge e respectively, p is the momentum, h is the Planck's constant and c is the speed of light. Equation (3.4) states that the energy is the sum of the kinetic energy ($E_{kin} = eV_{vac}$) and rest mass energy, while Eq. (3.5) is the relativistic expression for energy. Equation (3.6) is de Broglie's equation for the wavelength. When the electron passes through a potential field $V(\mathbf{r})$ (e.g. a solid) the change in potential energy $-eV(\mathbf{r})$ is converted to kinetic energy, so that eV_{vac} must be replaced by $e[V_{vac} + V(\mathbf{r})]$. This results in a shortening of the de Broglie wavelength for positive potentials, such as near an atom (Figure 3.3). The phase

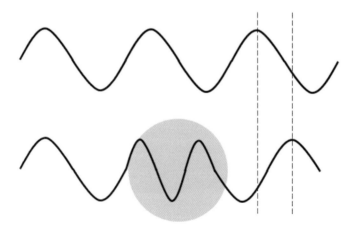

Figure 3.3 Schematic of the shortening of the de Broglie wavelength for an incident electron in a region of high potential, such as a single atom (depicted as a filled circle). The electron undergoes a phase shift compared to free space propagation. This is evident by comparing the relative positions of the first maximum in the electron wavefunction beyond the atom for the two cases (vertical dashed lines).

change $\varphi(\mathbf{r})$ relative to propagation in free space over an infinitesimal distance Δz is

$$\varphi(\mathbf{r}) = 2\pi \left(\frac{1}{\lambda'} - \frac{1}{\lambda} \right) \Delta z$$

$$= \frac{2\pi}{\lambda} \left\{ \sqrt{\left(1 + \frac{V(\mathbf{r})}{V_{\text{vac}}} \right) \cdot \left(1 + \frac{V(\mathbf{r})}{V_{\text{vac}} + (2m_o c^2/e)} \right)} - 1 \right\} \Delta z$$

$$\approx \frac{\pi e}{\lambda E_{\text{kin}}} V(\mathbf{r}) \left[1 + \frac{1}{1 + (2m_o c^2/e V_{\text{vac}})} \right] \Delta z \qquad (3.7)$$

where λ' is the modified electron wavelength in the potential field. In the above derivation it is reasonably assumed that the potential, which in our case is due to atoms in the solid, is small compared to the accelerating potential. Furthermore, if the relativistic correction $eV_{\text{vac}}/(2m_o c^2)$ to the electron wavelength is small the term within the square brackets in Eq. (3.7) can be taken as unity. Since only a phase change is involved in the electron wavefunction, it follows that the specimen transmission function $Q(\mathbf{R})$ must have the form:

$$Q(\mathbf{R}) = \exp[i\varphi(\mathbf{R})] = \exp \left\{ i \frac{\pi e}{\lambda E_{\text{kin}}} \int V(\mathbf{R}, z) dz \right\} = \exp \left[i\sigma V_{\text{p}}(\mathbf{R}) \right]$$

(3.8)

where $\sigma = (\pi e/\lambda E_{\text{kin}})$ is the interaction constant and $V_{\text{p}}(\mathbf{R}) = \int V(\mathbf{R}, z) dz$ is the projected potential ($\mathbf{r}$ is decomposed as $(\mathbf{R}, z)$ with z being along the slice normal). Consider an incident wavefunction $\psi_0(\mathbf{R})$ that undergoes scattering by a thin specimen with transmission function $Q(\mathbf{R})$. The new wavefunction $\psi(\mathbf{R})$ is given by

$$\psi(\mathbf{R}) = \psi_0(\mathbf{R}) Q(\mathbf{R}) = \psi_0(\mathbf{R}) \left[1 + i\sigma V_{\text{p}}(\mathbf{R}) + \cdots \right]$$

$$\approx \psi_0(\mathbf{R}) + i\sigma V_{\text{p}}(\mathbf{R}) \psi_0(\mathbf{R}) \qquad (3.9)$$

Here, the so-called *weak phase object approximation*, which assumes weak scattering by neglecting quadratic and higher order terms in V_{p}, has been applied. The higher the atomic number the thinner the specimen must be for the approximation to be valid. The first and second terms in Eq. (3.9) give the unscattered and scattered wavefunctions respectively; the latter is phase shifted by $\pi/2$ radians owing to the imaginary term i.

In order to calculate the transmission function the potential due to atomic nuclei and electrons in the solid must first be determined. One method is to construct the potential of the entire solid by superimposing the potential due to *free* atoms at their respective coordinates. Within the first Born approximation, which is valid for weak scattering, the atom scattering factor (f) for electrons is given by the Fourier transform of the atomic potential (Appendix A). Inverse Fourier transforming,

$$V(\mathbf{r}) = \frac{h^2}{2\pi m_o e} \int f(q) \exp(2\pi i \mathbf{q} \cdot \mathbf{r}) d\mathbf{q} \qquad (3.10)$$

where $q = 2\sin \theta / \lambda$ is the magnitude of the scattering vector, with scattering semi-angle θ. Atom scattering factors for various elements have been tabulated and parameterised in analytical form by a number of authors (see, e.g. Doyle and Turner, 1968; Peng *et al.*, 1996; Kirkland, 2010). These results are typically valid for single, isolated atoms and do not take into account bonding effects in the solid, such as electron transfer in ionic bonds. Since solid state bonding involves only valence electrons its effect on the final HREM or STEM image should in most cases be negligible. However, in low atomic number 2D materials such as graphene charge redistribution around nitrogen substitutional atoms has been reported in HREM images (Meyer *et al.*, 2011). In such cases, density functional theory is more appropriate for calculating the potential distribution and from that the specimen transmission function.

An alternative method is to express the crystal potential (V_{cryst}) as a Fourier series (Hirsch *et al.*, 1965):

$$V_{cryst}(\mathbf{r}) = \frac{h^2}{2\pi m_o e \Omega} \sum_{h,k,l} F_g(hkl) \exp(2\pi i \mathbf{g} \cdot \mathbf{r}) \qquad (3.11)$$

$$F_g(hkl) = \sum_i f_i(g) \exp(-2\pi i \mathbf{g} \cdot \mathbf{r}_i) \qquad (3.12)$$

where Ω is the unit cell volume. Just as the atomic potential is Fourier related to its atom scattering factor, the crystal potential is Fourier related to the structure factor $F_g(hkl)$, which is defined according to Eq. (3.12) (note that $\mathbf{g} = h\mathbf{a}^* + k\mathbf{b}^* + l\mathbf{c}^*$, where $\mathbf{a}^*$, $\mathbf{b}^*$ and $\mathbf{c}^*$ are the reciprocal lattice basis vectors; the h Miller index is not to be confused with Planck's constant in the pre-summation term of Eq. (3.11)). The summation in Eq. (3.12) is over all atoms i in the unit cell with atomic positions $\mathbf{r}_i$. Equations (3.11) and (3.12) are derived by summing Eq. (3.10) for all atoms in the crystal, taking into account their relative positions (see Appendix B for a derivation). The $\exp(-2\pi i g \cdot r_i)$ term represents the phase shifts due to scattering from atoms constituting the crystal motif (Hammond, 2009). For a perfect crystal only a relatively small number of Fourier components need to be summed to generate an accurate potential (the number of Fourier components increases with larger unit cell dimensions and higher atomic number), although this is not the case for defect crystals, where scattering between Bragg beams is permitted, so that a larger portion of Fourier space must be sampled.

For $r = x\hat{i} + y\hat{j} + z\hat{k}$, where $\hat{i}$, $\hat{j}$ and $\hat{k}$ are unit vectors along the crystal unit cell axes, $\exp(2\pi i g \cdot r) = 1$ if r is a lattice translation vector. Hence only a 'reduced' scheme, where r lies within the unit cell, is required for evaluating the crystal potential, similar to Brillouin zones in electronic structure theory. An alternative, but nevertheless equivalent, expression for V_{cryst} is therefore

$$V_{cryst}(r) = \frac{h^2}{2\pi m_o e\Omega} \sum_{h,k,l} F_g(hkl) \exp\left[2\pi i \left(\frac{hx}{a} + \frac{ky}{b} + \frac{lz}{c} \right) \right] \quad (3.13)$$

where a, b and c are the unit cell dimensions. The projected crystal potential (V_{cryst}^p) for a slice of thickness Δz located at a mean depth z_o within the overall crystal is therefore

$$V_{cryst}^p(R) = \frac{h^2}{2\pi m_o e\Omega} \sum_{h,k,l} F_g(hkl) \exp\left[2\pi i \left(\frac{hx}{a} + \frac{ky}{b} \right) \right]$$

$$\times \int_{z_o - \frac{\Delta z}{2}}^{z_o + \frac{\Delta z}{2}} \exp\left(\frac{2\pi i l z}{c} \right) dz = \frac{h^2 \Delta z}{2\pi m_o e\Omega}$$

$$\times \sum_{h,k,l} F_g(hkl) \left\{ \frac{\sin(\pi l \Delta z/c)}{(\pi l \Delta z/c)} \right\} \exp\left[2\pi i \left(\frac{hx}{a} + \frac{ky}{b} + \frac{lz_o}{c} \right) \right]$$

$$(3.14)$$

Equation (3.14) can be used to calculate the projected potential of any given slice in a multislice simulation. Consider the case where the unit cell basis vectors are defined such that c is parallel to the slice normal and a, b are in the plane of the slice. If the slice thickness Δz is equal to c (i.e. the lattice periodicity along the slice normal) then, due to the $[\sin(\pi l \Delta z/c)]/(\pi l \Delta z/c)$ term, only Fourier components of the form $hk0$ will contribute to the projected potential. To include the hkl Fourier components the slice thickness must be smaller than the lattice periodicity along the c-axis. The *hkl* Fourier components are essential for the correct simulation of HOLZ rings and HOLZ lines within the central disc of a CBED pattern (Kilaas *et al.*, 1987). By using a slice thickness smaller than the lattice repeat distance the incident electrons experience different projected potentials during propagation; if this criterion is not satisfied then the 3D crystal potential is effectively undersampled. Furthermore, if the repeat distance c is not an integer multiple of Δz, artefacts can result owing to beating between the crystal periodicity and slice thickness (Kirkland, 2010).

In the analysis thus far the atoms are assumed to be stationary, although in reality they are thermally vibrating about a mean position. In the Einstein approximation atoms are assumed to vibrate independently. The displacement follows a Gaussian distribution so that the time averaged potential $\langle V(\mathbf{r}) \rangle$ for a single atom at the origin is

$$\langle V(\mathbf{r}) \rangle = V(\mathbf{r}) \otimes \frac{1}{\left(\sqrt{2\pi \overline{x^2}} \right)^3} \exp \left(-\frac{r^2}{2\overline{x^2}} \right) \qquad (3.15)$$

where $\overline{x^2}$ is the mean square atomic displacement. The Gaussian distribution effectively smooths $V(\mathbf{r})$ to give $\langle V(\mathbf{r}) \rangle$. The time-averaged atom scattering factor $\langle f(\mathbf{q}) \rangle$ is given by the Fourier transform of $\langle V(\mathbf{r}) \rangle$, that is,

$$\langle f(\mathbf{q}) \rangle = f(\mathbf{q}) \exp \left(-2\pi^2 \overline{x^2} q^2 \right) = f(\mathbf{q}) \exp \left(-Bq^2 \right) \qquad (3.16)$$

Here, B is known as the Debye–Waller factor. Gao and Peng (1999) report calculated values of the Debye–Waller factor for a number of different elements and compounds. Scattering from a single atom is modelled as an outgoing distorted spherical wave, which has the form $f(\mathbf{q})e^{2\pi i k r}/r$ (Section 2.3.1). The intensity at the scattering vector $\mathbf{q}$ is therefore proportional to the square modulus of the atom scattering factor $f(\mathbf{q})$. From Eq. (3.16) it therefore follows that the effect of atomic vibrations is to reduce the intensity of the elastic scattered wave by a factor $\exp(-2Bq^2)$. The remaining intensity, that is, $|f(\mathbf{q})|^2[1 - \exp(-2Bq^2)]$, is the so-called *thermal diffuse scattered* (TDS) intensity. Figure 3.4 plots the elastic scattered intensity $|f(\mathbf{q})|^2 \exp(-2Bq^2)$ and the TDS intensity, $|f(\mathbf{q})|^2[1 - \exp(-2Bq^2)]$, as a function of the scattering vector magnitude $q = 2\sin \theta/\lambda$ for silicon at 200 kV. The elastic scattering decreases monotonically with q, while the behaviour of TDS is more complicated. Atomic vibrations dampen the central peak of the atomic

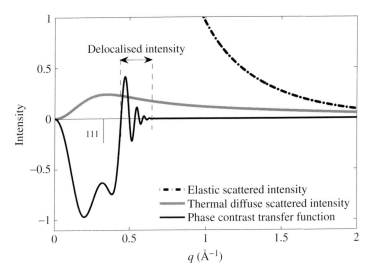

Figure 3.4 Elastic and thermal diffuse scattered intensity (TDS) of 200 kV electrons by a single silicon atom plotted as a function of scattering vector magnitude (q). The elastic scattered intensity increases rapidly at small q and extends beyond the intensity scale of the figure. The phase contrast transfer function for an objective lens of 1 mm spherical aberration coefficient at Scherzer defocus (-61 nm) is also superimposed (beam semi-convergence angle and defocus spread are 0.5 mrad and 6 nm respectively). The TDS intensity at spatial frequencies between the point resolution and information limit of the microscope is delocalised due to oscillations in the transfer function. The spatial frequency of the 111 zero order Laue zone reflection in silicon is also indicated.

potential, but have a weaker effect on the long 'tails' of the potential curve. TDS therefore increases at the expense of elastic scattering as q is increased from zero; this is the origin of the $[1 - \exp(-2Bq^2)]$ term in the TDS intensity. However, the total scattering $|f(\mathbf{q})|^2$ decreases with q (Section 2.1.1), so that at some finite scattering vector the TDS intensity starts decreasing.

For multislice simulations it is important to consider the effect of atomic vibrations on the specimen transmission function. The projected potential $V_p(\mathbf{R})$ in Eq. (3.8) is then the sum of a time-independent projected potential $\overline{V}_p(\mathbf{R})$ (derived from $\langle V(\mathbf{r}) \rangle$ in Eq. (3.15)) and a time-dependent projected potential $V'(\mathbf{R}, t)$. The time-averaged specimen transmission function is therefore

$$\langle Q(\mathbf{R}) \rangle = \langle \exp[i\sigma V_p(\mathbf{R})] \rangle = \exp\left[i\sigma \overline{V}_p(\mathbf{R})\right] \cdot \langle \exp\left[i\sigma V'(\mathbf{R}, t)\right] \rangle \quad (3.17)$$

Expanding $\langle \exp[i\sigma V'(\mathbf{R}, t)]\rangle$ only up to quadratic powers and noting that $\langle V'(\mathbf{R}, t)\rangle = 0$ gives

$$\langle Q(\mathbf{R})\rangle \approx \exp\left[i\sigma \overline{V}_\mathrm{p}(\mathbf{R})\right] \cdot \left(1 - \frac{1}{2}\sigma^2 \left\langle V'(\mathbf{R}, t)^2\right\rangle - \cdots\right)$$

$$\approx \exp\left[i\sigma \overline{V}_\mathrm{p}(\mathbf{R})\right] \cdot \exp\left(-\frac{1}{2}\sigma^2 \left\langle V'(\mathbf{R}, t)^2\right\rangle\right) \qquad (3.18)$$

$$\boxed{\langle Q(\mathbf{R})\rangle = \exp\left\{i\sigma\left[\overline{V}_\mathrm{p}(\mathbf{R}) + i\sigma\mu(\mathbf{R})\right]\right\}} \qquad (3.19)$$

where $\mu(\mathbf{R}) = \frac{1}{2}\langle V'(\mathbf{R}, t)^2\rangle$ is an 'absorption' term, since it introduces an amplitude decay term $\exp[-\sigma^2\mu(\mathbf{R})]$ in the specimen transmission function. A specimen with stationary atoms is a pure phase object, where scattering only changes the phase of the incident wavefunction and not its amplitude. With vibrating atoms, however, the specimen is a phase–amplitude object, that is, both the phase and amplitude of the incident electron wavefunction are modified. The absorption term has the effect of removing electrons from a multislice simulation, that is, the transmitted intensity decreases by a factor of $\exp[-2\sigma^2\mu(\mathbf{R})]$ through each slice. This 'anomaly' is due to the fact that the calculation is restricted to only elastically scattered electrons; in a more detailed calculation, which includes both elastic and inelastic scattering (see Chapter 5), the electrons 'lost' from the elastic wavefunction reappear in the wavefunction for the inelastically scattered electrons, so that the total intensity is conserved. In this case the inelastic wavefunction corresponds to the TDS and the energy transfer is due to phonon creation. From Figure 3.4 the Einstein approximated TDS intensity is weak compared to elastic scattering within the Scherzer passband and hence does not contribute significantly to the lattice image of a crystal (note, however, that if atom vibrations are correlated there can be peaks in the TDS intensity around Bragg reflections). Furthermore, TDS at larger q is spatially delocalised by the objective lens due to rapid oscillations in the contrast transfer function beyond the point resolution of the microscope. The delocalised intensity contributes a more or less uniform background, thus reducing the lattice contrast in an HREM image (van Dyck, 2011). This is likely to be one reason for the poor contrast of experimental HREM images compared to multislice simulations (the contrast is reduced by a factor of $\sim$3, the so-called 'Stobbs' factor). Note that the TDS intensity at large scattering angles can be quite significant, since the curve in Figure 3.4 must be multiplied by $2\pi q(dq)$ to take into account the total scattered area in reciprocal space; this is, however, not the case for elastic scattering in a crystal where Bragg peaks are observed at only discrete points in reciprocal space. As an example Boothroyd and Yeadon (2003) have estimated that 7.5% of the intensity scattered to less than 1.8 $\mathrm{\AA}^{-1}$ is due to TDS in a 25 nm thick silicon sample.

$\overline{V}_p(\mathbf{R})$ in Eq. (3.19) can be calculated for a given slice using an expression similar to Eq. (3.14). However, the Debye–Waller factor must be introduced into the atom scattering factors, and hence $F_g(hkl)$, through Eq. (3.16), to take into account the effects of thermal vibrations. Equation (3.19) has a mathematically similar form to Eq. (3.8); this similarity can be exploited by making the crystal potential a complex quantity, with the imaginary part equal to $\sigma\mu(\mathbf{R})$ in Eq. (3.19). Within this phenomenological model the atom scattering factors are also complex quantities, since they are derived from the atomic potential via a Fourier transform. More details of this so-called *optical potential* model can be found in Chapter 4.

3.1.2 Fresnel Propagator Function

The other important function in a multislice simulation is the free space propagator function $P(\mathbf{R})$, which describes an outgoing Hüygen wavelet. By Babinet's principle (see, e.g. Fowles, 1989) the diffracted intensity from a single atom is identical to that from an opaque screen containing a slit the size of the atom. If the source and observation points are located close to the screen then curvature of the incident and diffracted wavefronts must be taken into account. This is scattering in the near-field or Fresnel diffraction. On the other hand, in far-field diffraction, or Fraunhofer diffraction, the source and observation points are sufficiently far away that the wavefronts are effectively planar. An approximate criterion for Fraunhofer diffraction is (Fowles, 1989)

$$\frac{\delta^2}{2}\left(\frac{1}{d} + \frac{1}{d'}\right) \ll \lambda \tag{3.20}$$

where δ is the slit width, d, d' are the distances from the source and observation points to the opaque screen respectively and λ is the wavelength. For scattering by a single atom in a multislice simulation δ is the size of the atom (~ 1 $\mathrm{\AA}$) and (d, d') are equal to the slice thickness, which is of the order of the crystal inter-planar spacing (i.e. a few Angstroms). The left-hand side of Eq. (3.20) is therefore of the order of Angstroms and is several orders of magnitude larger than the wavelength of high energy electrons (i.e. a few picometers). Curvature of the wavefront is therefore important, so that the propagator function must be derived using Fresnel diffraction.

The 'optical disturbance' due to diffraction by a slit or aperture is given by the Fresnel–Kirchhoff integral formula (see, e.g. Fowles, 1989 or Born and Wolf, 2002). In one particular version of this formula the diffracted wavefunction ψ_{diff} is given by

$$\psi_{diff}(x, y, z) = \frac{-i}{2\lambda} \iint \psi_{inc}(x', y', z') \frac{\exp(2\pi i r/\lambda)}{r} \left[1 + \cos(\mathbf{n}, \mathbf{r})\right] dS_{apt}$$

(3.21)

The physical set up for which Eq. (3.21) applies is illustrated in Figure 3.5. The (fictitious) aperture is circular and integration is carried out over a spherical cap bounded by the aperture (the area element of this spherical cap is denoted by dS_{apt}). The aperture is located in the plane of the sample and is determined by the area of sample illuminated by the source S, which is located at the origin of the spherical cap. It can be reasonably assumed that the distance from the source to the sample is much larger than the area of illumination, so that the curvature

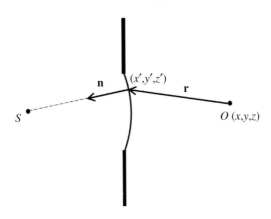

Figure 3.5 Special case of the Fresnel–Kirchhoff integral formula used for deriving the propagator function. A (fictitious) circular aperture is placed at the specimen plane; the source S is symmetrically located about the aperture and at the origin of the spherical cap. $\mathbf{n}$ is the surface normal to the spherical cap at the general position (x', y', z'). Each position along the spherical cap can potentially contribute to the wavefunction beyond the aperture, such as, at the arbitrary point O (x, y, z), which is related to (x', y', z') by the position vector $\mathbf{r}$.

of the spherical cap can be neglected. $\psi_{inc}(x', y', z')$ is the incident wavefunction at the aperture plane or equivalently the specimen. For a thin specimen it is equal to the wavefunction from the source multiplied by the specimen transmission function. The diffracted wavefunction $\psi_{diff}(x, y, z)$ is evaluated at the point O. The $[1 + \cos(\mathbf{n}, \mathbf{r})]$ term in Eq. (3.21) is known as the *obliquity factor*, with $\cos(\mathbf{n}, \mathbf{r})$ being the cosine of the angle between the position vector $\mathbf{r}$ and the surface normal $\mathbf{n}$ to the spherical cap (see Figure 3.5).

High energy electron diffraction is primarily in the forward direction, so that $|x - x'|$ and $|y - y'|$ are small compared to $\Delta z = z - z'$. Therefore, $\cos(\mathbf{n}, \mathbf{r}) \approx 1$ and

$$
\begin{aligned}
r &= \Delta z \sqrt{1 + (x - x')^2/\Delta z^2 + (y - y')^2/\Delta z^2} \\
&\approx \Delta z + \frac{(x - x')^2}{2(\Delta z)} + \frac{(y - y')^2}{2(\Delta z)}
\end{aligned}
\tag{3.22}
$$

Applying these approximations in Eq. (3.21) and neglecting the curvature of the spherical cap gives

$$
\begin{aligned}
\psi_{diff}(x, y, z) &\approx \frac{-i}{\lambda} \frac{\exp(2\pi i \Delta z/\lambda)}{\Delta z} \iint \psi_{inc}(x', y', z') \\
&\quad \times \exp\left\{\frac{i\pi}{\lambda \Delta z}\left[(x - x')^2 + (y - y')^2\right]\right\} dx' \, dy'
\end{aligned}
\tag{3.23}
$$

The above expression is valid when the domain of integration is bounded, such that $|x - x'|$ and $|y - y'|$ are small. If, however, the domain of integration is unbounded, the following result is obtained:

$$
\psi_{diff}(x, y, z) \approx \exp(2\pi i \Delta z/\lambda)\left\{\psi_{inc}(x, y, z') \otimes \frac{-i}{\lambda \Delta z} \exp\left[\frac{i\pi}{\lambda \Delta z}(x^2 + y^2)\right]\right\}
\tag{3.24}
$$

$\exp(2\pi i \Delta z/\lambda)$ is the phase change of a plane wave during free space propagation over a distance Δz and can be ignored as it is a constant. Comparing Eq. (3.24) with Eq. (3.1) gives an expression for the Fresnel free space propagator for a slice of thickness Δz (see Footnote[3]):

$$
\boxed{P(\mathbf{R}) = \frac{-i}{\lambda \Delta z} \exp\left[\frac{i\pi}{\lambda \Delta z}R^2\right]}
\tag{3.25}
$$

---

[3]Equation (3.25) is valid for illumination incident along the slice normal (Figure 3.6a). For a propagator function valid for small beam tilts, see Ishizuka (1982).

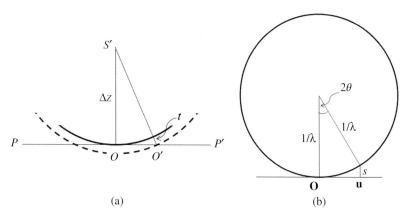

Figure 3.6 (a) Spreading of a Hüygen wavelet (solid arc) from the source S' in free space. The observation plane PP' is at a distance Δz from S'. The phase of the wave at O' is different to that at O due to the additional optical path length t along $S'O'$ compared to $S'O$. (b) Geometry of the deviation parameter (s) for the reciprocal vector $\mathbf{u}$ of scattering angle 2θ. For simplicity the specimen is assumed to be untilted, so that its surface normal, and hence s, is parallel to the incident wave vector (i.e. the line joining the Ewald sphere centre to the origin of the reciprocal lattice O). The equation of the Ewald sphere in the xy-plane is $x^2 + [y - (1/\lambda)]^2 = (1/\lambda)^2$. This gives $s = -\left(1 - \sqrt{1 - \lambda^2 u^2}\right)/\lambda$, which simplifies to $s = -\frac{1}{2}\lambda u^2$ for $\lambda u \ll 1$ (by convention s is negative if $\mathbf{u}$ lies outside the Ewald sphere).

The physical meaning of Eq. (3.25) is illustrated in Figure 3.6a. The propagator function models an outgoing Hüygen spherical wavelet over a distance Δz from the source S'. The Δz denominator in the pre-exponential constant of Eq. (3.25) conserves the total intensity of the expanding spherical wavefront. There is a phase change at the point O' compared to the origin O, owing to the wave travelling an additional distance t. Assuming again small angle scattering

$$t = \sqrt{R^2 + (\Delta z)^2} - \Delta z \approx \frac{R^2}{2\Delta z} \tag{3.26}$$

The relative phase change is therefore $(2\pi t/\lambda)$ or $(\pi R^2/\lambda \Delta z)$, which is the argument within the exponential term of Eq. (3.25). The intensity at O' is slightly smaller compared to the origin O, since the Hüygen wavelet would need to travel beyond Δz to reach O' (dashed arc in Figure 3.6a), but this minor difference is not reproduced in Eq. (3.25) owing to the approximations used in its derivation. Finally, the Fresnel–Kirchhoff formula predicts an additional phase change of $-\pi/2$ radians due to the $-i$

term in the pre-exponential factor (Eq. (3.25)). This is, however, a constant phase and affects all points **R** equally.

The Fourier transform of Eq. (3.25) is

$$\tilde{P}(\mathbf{u}) = \exp\left(-i\pi\lambda u^2 \Delta z\right) = \exp\left(2\pi i s \Delta z\right) \qquad (3.27)$$

where $s = -\frac{1}{2}\lambda u^2$ is the magnitude of the deviation parameter (otherwise known as the excitation error), defined as being negative for reflections lying outside the Ewald sphere (Figure 3.6b). The physical meaning of Eq. (3.27) can be interpreted as follows. From Eq. (3.2) $\tilde{P}(\mathbf{u})$ is a multiplicative function that alters the electron wavefunction in reciprocal space. The wave vector of a diffracted beam is equal to $(\mathbf{k}_{inc} + \mathbf{g} + \mathbf{s})$, where $\mathbf{k}_{inc}$ is the incident wave vector, $\mathbf{g}$ is the reciprocal lattice vector and $\mathbf{s}$ is the deviation parameter. At an arbitrary position $\mathbf{r}$ the phase of the unscattered wave is $(2\pi \mathbf{k}_{inc} \cdot \mathbf{r})$, while the phase of the diffracted wave is $[2\pi(\mathbf{k}_{inc} + \mathbf{g} + \mathbf{s}) \cdot \mathbf{r}]$. The important parameter is the phase difference between the two waves, which is $[2\pi(\mathbf{g} + \mathbf{s}) \cdot \mathbf{r}]$. The effect of using a projected potential, as in a multislice simulation, is that the slice of thickness Δz is effectively two dimensional, and hence the reciprocal vector $\mathbf{g}$ lies in the plane of the slice. The deviation parameter $\mathbf{s}$ is taken to be parallel to the slice normal. Therefore, when the electron wavefunction propagates a distance Δz the phase of the diffracted beam will shift by $\exp(2\pi i s \Delta z)$ relative to the unscattered beam. This is the origin of Eq. (3.27). Note that in diffraction from a thin foil the deviation parameter is parallel to the foil normal, and will therefore not be along the slice normal (i.e. electron beam direction) for a tilted sample. This point is, however, frequently overlooked in multislice simulations.

Although a phase shift of $\exp(2\pi i s \Delta z)$ is exact for free space propagation of a diffracted beam, the mathematical form of the deviation parameter, $s = -\frac{1}{2}\lambda u^2$, in Eq. (3.27) is an approximation. It is known as the *parabolic approximation*, since it (incorrectly) implies that the Ewald sphere is paraboloid in shape. This is a consequence of the small angle scattering assumption used in deriving Eq. (3.25) and hence its Fourier transform, Eq. (3.27). The exact form of the deviation parameter is $s = -\left(1 - \sqrt{1 - \lambda^2 u^2}\right)/\lambda$ (see Figure 3.6b), which is identical to the parabolic approximation at small scattering angles (i.e. $\lambda u \approx \sin 2\theta \ll 1$), but not when the scattering angle is large, such as, for example, in HOLZ reflections. For accurate simulation of HOLZ effects, it is therefore desirable to use the complete expression for the deviation parameter. A rigorous analysis by Chen and van Dyck (1997) has shown that apart from this modification to the *free space* propagator function, there is also a further 'mixed operator' term that models the distortion of the Ewald sphere within the electrostatic potential field of the specimen (recall that the Ewald sphere radius is $1/\lambda$ and that λ is a function of the specimen potential; Section 3.1.1).

3.1.3 Objective Lens Contrast Transfer Function and Partial Coherence

In electron microscopy the image formation process is often imperfect, although recently significant progress has been made in minimising arte-facts. First aberrations in the image-forming lenses produce blurring, that is, a single point in the object is distorted by the lens point spread func-tion $H(\mathbf{R})$. The aberrations are largest for the objective lens, owing to the larger fields of view and trajectory angles of the electrons used in imag-ing, so that the effect of any projector lens distortions can be ignored. Secondly illumination of the specimen may not be perfectly coherent. Here there are two effects: (i) a spread of illumination angles that leads to limited spatial coherence in the plane of the sample and (ii) a spread of electron energies, as well as any instability in the microscope voltage and objective lens current, which gives rise to a range of focal lengths and hence limited temporal coherence in the direction normal to the speci-men plane. These effects are extraneous to the limitations imposed by electron beam–specimen interactions, which is the subject of this book, and therefore will only be summarised here for completeness. The reader is referred to the many excellent books in the literature for more details (e.g. Williams and Carter, 1996; Spence, 2003; Brydson, 2011; Ernie, 2010).

A perfect lens transforms the illumination from a point object into a spherical wavefront that converges to a diffraction-limited focal 'point', that is, the Airy disc. Aberrations in the lens, however, cause deviations W of the wavefront from an ideal spherical surface (see Figure 3.7), such that the resulting phase change $\chi(\mathbf{u})$ is given by $(2\pi W/\lambda)$. The lens aber-ration function χ is a function of the transverse (i.e. perpendicular to the optic axis) wave vector component $\mathbf{u}$ of the incoming 'ray'. For an uncor-rected lens spherical aberration and defocus are the dominant residual aberrations. It can be shown that (Spence, 2003)

$$\chi(\mathbf{u}) = \frac{2\pi}{\lambda}\left[\frac{1}{2}\Delta f \lambda^2 u^2 + \frac{1}{4}C_s \lambda^4 u^4\right] \qquad (3.28)$$

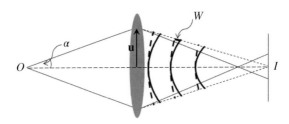

Figure 3.7 Distortion of the electron wavefunction due to lens aberrations, in this case spherical aberration. 'Rays' emitted at an angle α to the optic axis and transverse wave vector component $\mathbf{u}$ by a point O in the specimen converge to the point I for a perfect lens. In the ideal case, the electron wavefunction is a converging spherical wavefront (dashed arcs) and the ray paths are along the dashed lines. With spherical aberration the rays (solid lines) are brought to a premature focus. The image of O is therefore no longer a point, but blurred in the image plane. The wavefront (solid arcs) is distorted by a distance W compared to the ideal case.

where C_s is the spherical aberration coefficient and Δf is the defocus (negative for underfocus). More complete expressions for χ containing, for example, higher order aberrations can be found in Ernie (2010). In HREM the electron wavefunction exiting the specimen, $\psi_{\text{exit}}(\mathbf{R})$, is modified by objective lens aberrations. In real space this is expressed as a convolution with the lens point spread function (i.e. $\psi_{\text{exit}}(\mathbf{R}) \otimes H(\mathbf{R})$), while in reciprocal space the phase of the exit wavefunction at a given spatial frequency $\mathbf{u}$ is shifted by the aberration function (i.e. $\tilde{\psi}_{\text{exit}}(\mathbf{u}) \exp[-i\chi(\mathbf{u})]$). It therefore follows that

$$\tilde{H}(\mathbf{u}) = \exp[-i\chi(\mathbf{u})] \qquad (3.29)$$

Note that the phase change is expressed as $\exp[-i\chi(\mathbf{u})]$, rather than $\exp[i\chi(\mathbf{u})]$, since the *outgoing* wave from the lens is of the form $\exp(-2\pi ikr)$; with this notation there is a phase advance for $W > 0$ as required. If an objective aperture is inserted then $\tilde{H}(\mathbf{u})$ must be multiplied by an aperture function $A(\mathbf{u})$, which has a value of unity for spatial frequencies within the aperture and is zero outside the aperture. The HREM image intensity $I(\mathbf{R})$ is given by

$$I(\mathbf{R}) = \left| \psi_{\text{exit}}(\mathbf{R}) \otimes H(\mathbf{R}) \right|^2 \approx \left| \left(1 + i\sigma \overline{V}_p(\mathbf{R}) - \sigma^2 \mu(\mathbf{R}) \right) \otimes H(\mathbf{R}) \right|^2 \quad (3.30)$$

Here, the exit wavefunction has been replaced by the time-averaged specimen transmission function of Eq. (3.19) under the weak phase, weak amplitude approximation (i.e. only terms upto first order are retained in the exponential). Since the specimen is infinitesimally thin, it is assumed that the Fresnel propagator function does not have an effect on $\psi_{\text{exit}}(\mathbf{R})$. The above equation simplifies as

$$I(\mathbf{R}) \approx \left| 1 + (i\sigma \overline{V}_{\text{p}}(\mathbf{R}) \otimes H(\mathbf{R})) - (\sigma^2 \mu(\mathbf{R}) \otimes H(\mathbf{R})) \right|^2$$

$$= \left[1 + (i\sigma \overline{V}_{\text{p}}(\mathbf{R}) \otimes H(\mathbf{R})) - (\sigma^2 \mu(\mathbf{R}) \otimes H(\mathbf{R})) \right]$$

$$\cdot \left[1 - (i\sigma \overline{V}_{\text{p}}(\mathbf{R}) \otimes H^*(\mathbf{R})) - (\sigma^2 \mu(\mathbf{R}) \otimes H^*(\mathbf{R})) \right]$$

$$\approx 1 + i\sigma \overline{V}_{\text{p}}(\mathbf{R}) \otimes \left[H(\mathbf{R}) - H^*(\mathbf{R}) \right] - \sigma^2 \mu(\mathbf{R}) \otimes \left[H(\mathbf{R}) + H^*(\mathbf{R}) \right]$$

$$(3.31)$$

where the asterisk denotes the complex conjugate (note that the projected potential $\overline{V}_{\text{p}}$ and absorption μ are both real numbers). Taking Fourier transforms

$$\tilde{I}(\mathbf{u}) \approx \delta(\mathbf{u}) + i\sigma \widetilde{\overline{V}}_{\text{p}}(\mathbf{u}) \cdot [\tilde{H}(\mathbf{u}) - \tilde{H}(-\mathbf{u})] - \sigma^2 \tilde{\mu}(\mathbf{u}) \cdot [\tilde{H}(\mathbf{u}) + \tilde{H}(-\mathbf{u})]$$

$$= \delta(\mathbf{u}) + 2\sigma \widetilde{\overline{V}}_{\text{p}}(\mathbf{u}) \sin \chi(\mathbf{u}) - 2\sigma^2 \tilde{\mu}(\mathbf{u}) \cos \chi(\mathbf{u})$$

$$(3.32)$$

Here, $\sin \chi$ (or $2\sin \chi$ in some texts) is the phase contrast transfer function for a weak phase object and represents the role of the objective lens in transferring a given Fourier component of the specimen projected potential onto the HREM image. Ideally, $\sin \chi$ should be equal to ± 1 for all non-zero spatial frequencies. For an uncorrected lens only the $\sin \chi = -1$ criterion can be approximately satisfied, where an underfocus is used to compensate spherical aberration (i.e. the Scherzer imaging condition). In a C_{s}-'corrected' lens optimum imaging is for $\sin \chi \approx +1$, where an overfocus and small, negative C_{s} is used to compensate for the uncorrected fifth order spherical aberration (Ernie, 2010; note that a long hexapole corrector provides some flexibility in tuning the C_{s} value). Finally, $\cos \chi$ in Eq. (3.32) is the amplitude contrast transfer function for a weak amplitude object and represents the role of the objective lens in imaging the absorptive potential of the sample.

In the discussion thus far, the illumination has been assumed to be perfectly coherent (i.e. fixed illumination angle and focal point of the incident electrons). In practice, there will be a distribution of illumination wave vectors $p(\mathbf{k}_t)$ and defoci $p(\delta f)$, such that in normalised form

$$\int p\left(\mathbf{k}_t\right) d\mathbf{k}_t = 1; \quad \int p\left(\delta f\right) d(\delta f) = 1 \qquad (3.33)$$

where $\mathbf{k}_t$ is the transverse component of the incident wave vector and δf is the deviation from a given defocus Δf. If a weak phase, weak amplitude object is again assumed, then the imaging wavefunction for tilted illumination is given by $[\langle Q(\mathbf{R})\rangle \exp(2\pi i\mathbf{k}_t \cdot \mathbf{R})] \otimes H(\mathbf{R})$. The Fourier transform is $\langle \tilde{Q}(\mathbf{u} - \mathbf{k}_t)\rangle \tilde{H}(\mathbf{u})$ or equivalently $\langle \tilde{Q}(\mathbf{u})\rangle \tilde{H}(\mathbf{u} + \mathbf{k}_t)$, that is, the effect of an inclined beam can be taken into account purely by the point spread function provided the specimen is thin. The imaging wavefunction in real space is therefore equivalent to $\langle Q(\mathbf{R})\rangle \otimes [H(\mathbf{R}) \exp(-2\pi i\mathbf{k}_t \cdot \mathbf{R})]$. In an electron microscope the probe current and accelerating voltage are such that on average the incident electrons reach the specimen one at a time, rather than as multiple electrons.[4] Within the exposure time of the detector the signals from the individual electrons of varying illumination angle and defoci will therefore add incoherently to give the final HREM image, that is, there is no interference between individual signals. This can be expressed mathematically as

$$I(\mathbf{R}) = \iint \left|\psi_{\text{exit}}(\mathbf{R}) \otimes H(\mathbf{R}, \mathbf{k}_t, \delta f)\right|^2 p\left(\mathbf{k}_t\right) p\left(\delta f\right) d\mathbf{k}_t d\left(\delta f\right) \qquad (3.34)$$

where $H(\mathbf{R}, \mathbf{k}_t, \delta f)$ is a modified point spread function that takes into account the beam tilt and defocus change. This is easiest to evaluate in reciprocal space where the spatial frequency $\mathbf{u}$ in the lens aberration function (Eq. (3.28)) is replaced by $(\mathbf{u} + \mathbf{k}_t)$ and defocus Δf is replaced by $(\Delta f + \delta f)$. $\psi_{\text{exit}}(\mathbf{R})$ in Eq. (3.34) is evaluated assuming normal incidence and perfect focusing. One reason for a change in defocus δf is a fluctuation of the incident electron energy, owing to an intrinsic energy spread that is characteristic of the electron emitter. The spread in incident electron energies is however small compared to the primary energy and therefore does not have a significant effect on $\psi_{\text{exit}}(\mathbf{R})$. Assuming a weak phase, weak amplitude object and simplifying along the lines outlined in

---

[4] Assume a 200 kV electron beam with 100 pA current (typical operating conditions in a TEM). The average time between successive electrons is approximately 0.6 ns. In this time the two electrons will be separated by approximately 10 cm, that is, far enough to assume the electrons reach the specimen one at a time.

Eqs. (3.30)–(3.32) gives

$$\tilde{I}(\mathbf{u}) = \delta(\mathbf{u}) + 2\sigma \tilde{\overline{V}}_p(\mathbf{u}) \iint \sin \chi(\mathbf{u}, \mathbf{k}_t, \delta f) p(\mathbf{k}_t) p(\delta f) d\mathbf{k}_t d(\delta f)$$

$$-2\sigma^2 \tilde{\mu}(\mathbf{u}) \iint \cos \chi(\mathbf{u}, \mathbf{k}_t, \delta f) p(\mathbf{k}_t) p(\delta f) d\mathbf{k}_t d(\delta f) \qquad (3.35)$$

where $\chi(\mathbf{u}, \mathbf{k}_t, \delta f)$ is the lens aberration function modified for beam tilt and defocus change. The above equation has been evaluated assuming Gaussian distributions for $p(\mathbf{k}_t)$ and $p(\delta f)$. The result is (Kirkland, 2010)

$$\tilde{I}(\mathbf{u}) = \delta(\mathbf{u}) + 2\sigma \tilde{\overline{V}}_p(\mathbf{u}) \sin \chi(\mathbf{u}) E_s(u) E_t(u) - 2\sigma^2 \tilde{\mu}(\mathbf{u}) \cos \chi(\mathbf{u}) E_s(u) E_t(u)$$
$$(3.36)$$

where $E_s(u)$ and $E_t(u)$ are envelope (or damping) functions due to the spatial and temporal coherence respectively (Kirkland, 2010; de Jong and van Dyck, 1993):

$$E_s(u) = \exp\left[-\left(\frac{\pi\alpha}{\lambda}\frac{\partial\chi(u)}{\partial u}\right)^2\right] = \exp\left[-\left(\frac{\pi\alpha}{\lambda}\right)^2 (C_s\lambda^3 u^3 + \lambda\Delta f u)^2\right]$$

$$(3.37)$$

$$E_t(u) = \exp\left[-\frac{1}{2}(\pi\Delta_0\lambda u^2)^2\right]$$

$$\Delta_0 = C_c\sqrt{4\left(\frac{\Delta I}{I}\right)^2 + \left(\frac{\Delta E}{E}\right)^2 + \left(\frac{\Delta V}{V}\right)^2}$$

$$(3.38)$$

where α is the semi-convergence angle of illumination, Δ_0 is the defocus spread, C_c is the chromatic aberration coefficient, $(\Delta I/I)$ and $(\Delta V/V)$ are fractional instabilities in the objective lens current and accelerating voltage respectively and $(\Delta E/E)$ is the fractional energy spread at the electron emitter (since Eq. (3.36) applies to thin specimens inelastic scattering does not significantly affect ΔE). Note that damping by $E_s(u)$ at a given spatial frequency can be eliminated by adjusting the defocus to give zero aberration function gradient (i.e. $\partial\chi/\partial u = 0$) at the spatial frequency of interest. The information limit of the microscope is therefore determined by the temporal coherence envelope $E_t(u)$. The Fourier transform of the point spread function for partially coherent illumination ($\tilde{H}(\mathbf{u})^{\text{partial}}$) can therefore be expressed as

$$\tilde{H}(\mathbf{u})^{\text{partial}} = E_s(u)E_t(u)\exp\left[-i\chi(\mathbf{u})\right] \qquad (3.39)$$

From Eq. (3.36) it can be seen that the real and imaginary parts of $\tilde{H}(\mathbf{u})^{\text{partial}}$ represent the phase and amplitude contrast transfer functions for partially coherent illumination. Apart from spatial and temporal coherence, other factors, such as mechanical vibrations, drift, imaging detector response, etc., can also introduce damping envelopes; for more details, see de Jong and van Dyck (1993).

3.1.4 Implementation of the Multislice Algorithm

When running a multislice program due consideration must be given to potential artefacts, which will now be discussed. A computationally more efficient form of Eq. (3.1) is

$$\psi_{n+1}(\mathbf{R}) = \text{FT}^{-1}\left\{\text{FT}\left[\psi_n(\mathbf{R}) \cdot Q_{n+1}(\mathbf{R})\right] \cdot \tilde{P}_{n+1}(\mathbf{u})\right\} \qquad (3.40)$$

where FT and FT^{-1} represent Fourier and inverse Fourier transform operations respectively. In this form a convolution is reduced to the much simpler multiplication operation via Fourier transforms, which can be efficiently calculated using the fast Fourier transform (FFT) algorithm of Cooley and Tukey (1965). Since periodic repetition is involved in Fourier transforms the sample supercell must satisfy periodic boundary conditions in the plane of the slice. In the case of defects this requirement can be satisfied by 'padding' the defect with a perfect crystal region, or more realistically a long range elastic stress field region where the atomic displacements gradually approach zero. In this way, there is negligible overlap of stress fields between neighbouring supercells during periodic continuation. In Eq. (3.40) the size and sampling in real and reciprocal space must be identical for the specimen (strictly speaking specimen transmission function Q), electron wavefunction ψ and propagator function P respectively.

When constructing the projected potential for the specimen transmission function, either by Eq. (3.14) or more generally by superimposing projected potentials for individual atoms (Eq. (3.10)), the sampling has to be done at discrete points within the supercell. The maximum spatial frequency u_{max} is therefore limited to the Nyquist frequency

$(N/2a)$, where N is the number of sampling points along the dimension of the supercell (a). Multiplication of the specimen transmission function $Q_{n+1}(\mathbf{R})$ with the electron wavefunction $\psi_n(\mathbf{R})$ in Eq. (3.40) is a convolution in reciprocal space, so that spatial frequencies upto $2u_{max}$ will contribute to $Q_{n+1}(\mathbf{R})\psi_n(\mathbf{R})$. To avoid aliasing artefacts due to periodic continuation the specimen transmission function must therefore be bandwidth limited to $(2/3)u_{max}$ (Kirkland, 2010; Self and O'Keefe, 1988). Since from Eq. (3.8) $|Q(\mathbf{R})|^2 = 1$ for the case of static atoms (i.e. no absorption), a similar relationship must be obtained following bandwidth limiting. This is the so-called unitarity test and is used to examine the effect of bandwidth limiting on the accuracy of the specimen transmission function. The test is frequently carried out in reciprocal space where the Fourier transform of $|Q(\mathbf{R})|^2$ is compared to the Dirac delta function (i.e. the Fourier transform of unity). A deviation of no more than 1 part in 10^5 is sufficiently accurate for most multislice simulations (Self and O'Keefe, 1988). If required the accuracy can be improved by finer sampling of the supercell in real space (i.e. larger values of N).

The propagator function $\tilde{P}_{n+1}(\mathbf{u})$ in Eq. (3.40) is in reciprocal space, so that reciprocal space sampling is also important. Since the pixel size in reciprocal space is $(1/a)$ the supercell must be sufficiently large for good resolution. Equation (3.27) predicts that for reflections close to the Bragg orientation (i.e. $s \approx 0$) the propagator function is approximately unity. The Bragg scattering from a given slice will therefore be nearly in phase with that from the preceding slice and the diffracted beam intensity increases. This partly explains the typically large intensities observed for zero order Laue zone (ZOLZ) reflections in a diffraction pattern (note that the intensities cannot increase indefinitely since interference with the unscattered beam and other diffracted beams is also important). Alternatively, if the deviation parameter is large enough such that $s\Delta z$ is an integer, then those reflections could potentially have anomalously high intensities, owing to the propagator function being equal to unity. In such cases the number of reflections must be truncated (by setting the transmission function to zero beyond a certain spatial frequency) and/or the slice thickness decreased for accurate results.

Apart from the unitarity test for the specimen transmission function and maximum deviation parameter for the propagator function, the accuracy of the overall multislice simulation can be monitored by calculating the integrated intensity $\int |\psi(\mathbf{R})|^2 d\mathbf{R}$ as the electrons propagate through the crystal. This must be done in the absence of absorption. High angle scattering beyond the bandwidth limited spatial frequency leads to a gradual loss of intensity. For most purposes, a (normalised) integrated intensity of 0.9 or above at the specimen exit surface is sufficient. In Section 3.3.1, it will be shown that the multislice equation (3.1) is accurate to within first order of the slice thickness, so that the use of thin slices is desirable. Frequently, the effects of the objective lens and partial coherence of illumination are taken into account by multiplying $\psi_{\text{exit}}(\mathbf{u})$ by $\tilde{H}(\mathbf{u})^{\text{partial}}$ (Eq. (3.39)), even though this approach is only valid for thin specimens. For thicker specimens the method of transmission cross coefficients can be used (Kirkland, 2010), although it is computationally more expensive. The HREM image intensity is calculated by inverse Fourier transforming $\psi_{\text{exit}}(\mathbf{u})\tilde{H}(\mathbf{u})^{\text{partial}}$ and taking the square modulus. Similarly, Fourier transforming the exit wavefunction $\psi_{\text{exit}}(\mathbf{R})$ and taking the square modulus gives the selected area electron diffraction (SAED) pattern (since only intensities are measured in a diffraction pattern the phase shifts of the diffracted beams by lens aberrations can be ignored).

3.2 APPLICATIONS OF MULTISLICE SIMULATIONS

3.2.1 HREM Imaging and Electron Crystallography

HREM imaging is a powerful tool for solving (or confirming) the crystal structure of a given material. There are, however, artefacts from dynamic diffraction and lens aberrations that need to be considered during interpretation. The electron wavefunction $\psi_{\text{HREM}}(\mathbf{R})$ in HREM is due to interference of the unscattered and diffracted beams and can be expressed as

$$\psi_{\text{HREM}}(\mathbf{R}) = \sum_g \Psi_g(t) \exp\left[2\pi i(\mathbf{k}_t + \mathbf{g} + \mathbf{s}) \cdot \mathbf{R}\right] \exp\left[-i\chi(\mathbf{u}_g)\right] \quad (3.41)$$

$\Psi_g(t)\exp[2\pi i(\mathbf{k}_t + \mathbf{g} + \mathbf{s})\cdot\mathbf{R}]$ is the diffracted beam wavefunction at the exit surface of a specimen of thickness t, and $-\chi(\mathbf{u}_g)$ is the lens aberration phase shift for the $\mathbf{g}$ diffracted beam with transverse wave vector component $\mathbf{u}_g$. Apart from the two-beam case (Chapter 4) there is no simple expression for $\Psi_g(t)$ under dynamic diffraction conditions, but much insight can be gained by deriving $\Psi_g(t)$ assuming weak (i.e. kinematic) scattering. The derivation is illustrated in Figure 3.8. Since high energy electron diffraction is primarily in the forward direction the diffracted beam **g** at the exit surface point O is largely due to scattering events taking place within a narrow column centred about the diffracted beam. This is known as the column approximation. The diffracted wave for a weak phase object of thickness dz is obtained by Fourier transforming Eq. (3.9). For an incident plane wave $\psi_0(\mathbf{R}) = 1$, the **g** diffracted wave is $i\sigma V_g\, dz$, where V_g is the **g** Fourier component of the crystal potential (from Eq. (3.11) equal to $h^2 F_g / 2\pi m_o e\Omega$). Consider scattering from the point P within a thin slice located at depth z (Figure 3.8). The contribution to $\Psi_g(t)$ at the exit surface point O of the column is $i\sigma V_g\, dz\{\exp(2\pi i k_{inc}\cdot z)\exp[2\pi i(k_{inc}+\mathbf{g}+\mathbf{s})\cdot\mathbf{r}_0]\}$. The first phase factor $\exp(2\pi i k_{inc}\cdot z)$ takes into account the phase shift of the plane wave incident on the thin slice with respect to the beam entrance surface. The second phase factor is due to propagation of the diffracted beam of wave vector $(k_{inc}+\mathbf{g}+\mathbf{s})$ along $\mathbf{r}_0$ before reaching the point O (Figure 3.8). Since $\mathbf{r}_0$ is nearly parallel to the incident beam direction the above expression simplifies to $i\sigma V_g\, dz\exp(-2\pi i s z)\exp[2\pi i(k_{inc}+s)t]$, where normal incidence and an untilted specimen have been assumed, so that $\mathbf{g}\cdot\mathbf{r}_0 \approx 0$ and $\mathbf{s}\cdot\mathbf{r}_0 \approx s(t-z)$. The $\exp[2\pi i(k_{inc}+s)t]$ term represents a constant phase and can therefore be omitted. In the kinematic theory, scattering of a

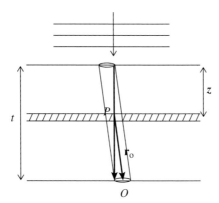

Figure 3.8 Illustration of the column approximation used for calculating the kinematic electron diffraction intensity for a specimen of thickness t. Owing to small scattering angles the diffracted intensity from a high energy incident plane wave is confined to a narrow column centred about the diffracted beam observed at the exit surface point O. Scattering at different depths along the column, such as at the point P, will contribute to the net diffracted beam intensity observed at O.

diffracted beam, once generated, is assumed to be negligible. $\Psi_g(t)$ is therefore due to interference of the diffracted beams originating from all depths within the specimen, that is,

$$\Psi_g(t) = i\sigma V_g \int_0^t e^{-2\pi i s z}\, dz = i\sigma t V_g \left(\frac{\sin \pi s t}{\pi s t}\right) e^{-i\pi s t} = \phi_g \exp\left[i\left(\frac{\pi}{2} - \pi s t\right)\right]$$

$$\phi_g = \frac{\pi t}{\xi_g}\left(\frac{\sin \pi s t}{\pi s t}\right) \tag{3.42}$$

where ξ_g is known as the extinction distance ($=\pi/\sigma V_g$; see also Section 4.1.2). Consider a three-beam HREM image, consisting of the unscattered beam and two $\pm g$ diffracted beams from crystal planes that are parallel to the incident electron beam. For normal incidence and an untilted specimen the one-dimensional HREM image intensity profile, $I_{\mathrm{HREM}}(x) = |\psi_{\mathrm{HREM}}(x)|^2$, along a direction normal to the scattering planes is determined by substituting Eq. (3.42) in Eq. (3.41):

$$I_{\mathrm{HREM}}(x) = \left[\phi_o^2 + 2\phi_g^2\right] + 4\phi_o\phi_g \cos(2\pi gx)\sin(\chi(\mathbf{g}) + \pi s t)$$
$$+ 2\phi_g^2 \cos(4\pi gx) \tag{3.43}$$

where ϕ_o, ϕ_g are the amplitudes of the unscattered and diffracted beams respectively (note that $\phi_g = \phi_{-g}$ for end-on planes in a centrosymmetric crystal). The first bracketed term represents a constant background intensity; selecting either ϕ_o^2 or ϕ_g^2 by inserting a small objective aperture in the back focal plane gives the standard bright-field and dark-field images used in diffraction contrast analysis. The second term is the sum of two interference terms between the unscattered beam and $\pm g$ diffracted beams. Its intensity modulates as a cosine wave with periodicity equal to the inter-planar spacing; this is the origin of lattice fringes in an HREM image. Lattice fringe contrast is greatest when $\sin(\chi(\mathbf{g}) + \pi s t) = \pm 1$, or equivalently when $\chi(\mathbf{g}) + \pi s t = (2n+1)\pi/2$, where n is an integer. Note that this is slightly different from the Scherzer condition, $\sin(\chi(\mathbf{g})) = -1$, due to the additional $-\pi s t$ phase shift of the diffracted beam for a specimen of thickness t (Eq. (3.42)). From the lens aberration function $\chi(\mathbf{g})$ (Eq. (3.28)), the lattice fringe profile for two defoci Δf_1 and Δf_2 are identical provided that $|\Delta f_1 - \Delta f_2| = 2n/\lambda g^2$. These are known as Fourier images.[5] However, if $|\Delta f_1 - \Delta f_2|$ is half this

---

[5] Experimentally, there will only be a limited focus range over which Fourier images may be observed (Spence, 2003). This is because of damping by the spatial coherence envelope for those defocus values where $\partial \chi/\partial u$ is not zero (Eq. (3.37)). The larger the beam semi-convergence angle α the narrower the focus range for observable Fourier images.

value the lattice fringes shift by half the inter-planar spacing. Specimen thickness (t) and orientation (s) also affect the HREM image via the amplitude ϕ_g and phase $-\pi st$ (Eq. (3.42)). Finally the third, so-called *non-linear*, term in Eq. (3.43) is due to interference between the two $\pm g$ diffracted beams. The periodicity of these fringes is half the inter-planar spacing (hence also called half-period fringes). Since scattering of the $+g$ beam along $-g$ (and vice versa) results in a beam along the optic axis (Figure 3.2) the non-linear term in this case is not affected by lens aberrations.

The above effects are readily observed in multislice simulated HREM images of [110]-Si. The structure along [110] consists of an 'AB' periodic stacking sequence of atomic layers. The projected potential, shown in Figure 3.9a, reveals 'dumbbell' atom columns separated by only 1.4 Å (a single dumbbell is circled for reference). A multislice simulated diffraction pattern for a 50 nm thick specimen is also shown in the inset. The 002 reflection is kinematically forbidden, but arises through double diffraction of the $\bar{1}11$ and $1\bar{1}1$ beams. To simplify the image analysis an objective aperture is placed such that only the 111-type beams can contribute to the HREM image; see Figure 3.4 for a superimposition of the 111 spatial frequency within the phase contrast transfer function. Figure 3.9b is the simulated image at 200 kV and Scherzer defocus for a specimen only 3.8 Å thick (i.e. a single 'AB' stacking sequence). The specimen can therefore reasonably be assumed to be a weak phase object. Comparing with Figure 3.9a it is clear that the dumbbell atom columns have not been resolved owing to the limited spatial resolution (2.9 Å) imposed by the objective aperture. Furthermore, the atom columns appear dark against a white background. From Eq. (3.32) the image Fourier transform for a weak phase object is $\delta(\mathbf{u}) + 2\sigma \overline{V}_p(\mathbf{u}) \sin \chi(\mathbf{u})$. For the ideal lens, $\sin \chi(\mathbf{u}) = -1$ for $u \neq 0$ and $\sin \chi(0) = 0$, that is, the diffracted beams are phase shifted by $\pi/2$ radians compared to the unscattered beam (Eq. (3.29)). Inverse Fourier transforming $\delta(\mathbf{u}) - 2\sigma \tilde{V}_p(\mathbf{u})$ gives the ideal image $[1 - 2\sigma \overline{V}_p(\mathbf{R})]$ for a weak phase object, that is, positive potential atom columns show dark contrast, consistent with Figure 3.9b. In other words, the $\pi/2$ radian phase shift due to scattering (Eq. (3.9)) together with further $\pi/2$ radians from lens aberrations means that the diffracted beams interfere destructively with the unscattered beam to give reduced image intensity at the atom column positions.

The defocus change $(2/\lambda g^2)$ for Fourier images of the 111 lattice fringes is 78 nm. Increasing the underfocus by half this value from

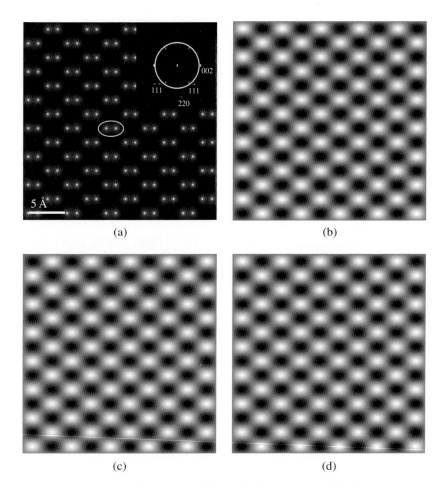

(a)

(b)

(c)

(d)

Figure 3.9 (a) Projected potential of [110]-Si. A pair of dumbbell atom columns is circled for reference. The inset shows the multislice simulated diffraction pattern for a 50 nm thick specimen; the circle is the outline of the objective aperture used in HREM image simulations. (b) Simulated HREM image for a 3.8 Å thick specimen (i.e. a single AB stacking sequence) at −61 nm Scherzer defocus. (c) and (d) are HREM images for the same specimen but at −100 and −140 nm defocus respectively. (e) Variation of $\bar{1}11$ and 000 beam intensities as a function of depth. In (f) the variation in phase for the 000, $\bar{1}11$ and $\bar{1}1\bar{1}$ beams are plotted with respect to depth. The Scherzer defocus HREM images for 12.6, 5 and 20 nm thick specimens are shown in (g), (h) and (i) respectively. Traces of the $(\bar{1}11)$ and $(\bar{1}1\bar{1})$ planes have been superimposed (solid lines) on the left-hand corner of each image. For all simulations the following parameters were assumed: 200 kV accelerating voltage, 1 mm spherical aberration coefficient, 0.5 mrad beam semi-convergence angle, 6 nm defocus spread and 8.5 mrad objective aperture radius. Absorption is not included. The supercell structure and scale bar in (a) are common to all HREM images. Kirkland's (2010) multislice code was used in the simulations.

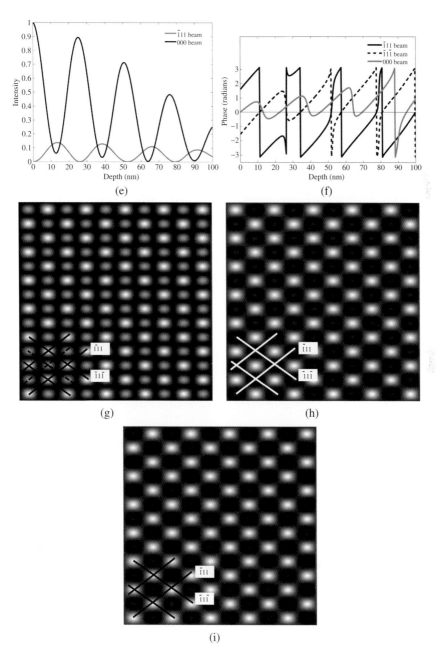

Figure 3.9 (*Continued*)

−61 nm (Scherzer defocus) to −100 nm should therefore result in a contrast reversal. This is indeed apparent in Figure 3.9c where atom columns appear bright on a dark background (the specimen is identical to that in Figure 3.9b). Further underfocusing to −140 nm gives Figure 3.9d, which is the Fourier image of Figure 3.9b. To examine the role of specimen thickness the multislice simulated intensities of the 000 and $\bar{1}11$ beams are plotted as a function of depth in Figure 3.9e (the intensity of $\bar{1}1\bar{1}$ is identical to $\bar{1}11$). The intensities show an oscillatory behaviour. Note that the 000 beam intensity *in general* decreases with depth. This is because of scattering to other higher order reflections and to a lesser extent loss of intensity in the simulation due to large angle scattering (Section 3.1.4; note that absorption was not included and the integrated intensity was >0.95 at 100 nm depth). At 12.6 nm depth, the 000 beam intensity is at a minimum and is less than the $\bar{1}11$ beam intensity. From Eq. (3.43) these are ideal conditions for the observation of half-period fringes due to the non-linear term. The simulated image at Scherzer defocus is shown in Figure 3.9g. The trace (solid lines) of $(\bar{1}11)$ and $(\bar{1}1\bar{1})$ lattice planes derived from the supercell (Figure 3.9a) are superimposed, from which it can be seen that the lattice fringe periodicity is half the inter-planar spacing (dashed lines) as expected. The overall symmetry of the image is clearly different to the projected structure and highlights some of the pitfalls in structure determination using HREM.

Apart from intensity, the phase of the diffracted beams is also important in HREM. In Figure 3.9f the multislice simulated phases of 000, $\bar{1}11$ and $\bar{1}1\bar{1}$ beams are shown as a function of depth. The phase is plotted within the range $[-\pi, \pi]$ and the additional phase change due to lens aberrations is not included. From Eq. (3.42) the phase of a diffracted beam must be $\pi/2$ radians at zero depth, which is consistent with the $\bar{1}11$ beam. The $\bar{1}1\bar{1}$ beam, however, has an additional phase shift of π radians due to a change in sign of V_g, or equivalently F_g, in Eq. (3.42). This is due to the origin of the supercell being at the top left-hand corner of Figure 3.9a rather than at the crystal centre of inversion (i.e. the mid-point of a dumbbell). The kinematic theory also predicts the phase of a diffracted beam to vary linearly with depth (Eq. (3.42)), which is the case in Figure 3.9f, although beyond ∼24 nm the linearity breaks down due to dynamic diffraction. Following a further phase shift of the diffracted beam due to lens aberrations destructive- or constructive-type interference with the 000 unscattered beam is possible depending on the relative phases of the two beams.[6] This leads to dark and bright atom column contrast in the HREM image respectively. Consider, for example, specimen thicknesses of 5 and 20 nm, which are equally spaced on either side of the first minimum in 000 beam intensity at 12.6 nm and therefore have roughly equal intensities for the diffracted beam as well as appreciable intensity in the unscattered beam (Figure 3.9e). At 5 nm the $\bar{1}11$ beam has phase advanced by 0.58π radians compared to the 000 beam (Figure 3.9f), so that with an additional 0.75π radian phase shift due to lens distortions at Scherzer defocus destructive-type interference takes place. On the other hand, at 20 nm the $\bar{1}11$ beam phase lags the 000 beam by 0.56π radians and with lens distortions constructive-type interference takes place. Similar arguments can be applied to the $11\bar{1}$ beam once the extraneous phase shift of π radians has been taken into account. The contrast reversal is confirmed from the simulated HREM images for 5 and 20 nm thick specimens at Scherzer defocus (Figure 3.9h and i respectively) by observing the intensity along the traces of $(\bar{1}11)$ and $(\bar{1}1\bar{1})$ lattice planes. Note that half-period fringes due to the non-linear term are also weakly present in both images and as predicted by Eq. (3.43) they do not undergo a contrast reversal.

Given that lens aberrations distort the image from the true projected crystal structure it is desirable to remove or at least minimise its effect. Lens aberrations can be removed by one of two ways: either through the use of multipole lenses to directly correct aberrations or by using exit wave reconstruction algorithms to process the data post-acquisition. The latter involves acquiring a focal series of images; varying the defocus shifts the passband of the phase contrast transfer function over the spatial frequency axis, thereby mitigating damping by the spatial coherence envelope and zeros in the phase contrast transfer function. Iterative algorithms have been developed to reconstruct the specimen exit wavefunction from the image focus series (see, e.g. Coene et al., 1996; Op de Beeck et al., 1996). In particular exit wave reconstruction is useful for crystallography studies of intermetallic superconductors, such as LuNiBC, which contain both heavy and light atoms (Zandbergen et al., 1996). The heavy atoms are visible in the amplitude image of the exit wavefunction due to stronger absorption, while under the right conditions light atoms are visible in the phase image (the phase can only vary over 2π radians, so that a light atom is potentially visible when a heavy atom is at the start of

---

[6] If the phase difference is between $-\pi/2$ and $\pi/2$ radians constructive-type interference can take place; otherwise, it is destructive.

a new phase cycle). Spatial frequencies up to the microscope information limit are potentially recovered through focal series reconstruction, but this limit can be surpassed by combining a focus series with a beam tilt series, the beam tilt having the effect of shifting the phase contrast transfer function along the tilt direction in reciprocal space (Section 3.1.3; Kirkland *et al.*, 1995). Application of this technique to [112]-Si is shown in Figure 3.10a (Haigh *et al.*, 2009); the two dumbbell atom columns separated by only 78 pm are clearly resolved in the phase image, despite the information limit for axial illumination being larger than an Angstrom. The advent of aberration corrected hardware has also produced similarly impressive results. An example is shown in Figure 3.10b of oxygen atom imaging in a 4 nm thick $SrTiO_3$ sample (Jia *et al.*, 2003). The image was acquired with negative C_s and small overfocus; under these conditions the diffracted beams are in-phase with the unscattered beam and therefore interfere constructively to produce white atom contrast. From Figure 3.10b it is also evident that the intensity of some of the oxygen atom columns differs noticeably from the average value due to variations in oxygen stoichiometry. The oxygen stoichiometry can be estimated by

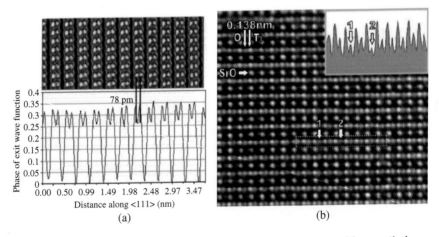

Figure 3.10 (a) Phase image of the [112]-Si exit wave reconstructed from a tilt-focus series. The dumbbell atom columns, separated by only 0.78 Å, are clearly resolved. From Haigh *et al.* (2009). Reproduced with permission; copyright American Physical Society. (b) Negative spherical aberration image of a 4 nm thick $SrTiO_3$ sample. The intensity of an oxygen atom column is sensitive to its occupancy, as revealed by the columns labelled 1 and 2 in the image, which are oxygen deficient compared to their neighbours. This can more easily be seen in the intensity profile acquired across the atom columns of interest (figure inset). From Jia *et al.* (2003). Reproduced with permission; copyright The American Association for the Advancement of Science.

comparing the experimental intensity profile with multislice simulations (Jia *et al.*, 2003).

SAED patterns are also important for solving crystal structures. Since only intensities are measured the phase shifts of individual diffracted beams due to lens aberrations are not important. Furthermore, unlike HREM, where partial coherence limits information transfer, in SAED the Bragg peaks are only limited by a curved Ewald sphere and TDS damping at large scattering angles (Eq. (3.16)), so that there is less demand on instrumentation. Jansen *et al.* (1998) have developed a least-squares algorithm for crystal structure refinement, where the experimental electron diffraction intensities are compared to multislice simulated values. Specimen parameters such as atom coordinates, Debye–Waller factors and thickness are treated as variables and are deemed accurate when there is close agreement between simulation and experiment. A basic knowledge of the crystal structure is required for this method to work. In practice, a better approach would be to measure kinematic electron diffraction intensities, which can then be used to extract the crystal potential and hence atom coordinates. However, even for thin specimens electron diffraction is typically dynamic in nature, but the recent emergence of precession electron diffraction can partly mitigate its effects. Here, the incident beam is precessed at a fixed angle about the optic axis and the diffraction patterns at each beam tilt are cumulatively measured after post-specimen de-scanning (Vincent and Midgley, 1994). At any given beam tilt, only a few reflections are close to the Bragg angle, so that dynamic scattering between diffracted beams is reduced, although dynamic diffraction between a Bragg beam and unscattered beam may still occur. Multislice simulations (Own *et al.*, 2006; White *et al.*, 2010) have shown that precession electron diffraction patterns agree with the kinematic result more closely as the precession angle is increased, although eventually the similarity breaks down for larger specimen thicknesses due to dynamic diffraction.

3.2.2 CBED and STEM Applications: Frozen Phonon Model

The previous sections focused on parallel beam illumination of the specimen, although convergent beam illumination is also important for applications such as CBED and STEM. For parallel illumination the (normalised) incident wavefunction $\psi_0(\mathbf{R})$ is given by $\exp(2\pi i \mathbf{k}_t \cdot \mathbf{R})$. For convergent illumination the incident wavefunction is calculated by coherently summing all partial plane waves constituting the electron probe, taking due account of lens aberrations, that is,

$$\psi_0 \left(\mathbf{R}, \mathbf{R}_p\right) = \beta \int A \left(\mathbf{k}_t\right) \exp\left[-i\chi\left(\mathbf{k}_t\right)\right] \exp\left[2\pi i\mathbf{k}_t \cdot \left(\mathbf{R} - \mathbf{R}_p\right)\right] d\mathbf{k}_t$$

(3.44)

where $\mathbf{R}_p$ is the probe incident position on the sample expressed in $\mathbf{R}$-coordinates and $A(\mathbf{k}_t)$ is the aperture function, which is equal to unity within the probe defining aperture and zero outside it. For an aberration-free lens the probe wavefunction is the inverse Fourier transform of the aperture function. The normalisation constant β is defined such that

$$\int \left|\psi_0 \left(\mathbf{R}, \mathbf{R}_p\right)\right|^2 d\mathbf{R} = 1$$

(3.45)

In multislice simulations of CBED patterns Eqs. (3.44) and (3.45) are used for the incident wavefunction. The specimen exit wavefunction is calculated via the previously described method of transmission and propagation for individual slices within the specimen. The CBED pattern is generated by Fourier transforming the exit wavefunction and taking the square modulus. A similar procedure is adopted for high angle annular dark-field (HAADF) imaging in STEM, except here the diffraction pattern intensity is integrated over the angular range of the detector to give the HAADF signal intensity for that probe position. The HAADF image is built up serially, pixel by pixel, by 'scanning' the probe over the specimen region of interest and repeating the above calculation for each scan position (Kirkland, 2010). In a CBED or STEM multislice simulation the size of the specimen supercell in real space (a) must be large enough for adequate sampling of the diffraction pattern, since the resolution in reciprocal space is ($1/a$). Furthermore, the number of sampling points (N) must be chosen such that the maximum spatial frequency ($N/2a$) is large enough to simulate either the HOLZ rings in CBED or the angular range of the HAADF detector.

Accurate simulations of CBED patterns and STEM HAADF images must also include thermal diffuse scattering due to vibrating atoms. For the former TDS gives rise to Kikuchi lines (i.e. phonon scattered electrons that are subsequently Bragg diffracted; Hirsch *et al.*, 1965) and reduces the intensity of Bragg reflections via the Debye–Waller factor. Furthermore, TDS is dominant at large scattering angles (Figure 3.4) and is therefore an important contribution to the HAADF signal. In Section 3.1.1, it was shown that the time-averaged specimen transmission function for a crystal with vibrating atoms included an absorption term, $\exp[-\sigma^2\mu(\mathbf{R})]$ (Eq. (3.19)), which depletes intensity from the (elastic) electron wavefunction. The lost intensity is the TDS intensity, but the simulation does not predict the redistribution of this intensity throughout reciprocal space, as required for CBED and HAADF simulations. This difficulty is overcome in the 'frozen phonon' model, which is based on the fact that the scattering time for an incident electron by an atom is significantly shorter than the phonon oscillation period ($\sim 10^{-13}$ s), so that the electron effectively sees a 'snapshot' of the vibrating atom frozen in time (Hall and Hirsch, 1965; Loane *et al.*, 1991). The scattering time for an electron with velocity v and impact parameter b is $\sim b/v$ (Section 2.1.2). As an example if b is taken as the square root of the total elastic scattering cross-section for copper (Eq. (2.34), Section 2.1.3), then for a 200 kV electron beam the scattering time is only $\sim 10^{-19}$ s, which is significantly shorter than the phonon oscillation period. Furthermore, the time required for a 200 kV electron to propagate through a 100 nm thin foil is only $\sim 5 \times 10^{-16}$ s, so that decay of a given phonon configuration during electron transmission is also negligible.

In the frozen phonon model multislice simulations are carried out for several snapshots of vibrating atoms frozen in time and the resulting diffraction patterns are added incoherently to give the final result. The rationale behind this approach is that the electrons are incident on the specimen one at a time, with typically a sub-nanosecond delay between successive electrons (see Footnote 4, Section 3.1.3). The time between successive electrons is long enough for the simulated phonon configurations to be (i) uncorrelated and (ii) governed by the equilibrium phonon distribution, owing to the crystal reaching thermal equilibrium following (potential) excitation by a previously incident electron. Within the acquisition time of a CBED pattern or HAADF image the diffraction patterns from a large number of incident electrons are cumulatively measured, so that many phonon configurations must be simulated to match theory with experiment. Since the specimen is a 'snapshot' of the vibrating atoms the transmission function does not contain the absorption term in Eq. (3.19). However, the transmission function is also not periodic, owing to the slight displacement of the atoms from their equilibrium positions. Scattering to reciprocal vectors between the Bragg reflections is therefore possible and gives rise to the TDS intensity. Note however that electron–phonon interactions are inelastic collisions. Although the energy transfer is small (phonon energies are only a few million electron volts) the phase relationship between the phonon scattered and incident wave is nevertheless lost due to a change in the crystal state, that is electron–phonon scattering is incoherent.

The frozen phonon model is however a fully elastic calculation, where there is a phase relationship between the scattered and unscattered wave, so that interference between the two waves is possible (i.e. coherent scattering). This discrepancy is discussed in greater detail in Section 3.3.2.

An exact calculation of the normal modes of vibration of atoms in a crystal can be quite demanding. Hence the Einstein approximation, where atoms are assumed to vibrate independently of one another, is often used. While this is computationally convenient it cannot reproduce those features of the TDS intensity that are due to correlated atom movement, such as, streaking around ZOLZ reflections. For silicon, correlations between nearest neighbour atoms are sufficient to reproduce the experimentally observed TDS intensity in a CBED pattern (Muller *et al.*, 2001). In the Einstein approximation atomic displacements follow a Gaussian distribution, with a mean square displacement $\overline{x^2}$ that is related to the Debye–Waller factor via Eq. (3.16). To generate a frozen phonon configuration a Gaussian distributed random number with variance $\overline{x^2}$ is used to calculate atomic displacements along the three Cartesian axes (i.e. atomic vibration is modelled as three independent harmonic oscillators; note also the similarity between the frozen phonon model and Monte Carlo simulations discussed in Chapter 2). Figure 3.11a–c shows CBED results for a 50 nm thick, [112]-Si specimen

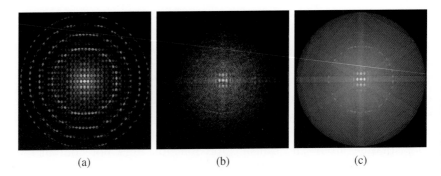

(a)   (b)   (c)

Figure 3.11 (a) Multislice simulated CBED pattern for a 50 nm thick, [112]-Si specimen at 120 kV and 4 mrad probe semi-convergence angle. Atomic vibrations are not included. (b) The CBED pattern obtained from a single frozen phonon configuration and (c) after averaging over 20 frozen phonon configurations. The Kikuchi pattern is clearly visible in (c), including the [223] minor zone-axis to the left of the 'image'. The sharp cut-off in intensity at large spatial frequencies is due to bandwidth limiting in the simulation to avoid aliasing artefacts. All CBED patterns are plotted on a logarithmic intensity scale. Kirkland's (2010) multislice code was used in the simulations.

and 120 kV electron beam. In Figure 3.11a atomic vibrations are not included; many ZOLZ reflections are visible along with HOLZ rings. There is a lot of 'noise' for a simulation of only one frozen phonon configuration (Figure 3.11b), but some indication of a HOLZ ring and Kikuchi lines are observed. Furthermore, the ZOLZ reflections further out in reciprocal space are strongly damped. Averaging over 20 frozen phonon configurations produces a more realistic CBED pattern (Figure 3.11c). The Kikuchi lines are now clearly visible, but ZOLZ reflections and HOLZ rings have lower contrast compared to the stationary atom case (Figure 3.11a).

In HAADF imaging the electron probe is required to propagate or channel along the atom columns, so that a larger proportion of the beam intensity can be scattered by the atomic nuclei out to the high angles intercepted by the annular detector. The atom columns thereby appear brighter compared to the background in a HAADF image. Figure 3.12a plots the variation in electron beam intensity as a function of depth (i.e. pendellösung) for the case of a 20 mrad, 200 kV STEM probe focused on an atom column of [110]-Si. The intensity here refers to that measured along the atom column on which the probe is incident, so that any increase in intensity is due to channelling, while any decrease is due to beam spreading. To evaluate the role of TDS, pendellösung plots were multislice simulated with and without atomic vibrations. The results show that atomic vibrations reduce the ability of an atom column to channel the beam along its length. Channelling is caused by Coulomb attraction of the electron beam intensity by the atomic nuclei. If the atoms are more randomly arranged in space due to thermal motion the channelling effect, and hence the HAADF signal, should be weaker. This is confirmed from the multislice simulated HAADF intensity profiles shown in Figure 3.12b and c, which were extracted across a dumbbell pair of atom columns in 10 and 50 nm thick, [110]-Si specimens respectively. The HAADF intensity for vibrating atoms is less than that for stationary atoms, as expected. Notice also the increase in background intensity with specimen thickness, due to increased generation of high angle scattered electrons in the thicker sample. Furthermore, as the specimen thickness increases the peak-to-background ratio of the dumbbell decreases, thereby making it harder to resolve the lattice image. Klenov and Stemmer (2006) have analysed the variation in background and atom column HAADF intensities for $PbTiO_3$ and $SrTiO_3$ as a function of specimen thickness as well as other imaging parameters. Frozen phonon multislice simulations were compared with experiment. Dwyer and Etheridge (2003) have reported frozen phonon

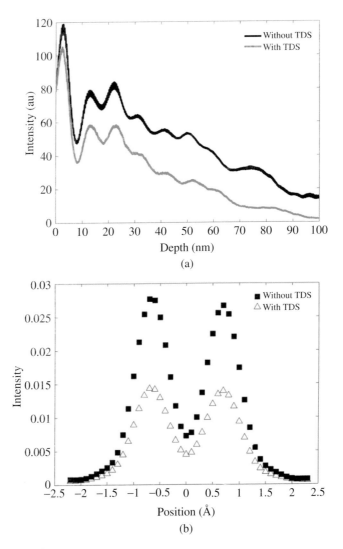

Figure 3.12 (a) Pendellösung plots for a STEM probe incident on an atom column of [110]-Si determined from multislice simulations with and without atomic vibrations (i.e. with and without TDS). The pendellösung intensity was calculated along the atom column on which the probe is incident. The probe parameters were 200 keV energy, 20 mrad semi-convergence angle and zero electron-optic aberrations. Twenty frozen phonon configurations were averaged for the pendellösung plot with atomic vibrations. (b) and (c) are the HAADF intensity profiles across the dumbbell atom columns, calculated with and without TDS, for 10 and 50 nm thick specimens respectively. Note the change in intensity scale between the two figures. Kirkland's (2010) multislice code was used in the simulations.

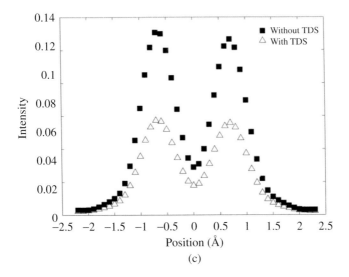

Figure 3.12   (*Continued*)

calculations of probe spreading in [111]- and [100]-oriented silicon. Further details of the STEM probe pendellösung (using Bloch waves) can be found in Chapter 4.

3.3 FURTHER TOPICS IN MULTISLICE SIMULATION

3.3.1 Accuracy of Multislice Algorithms

The multislice method is built purely on concepts borrowed from physical optics and historically this is how it was first proposed (Cowley and Moodie, 1957). Strictly speaking, however, the electron wavefunction within the specimen must also satisfy Schrödinger's equation. The quantum mechanical and multislice solutions can be shown to be equivalent under certain conditions. In this section, this equivalence is explored in more detail, since it gives an estimate on the accuracy of the multislice method and suggests ways to improve it.

Schrödinger's equation for an incident electron of wave number k $(=1/\lambda)$ is given by

$$\nabla^2 \psi(\mathbf{r}) + \left[4\pi^2 k^2 + \frac{8\pi^2 me}{h^2} V(\mathbf{r}) \right] \psi(\mathbf{r}) = 0 \qquad (3.46)$$

In our case, the specimen potential energy $-eV(\mathbf{r})$ is small compared to the electron kinetic energy, so that the wavefunction in the solid will undergo a relatively small change compared to the wavefunction in vacuum. In the most general scenario the wavefunction ψ can be written as the sum of a forward propagating wave and backscattered wave, that is,

$$\psi(\mathbf{r}) = \Phi^f(\mathbf{r})\exp(2\pi i\mathbf{k}\cdot\mathbf{r}) + \Phi^b(\mathbf{r})\exp(-2\pi i\mathbf{k}\cdot\mathbf{r}) \qquad (3.47)$$

Here, $\exp(2\pi i\mathbf{k}\cdot\mathbf{r})$ and $\exp(-2\pi i\mathbf{k}\cdot\mathbf{r})$ are the wavefunctions for forward and backscattered plane waves in vacuum and Φ^f, Φ^b represent the 'perturbations' due to interactions with the solid. For large electron energies scattering is primarily in the forward direction (i.e. $\Phi^b = 0$), so that for electrons incident along the z-axis (i.e. perpendicular to the sample plane), Eq. (3.47) simplifies to $\psi(\mathbf{r}) = \Phi^f(\mathbf{r})\exp(2\pi ikz)$. Substituting in Eq. (3.46) gives

$$\frac{\partial^2\Phi^f}{\partial z^2} + 4\pi ik\frac{\partial\Phi^f}{\partial z} + \nabla^2_{xy}\Phi^f + \frac{8\pi^2 me}{h^2}V(\mathbf{r})\Phi^f = 0$$

$$\nabla^2_{xy} = \frac{\partial^2}{\partial x^2} + \frac{\partial^2}{\partial y^2} \qquad (3.48)$$

For weak scattering, or equivalently small angle scattering, the term $\partial^2\Phi^f/\partial z^2$ in Eq. (3.48) is smaller than $4\pi ik(\partial\Phi^f/\partial z)$ and therefore can be neglected. Equation (3.48) then reduces to a first order linear differential equation:

$$\boxed{\frac{\partial\Phi^f(\mathbf{r})}{\partial z} = \left[\frac{i\lambda}{4\pi}\nabla^2_{xy} + i\sigma V(\mathbf{r})\right]\Phi^f(\mathbf{r})} \qquad (3.49)$$

where the substitution for the interaction constant $\sigma = \pi e/\lambda E_{\text{kin}} = 2\pi me\lambda/h^2$ has been made. The assumption of Φ^f varying slowly with z is often referred to in the literature as being equivalent to neglecting backscattering (where there is a sudden reversal of z), although strictly speaking backscattering is modelled through the $\Phi^b\exp(-2\pi i\mathbf{k}\cdot\mathbf{r})$ term in Eq. (3.47). The solution for Φ^f over a slice extending from z to $z + \Delta z$ is

$$\Phi^f(\mathbf{R}, z + \Delta z) = \exp\left[\int_z^{z+\Delta z}\left(\frac{i\lambda}{4\pi}\nabla^2_{xy} + i\sigma V(\mathbf{r})\right)dz\right]\Phi^f(\mathbf{R}, z)$$

$$= \exp\left(\frac{i\lambda\Delta z}{4\pi}\nabla^2_{xy} + i\sigma V_p(\mathbf{R})\right)\Phi^f(\mathbf{R}, z) \qquad (3.50)$$

It is important to realise that the exponential term in Eq. (3.50) is an operator acting on $\Phi^f(\mathbf{R}, z)$. An operator of the form $\exp[A + B]$, where A and B are non-commutative (i.e. $AB \neq BA$), can be expanded in the usual manner:

$$\exp[A + B] = \sum_{n=0}^{\infty} \frac{1}{n!}(A + B)^n$$

$$= 1 + (A + B) + \frac{1}{2}\left(A^2 + B^2 + AB + BA\right) + \cdots \quad (3.51)$$

Consider the two alternative expansions:

$$\exp[A]\exp[B] = \left(\sum_{n=0}^{\infty} \frac{A^n}{n!}\right) \cdot \left(\sum_{m=0}^{\infty} \frac{B^m}{m!}\right)$$

$$= 1 + (A + B) + \left(\frac{A^2 + B^2}{2} + AB\right) + \cdots \quad (3.52)$$

$$\exp[B]\exp[A] = \left(\sum_{n=0}^{\infty} \frac{B^n}{n!}\right) \cdot \left(\sum_{m=0}^{\infty} \frac{A^m}{m!}\right)$$

$$= 1 + (A + B) + \left(\frac{A^2 + B^2}{2} + BA\right) + \cdots \quad (3.53)$$

These differ from Eq. (3.51) in the quadratic and higher order terms. In a conventional multislice simulation one of the above operator approximations is adopted, that is,

$$\exp\left(\frac{i\lambda\Delta z}{4\pi}\nabla^2_{xy} + i\sigma V_p(\mathbf{R})\right) \approx \exp\left(\frac{i\lambda\Delta z}{4\pi}\nabla^2_{xy}\right)\exp[i\sigma V_p(\mathbf{R})] \quad \text{or}$$

$$\exp[i\sigma V_p(\mathbf{R})]\exp\left(\frac{i\lambda\Delta z}{4\pi}\nabla^2_{xy}\right) \quad (3.54)$$

The error of the operator is then of the order of $(\Delta z)^2$. The number of slices in the sample is however proportional to $1/\Delta z$ (assuming uniform slice thickness), so that the global error is Δz, that is, first order with respect to slice thickness. More accurate operator expansions are given in van Dyck (1979). In Eq. (3.54) the term $\exp[i\sigma V_p(\mathbf{R})]$ is the transmission function (Eq. (3.8)) and represents scattering by the specimen projected potential. To interpret the meaning of $\exp\left(\frac{i\lambda\Delta z}{4\pi}\nabla^2_{xy}\right)$ consider the case where $V_p(\mathbf{R}) = 0$. Since only free space propagation is involved in transforming $\Phi^f(\mathbf{R}, z)$ to $\Phi^f(\mathbf{R}, z + \Delta z)$ (Eq. (3.50)), the operator must represent convolution by the Fresnel propagator function (Eq. (3.25)), that is,

$$\exp\left(\frac{i\lambda\Delta z}{4\pi}\nabla_{xy}^2\right) = \frac{-i}{\lambda\Delta z}\exp\left[\frac{i\pi}{\lambda\Delta z}(x^2+y^2)\right]\otimes \qquad (3.55)$$

The derivation is easily carried out in reciprocal space. Since the two operators $\exp[(i\lambda\Delta z/4\pi)\partial^2/\partial x^2]$ and $\exp[(i\lambda\Delta z/4\pi)\partial^2/\partial y^2]$ are commutative (due to a similar property of mixed derivatives) the following Fourier transform is obtained:

$$\mathrm{FT}\left[\exp\left(\frac{i\lambda\Delta z}{4\pi}\nabla_{xy}^2\right)\Phi^f(\mathbf{R},z)\right] = \iint \sum_{n=0}^{\infty}\frac{1}{n!}\left(\frac{i\lambda\Delta z}{4\pi}\frac{\partial^2}{\partial x^2}\right)^n$$

$$\times \sum_{m=0}^{\infty}\frac{1}{m!}\left(\frac{i\lambda\Delta z}{4\pi}\frac{\partial^2}{\partial y^2}\right)^m \Phi^f(\mathbf{R},z)e^{-2\pi i(u_x x + u_y y)}dxdy \qquad (3.56)$$

where u_x, u_y are the components of the two-dimensional Fourier space vector $\mathbf{u}$. To evaluate the right-hand side note that integration by parts gives the following result, assuming Φ^f to be bounded and/or periodic:

$$\int_{-\infty}^{\infty}\frac{1}{m!}\left(\frac{i\lambda\Delta z}{4\pi}\right)^m\frac{\partial^{2m}\Phi^f}{\partial y^{2m}}e^{-2\pi i u_y y}\,dy = \frac{(-i\pi\lambda\Delta z u_y^2)^m}{m!}\int_{-\infty}^{\infty}\Phi^f e^{-2\pi i u_y y}\,dy$$

$$(3.57)$$

Equation (3.57) can be generalised to evaluate Eq. (3.56) with respect to the y- and x-integrals successively. The result is

$$\mathrm{FT}\left[\exp\left(\frac{i\lambda\Delta z}{4\pi}\nabla_{xy}^2\right)\Phi^f(\mathbf{R},z)\right] = \left\{\sum_{n=0}^{\infty}\frac{1}{n!}[-i\pi\lambda\Delta z(u_x^2 + u_y^2)]^n\right\}$$

$$\times \iint \Phi^f e^{-2\pi i(u_x x + u_y y)}dx\,dy = \tilde{P}(\mathbf{u})\tilde{\Phi}^f(\mathbf{u})$$

$$(3.58)$$

where $\tilde{P}(\mathbf{u})$ is the Fresnel propagator of Eq. (3.27). Inverse Fourier transforming Eq. (3.58) then gives Eq. (3.55), the desired result.

Note that the operator $\frac{1}{2}[\exp(A)\exp(B) + \exp(B)\exp(A)]$ has an error of $(\Delta z)^3$. This is, however, equivalent to carrying out two multislice simulations (i.e. in the first scattering followed by propagation and in the second propagation followed by scattering or vice versa), so that the computing time is doubled. A more efficient second order method is the 'real space' multislice method of van Dyck (1980), so called because the calculation is done entirely in real space without using any Fourier transforms. It is also better for simulating HOLZ effects, owing to the use of a 'potential eccentricity' term (van Dyck and Coene, 1984; Kilaas *et al.*, 1987). Finally, Chen and van Dyck (1997) have developed an alternative multislice method that does not assume that $\partial^2 \Phi^f / \partial z^2$ is negligible, so that even backscattering can be simulated. This is especially important for surface diffraction techniques, such as LEED and RHEED.

3.3.2 Is the Frozen Phonon Model Physically Realistic?

In Section 3.2.2, it was noted that the frozen phonon model is a fully elastic calculation, although electron–phonon scattering is an inelastic process. In this section, it will be demonstrated that the two approaches produce equivalent results, provided that certain conditions are satisfied. A rigorous analysis based on the Green's function method was first provided by Wang (1998), but here a simpler analysis of van Dyck (2009), which utilises the multislice equations for forward scattering, will be presented. The starting point of the analysis is Eq. (3.49), with the crystal potential $V(\mathbf{r})$ replaced by $[V_o(\mathbf{r}) + \Delta V(\mathbf{r}, t)]$, where $V_o(\mathbf{r})$ is a time-independent potential (Eq. (3.15)) and $\Delta V(\mathbf{r}, t)$ is a time (t)-dependent potential due to atomic vibrations. Similarly, $\Phi^f(\mathbf{r})$ can also be replaced by $[\Phi_o^f(\mathbf{r}) + \Delta \Phi^f(\mathbf{r}, t)]$, where $\Phi_o^f(\mathbf{r})$ is the elastic wavefunction and $\Delta \Phi^f(\mathbf{r}, t)$ is the phonon scattered TDS wavefunction. By definition $\langle \Delta V \rangle = 0$ and $\langle \Delta \Phi^f \rangle = 0$, where the $\langle \, \rangle$ symbol represents the time average or equivalently the average over phonon configurations. Substituting in Eq. (3.49) and (for convenience) dropping the superscript in Φ^f,

$$\frac{\partial \Phi_o(\mathbf{r})}{\partial z} + \frac{\partial \Delta \Phi(\mathbf{r}, t)}{\partial z} = \left[\frac{i\lambda}{4\pi} \nabla_{xy}^2 + i\sigma V_o(\mathbf{r}) + i\sigma \Delta V(\mathbf{r}, t) \right] (\Phi_o(\mathbf{r}) + \Delta \Phi(\mathbf{r}, t))$$

(3.59)

Averaging over phonon configurations,

$$\frac{\partial \Phi_o(\mathbf{r})}{\partial z} = \left[\frac{i\lambda}{4\pi} \nabla_{xy}^2 + i\sigma V_o(\mathbf{r}) \right] \Phi_o(\mathbf{r}) + i\sigma \langle \Delta V(\mathbf{r}, t) \cdot \Delta \Phi(\mathbf{r}, t) \rangle$$

(3.60)

Subtracting Eq. (3.60) from Eq. (3.59) gives,

$$\frac{\partial \Delta \Phi(\mathbf{r}, t)}{\partial z} = \left[\frac{i\lambda}{4\pi} \nabla_{xy}^2 + i\sigma V_o(\mathbf{r}) \right] \Delta \Phi(\mathbf{r}, t) + i\sigma \Delta V(\mathbf{r}, t) \Phi_o(\mathbf{r})$$

$$+ i\sigma \left[\Delta V(\mathbf{r}, t) \Delta \Phi(\mathbf{r}, t) - \langle \Delta V(\mathbf{r}, t) \cdot \Delta \Phi(\mathbf{r}, t) \rangle \right]$$

(3.61)

Equations (3.60) and (3.61) are coupled differential equations. In the latter, the first term represents propagation and elastic scattering of the

TDS wave, while the second term represents generation of TDS intensity by the elastic wave. The third term does not appear in van Dyck (2009); however for a thin specimen, where $\Phi_o(\mathbf{r})$ is large, it will be small compared to the second term. Ignoring the first and third terms gives

$$\Delta\Phi(\mathbf{R}, z, t) = i\sigma \int_0^z \Delta V(\mathbf{R}, z', t)\, \Phi_o(\mathbf{R}, z')\, dz' \tag{3.62}$$

Substituting in Eq. (3.60)

$$\frac{\partial\Phi_o(\mathbf{r})}{\partial z} = \left[\frac{i\lambda}{4\pi}\nabla_{xy}^2 + i\sigma V_o(\mathbf{r})\right]\Phi_o(\mathbf{r}) - \sigma^2$$

$$\times \int_0^z \langle\Delta V(\mathbf{R}, z, t)\Delta V(\mathbf{R}, z', t)\rangle\Phi_o(\mathbf{R}, z')dz' \tag{3.63}$$

If it is assumed that the atoms vibrate independently (i.e. Einstein approximation), then $\langle\Delta V(\mathbf{R}, z, t)\Delta V(\mathbf{R}, z', t)\rangle = \langle\Delta V(\mathbf{R}, z, t)\rangle \cdot \langle\Delta V(\mathbf{R}, z', t)\rangle$, which is equal to zero unless $z' = z$. Therefore,

$$\frac{\partial\Phi_o(\mathbf{r})}{\partial z} = \left[\frac{i\lambda}{4\pi}\nabla_{xy}^2 + i\sigma V_o(\mathbf{r})\right]\Phi_o(\mathbf{r}) - \sigma^2\langle\Delta V(\mathbf{r}, t)^2\rangle\Phi_o(\mathbf{r}) \tag{3.64}$$

Note that the last term in Eq. (3.64) is similar to the absorption term in Eqs. (3.18) and (3.19)[7] and represents depletion of the elastic wave due to TDS scattering. Eqs. (3.62) and (3.64) are the TDS and elastic wavefunctions simulated by the frozen phonon model, under two main assumptions: (i) there is negligible scattering and spreading of the TDS wave once generated and (ii) the atoms vibrate independently. It is now required to show that these results are equivalent to a more comprehensive analysis that includes both elastic and inelastic scattering. The theory is due to Yoshioka (1957) and is described in detail in Chapter 5, but here we shall use only the relevant results. The Schrödinger equation for this theory is

$$\nabla^2\psi_m(\mathbf{r}) + \left[4\pi^2 k_m{}^2 + \frac{8\pi^2 me}{h^2}H_{mm}(\mathbf{r})\right]\psi_n(\mathbf{r}) = -\frac{8\pi^2 me}{h^2}\sum_{m\neq n}H_{mn}(\mathbf{r})\psi_n(\mathbf{r}) \tag{3.65}$$

The wavefunction ψ_m corresponds to the elastic wave when $m = 0$ and an inelastic wave when $m \neq 0$ (the index m refers to the mth excited state of

---

[7]Equation (3.64) differs from Eqs. (3.18) and (3.19) by a factor of ½. This is because the latter is valid in the projection approximation, that is, the potential is integrated over z. The two expressions can nevertheless be shown to be equivalent (van Dyck, 2009).

the crystal of energy E_m, with the ground state being arbitrarily assigned zero energy, i.e. $E_0 = 0$). The wave number $k_m^2 = k^2 - (2mE_m/h^2)$, where k is the wave number of the incident electron in vacuum. Clearly, inelastic waves will have longer de Broglie wavelengths due to E_m energy transfer to the crystal. The excitation matrix element $H_{mn}(\mathbf{r})$ is the interaction term for the crystal transitioning from the n-state to the m-state via $(E_m - E_n)$ energy transfer from the incident electron (if $m = n$ then $H_{mm}(\mathbf{r})$ represents the interaction term for elastic scattering by the crystal of energy E_m), and is expressed as

$$H_{mn}(\mathbf{r}) = \langle a_m \left| V(\mathbf{r}, \mathbf{r}_i) \right| a_n \rangle = \int a_m^*(\mathbf{r}_i) V(\mathbf{r}, \mathbf{r}_i) a_n(\mathbf{r}_i) d\mathbf{r}_i \qquad (3.66)$$

$a_m(\mathbf{r}_i)$ is the eigenfunction of the crystal in the mth state expressed in terms of the nuclei and atomic electron coordinates $\mathbf{r}_i$. The instantaneous potential $V(\mathbf{r}, \mathbf{r}_i)$ is equal to $[V_o(\mathbf{r}) + \Delta V(\mathbf{r}, t)]$ encountered earlier. The $\mathbf{r}_i$ dependence is due to the potential being a function of time or equivalently phonon configuration. It is reasonably assumed that $H_{mm}(\mathbf{r}) \approx V_o(\mathbf{r})$, since phonon excitation is not expected to significantly change the time-independent potential V_o and because ΔV does not contribute to $H_{mm}(\mathbf{r})$ once integrated over all phonon configurations $\mathbf{r}_i$ (Eq. (3.66)). Following a similar approach to that outlined in Section 3.3.1 and substituting $\psi_m(\mathbf{r}) = \varphi_m(\mathbf{r})\exp(2\pi i k z)$ for the forward scattered wave in Eq. (3.65) gives, under the reasonable assumption of small energy loss (i.e. $k_m \approx k$),

$$\frac{\partial \varphi_0(\mathbf{r})}{\partial z} = \left[\frac{i\lambda}{4\pi} \nabla_{xy}^2 + i\sigma V_o(\mathbf{r}) \right] \varphi_0(\mathbf{r}) + i\sigma \sum_{m \neq 0} H_{0m}(\mathbf{r})\varphi_m(\mathbf{r}) \quad (3.67)$$

$$\left(\frac{\partial \varphi_m(\mathbf{r})}{\partial z} \right)_{m \neq 0} = \left[\frac{i\lambda}{4\pi} \nabla_{xy}^2 + i\sigma V_o(\mathbf{r}) \right] \varphi_m(\mathbf{r}) + i\sigma \sum_{m \neq n} H_{mn}(\mathbf{r})\varphi_n(\mathbf{r}) \quad (3.68)$$

First consider Eq. (3.68). Similar to the frozen phonon case we ignore the first term, which describes scattering and spreading of the inelastic wave φ_m, and concentrate solely on the second term relating to generation of φ_m by φ_n. At this stage the so-called *ground state* approximation (Wang, 1998) is introduced. This states that only $H_{m0}(\mathbf{r})\varphi_0(\mathbf{r})$ is significant in the summation term of Eq. (3.68), owing to the elastic wave having the highest intensity in thin crystals and because the incident electron largely interacts with the crystal in the ground state. This assumption will be discussed in more detail later on. Eq. (3.68) can then be simplified to

$$\varphi_m(\mathbf{R}, z) = i\sigma \int_0^z H_{m0}(\mathbf{R}, z')\varphi_0(\mathbf{R}, z')dz' \qquad (3.69)$$

Substituting in Eq. (3.67),

$$\frac{\partial \varphi_0(\mathbf{r})}{\partial z} = \left[\frac{i\lambda}{4\pi}\nabla_{xy}^2 + i\sigma V_o(\mathbf{r})\right]\varphi_0(\mathbf{r}) - \sigma^2$$
$$\times \sum_{m \neq 0}\left(\int_0^z H_{0m}(\mathbf{R}, z)H_{m0}(\mathbf{R}, z')\varphi_0(\mathbf{R}, z')dz'\right) \qquad (3.70)$$

The summation term is easiest to solve using bra–ket notation:

$$\sum_{m \neq 0} H_{0m}(\mathbf{R}, z)H_{m0}(\mathbf{R}, z') = \langle a_0|V(\mathbf{R}, z, \mathbf{r}_i)|\left(\sum_{m \neq 0}|a_m\rangle\langle a_m|\right)|V(\mathbf{R}, z', \mathbf{r}_i)|a_0\rangle$$
$$(3.71)$$

From the closure relation of eigenstates,

$$\sum_m |a_m\rangle\langle a_m| = \mathbf{I} \quad \text{or} \quad \sum_{m \neq 0}|a_m\rangle\langle a_m| = \mathbf{I} - |a_0\rangle\langle a_0| \qquad (3.72)$$

where $\mathbf{I}$ is the identity matrix. Substituting in Eq. (3.71),

$$\sum_{m \neq 0} H_{0m}(\mathbf{R}, z)H_{m0}(\mathbf{R}, z') = \langle a_0\left|V(\mathbf{R}, z, \mathbf{r}_i) \cdot V(\mathbf{R}, z', \mathbf{r}_i)\right|a_0\rangle$$
$$- \langle a_0|V(\mathbf{R}, z, \mathbf{r}_i)|a_0\rangle\langle a_0|V(\mathbf{R}, z', \mathbf{r}_i)|a_0\rangle$$
$$(3.73)$$

Each bra–ket term on the right-hand side of Eq. (3.73) is an expectation value of the instantaneous potential for the crystal in the ground state. The result is the potential averaged with respect to time or equivalently phonon configuration. Assuming independent atomic vibrations (i.e. Einstein approximation) the bra–ket terms simplify as

$$\langle a_0|V(\mathbf{R}, z, \mathbf{r}_i) \cdot V(\mathbf{R}, z', \mathbf{r}_i)|a_0\rangle$$
$$= \langle[V_o(\mathbf{R}, z) + \Delta V(\mathbf{R}, z, t)] \cdot [V_o(\mathbf{R}, z') + \Delta V(\mathbf{R}, z', t)]\rangle$$
$$= V_o(\mathbf{R}, z)V_o(\mathbf{R}, z') + \langle\Delta V(\mathbf{R}, z, t)\Delta V(\mathbf{R}, z', t)\rangle$$
$$= V_o(\mathbf{R}, z)V_o(\mathbf{R}, z') + <\Delta V(\mathbf{R}, z, t)^2 > \delta(z - z')$$

and

$$\langle a_0 | V(\mathbf{R}, z, \mathbf{r}_i) | a_0 \rangle = \langle [V_o(\mathbf{R}, z) + \Delta V(\mathbf{R}, z, t)] \rangle = V_o(\mathbf{R}, z) \quad \text{etc.} \quad (3.74)$$

where $\delta(z - z')$ is the Dirac delta function. The final form for Eq. (3.70) is therefore

$$\frac{\partial \varphi_0(\mathbf{r})}{\partial z} = \left[\frac{i\lambda}{4\pi} \nabla^2_{xy} + i\sigma V_o(\mathbf{r}) \right] \varphi_0(\mathbf{r}) - \sigma^2 \left\langle \Delta V(\mathbf{r}, t)^2 \right\rangle \varphi_0(\mathbf{r}) \quad (3.75)$$

Equations (3.75) and (3.69) represent the elastic and TDS wavefunctions for a full elastic–inelastic calculation. These should be compared with Eqs. (3.64) and (3.62) for the frozen phonon case respectively. The two elastic wavefunction equations are identical and furthermore it can be shown that the TDS intensities are also identical (van Dyck, 2009). Therefore, the frozen phonon model is physically realistic provided the following assumptions are satisfied: (i) scattering and spreading of the TDS wave is negligible, (ii) the atoms vibrate independently (Einstein approximation) and (iii) ground state approximation. The final assumption is valid because the time between successive incident electrons is much longer than the phonon oscillation period (Section 3.2.2), so that any phonons excited by a high energy electron have sufficient time to thermalise. Multiple electron–phonon scattering events for a single incident electron are also unlikely to alter the crystal significantly from its ground state, owing to the large number of phonons already present at room temperature thermal equilibrium. The more rigorous Green's function analysis by Wang (1998) also requires the ground state approximation to be valid for the frozen phonon method to be physically realistic. A second requirement is that multiple electron–phonon scattering events must be incoherent. The frozen phonon method is however a fully elastic calculation where scattering is coherent, so that it can only be accurate if the specimen is thin enough to avoid multiple electron–phonon scattering. Finally the condition of negligible scattering and spreading of the TDS wave is not a requirement in Wang's (1998) analysis. It may not even be a strict requirement in the above analysis, although its introduction does considerably simplify the mathematics. A similar argument applies to the Einstein approximation; without it the elastic wave has the less intuitive form given by Eq. (3.63), which is nevertheless valid for both the frozen phonon and Yoshioka elastic–inelastic scattering models.

3.4 SUMMARY

The multislice method is based on the physical optics concepts of scattering and free space propagation. Scattering is modelled by treating the specimen as a phase object, that is, the local phase of the incident wave is a function of the specimen potential in that region. Free space propagation models the spreading of the scattered wave as a Hüygen wavelet. A high energy electron within a crystal will repeatedly undergo scattering and propagation, and in a multislice simulation this behaviour is taken into account by dividing the specimen into thin slices. Comparing the conventional multislice method to the more rigorous quantum mechanical theory shows that it is accurate to first order of the slice thickness, although more accurate schemes are available. The advantage of the multislice method is that it can handle crystalline (including crystal defects) as well as amorphous specimens in many detection modes, such as imaging and diffraction in TEM and STEM. Multislice methods are typically elastic calculations, but an absorption term can be introduced to simulate the loss of intensity from the elastic wave into the inelastic wave, particularly the TDS from electron–phonon interactions. If the TDS distribution in reciprocal space is important, for example, in CBED or STEM HAADF measurements, then frozen phonon calculations must be carried out. The frozen phonon method gives physically realistic results provided that multiple electron–phonon scattering events are treated incoherently and the incident electron interacts with the crystal in the ground state. The latter is nearly always satisfied, while the former is never satisfied, although multiple scattering can be avoided if the specimen is sufficiently thin.

REFERENCES

Boothroyd, C.B. and Yeadon, M. (2003) *Ultramicroscopy* **96**, 361.

Born, M. and Wolf, E. (2002) *Principles of Optics*, 7th edition, Cambridge University Press, Cambridge.

Brydson, R. (2011) editor, *Aberration-Corrected Analytical Transmission Electron Microscopy*, John Wiley & Sons, UK.

Chen, J.H. and van Dyck, D. (1997) *Ultramicroscopy* **70**, 29.

Coene, W.M.J., Thust, A., Op de Beeck, M. and van Dyck, D. (1996) *Ultramicroscopy* **64**, 109.

Cooley, J.W. and Tukey, J.W. (1965) *Math Comput.* **19**, 297.

Cowley, J.M. and Moodie, A.F. (1957) *Acta Cryst.* **10**, 609.

Doyle, P.A. and Turner, P.S. (1968) *Acta Cryst. A* **24**, 390.

Dwyer, C. and Etheridge, J. (2003) *Ultramicroscopy* **96**, 343.

van Dyck, D. (1980) *J. Microscopy* **119**, 141.

van Dyck, D. (1979) *Phys. Status Solidi (a)* **52**, 283.

van Dyck, D. (2009) *Ultramicroscopy* **109**, 677.

van Dyck, D. (2011) *Ultramicroscopy* **111**, 894.

van Dyck, D. and Coene, W. (1984) *Ultramicroscopy* **15**, 29.

Ernie, R. (2010) *Aberration-Corrected Imaging in Transmission Electron Microscopy*, Imperial College Press, London.

Fowles, G.R. (1989) *Introduction to Modern Optics*, Dover edition, USA.

Gao, H.X. and Peng, L.-M. (1999) *Acta Cryst. A* **55**, 926.

Goodman, P. and Moodie, A.F. (1974) *Acta Cryst. A* **30**, 280.

Haigh, S.J., Sawada, H. and Kirkland, A.I. (2009) *Phys. Rev. Lett.* **103**, 126101.

Hall, C.R. and Hirsch, P.B. (1965) *Proc. Roy. Soc. A* **286**, 158.

Hammond, C. (2009) *The Basics of Crystallography and Diffraction*, Oxford University Press, Oxford.

Hirsch, P.B., Howie, A., Nicholson, R.B., Pashley, D.W. and Whelan, M.J. (1965) *Electron Microscopy of Thin Crystals*, Butterworths, Great Britain.

Ishizuka, K. (1982) *Acta Cryst. A* **38**, 773.

Jansen, J., Tang, D., Zandbergen, H.W. and Schenk, H. (1998) *Acta Cryst. A* **54**, 91.

Jia, C.L., Lentzen, M. and Urban, K. (2003) *Science* **299**, 870.

de Jong, A.F. and van Dyck, D. (1993) *Ultramicroscopy* **49**, 66.

Kilaas, R., O'Keefe, M.A. and Krishnan, K.M. (1987) *Ultramicroscopy* **21**, 47.

Kirkland, E.J. (2010) *Advanced Computing in Electron Microscopy*, 2nd edition, Springer, USA.

Kirkland, A.I., Saxton, W.O., Chau, K.-L., Tsuno, K. and Kawasaki, M. (1995) *Ultramicroscopy* **57**, 355.

Klenov, D.O. and Stemmer, S. (2006) *Ultramicroscopy* **106**, 889.

Loane, R.F., Xu, P. and Silcox, J. (1991) *Acta Cryst. A* **47**, 267.

Meyer, J.C., Kurasch, S., Park, H.J., Skakalova, V., Künzel, D., Groß, A., Chuvilin, A., Algara-Siller, G., Roth, S., Iwasaki, T., Starke, U., Smet, J.H. and Kaiser, U. (2011) *Nature Mat.* **10**, 209.

Muller, D.A., Edwards, B., Kirkland, E.J. and Silcox, J. (2001) *Ultramicroscopy* **86**, 371.

Op de Beeck, M., van Dyck, D. and Coene, W. (1996) *Ultramicroscopy* **64**, 167.

Own, C.S., Marks, L.D. and Sinkler, W. (2006) *Acta Cryst. A* **62**, 434.

Peng, L.-M., Ren, G., Dudarev, S.L. and Whelan, M.J. (1996) *Acta Cryst. A* **52**, 257.

Self, P.G. and O'Keefe, M.A. (1988) in *High-Resolution Transmission Electron Microscopy and Associated Techniques*, Buseck, P.R., Cowley, J.M. and Eyring, L. eds., Oxford University Press, Oxford.

Spence, J.C.H. (2003) *High-Resolution Electron Microscopy*, 3rd edition, Oxford University Press, Oxford.

Vincent, R. and Midgley, P.A. (1994) *Ultramicroscopy* **53**, 271.

Wang, Z.L. (1998) *Acta Cryst. A* **54**, 460.

White, T.A., Eggeman, A.S. and Midgley, P.A. (2010) *Ultramicroscopy* **110**, 763.

Williams, D.B. and Carter, C.B. (1996) *Transmission Electron Microscopy, Vol. III-Imaging*, Springer, New York.

Yoshioka, H. (1957) *J. Phys. Soc. Japan* **12**, 618.

Zandbergen, H.W., Tang, D., Jansen, J. and Cava, R.J. (1996) *Ultramicroscopy* **64**, 231.

4

Bloch Waves

The second of the dynamic diffraction theories, which is also widely used for analysing electron beam–specimen interactions, is the Bloch wave method. This involves solving Schrödinger's equation for the incident electron within the electrostatic potential field of the specimen. Solutions to Schrödinger's equation are known as Bloch waves or Bloch states (here the two terms are used interchangeably). There are many Bloch wave solutions, but in most cases only a few of them are 'excited' by the electron beam. For a crystalline specimen oriented along a zone-axis the Bloch waves excited typically have a direct relationship to the projected potential, which is also periodic. For example, the so-called 1s-*states* have intensity maxima at the atom columns, which gives rise to highly localised scattering events, such as inner-shell ionisation or high angle thermal diffuse scattering (TDS). On the other hand, Bloch states with intensity maxima between the atom columns are strongly transmitted and enable imaging of specimens that are moderately thick, typically up to a few hundred nanometres in thickness. Both 1s- and non-1s states may be simultaneously excited, depending on the crystal orientation and operating voltage of the microscope. The phase difference between these Bloch states gives rise to interference or 'beating' effects (i.e. the Bloch waves go in and out of phase as the depth within the crystal is varied), which is the origin of dynamic diffraction.

This relatively simple picture of dynamic diffraction makes the Bloch wave method physically intuitive compared to multislice (Chapter 3). Computationally it is suitable for simulating materials with small unit cells, such as metals and semiconductors. For larger unit cells, however, the number of Fourier terms required to accurately represent a given Bloch state increases significantly, so that the Bloch wave method

Electron Beam-Specimen Interactions and Simulation Methods in Microscopy,
First Edition. Budhika G. Mendis.
© 2018 John Wiley & Sons Ltd. Published 2018 by John Wiley & Sons Ltd.

becomes computationally too expensive. Similarly, it is also unsuitable for defect crystals with rapidly varying strain fields, such as the core region of a dislocation, although slowly varying strain fields (e.g. the long-range elastic stress field of a dislocation) can be analysed by invoking the column approximation (Section 3.2.1). The method therefore lacks the versatility of the multislice method, but nevertheless provides much insight into the fundamental physics of dynamic diffraction.

In this chapter, the Bloch wave method is discussed for a thin specimen where elastic scattering is dominant, with TDS being treated phenomenologically (i.e. as a depletion of the elastic wavefield intensity). In Section 4.1, the mathematical background is presented, from which results for the so-called *two beam* case, that is, transmitted beam and only one diffracted beam, are derived. This is by far one of the simplest cases encountered in electron diffraction, but is an extremely useful starting point for analysing more complex cases. Application of Bloch wave theory to high resolution electron microscopy (HREM) and high angle annular dark field (HAADF) imaging, as well as diffraction contrast imaging of crystal defects, are presented in Section 4.2. The chapter concludes with some further topics, namely scanning transmission electron microscopy (STEM) methods for imaging dopant atoms as well as the effect of channelling on chemical microanalysis and electron backscattering.

4.1 BASIC PRINCIPLES

4.1.1 Mathematical Background

The theory of Bloch waves, as presented in this section, is based on the texts by Hirsch *et al.* (1965) and Spence and Zuo (1992). Schrödinger's equation for a high energy electron in the crystal is given by

$$\nabla^2 \psi(\mathbf{r}) + \left[4\pi^2 k_{vac}^2 + \frac{8\pi^2 m e}{h^2} V(\mathbf{r}) \right] \psi(\mathbf{r}) = 0 \qquad (4.1)$$

Here, k_{vac} is the wave number of the electron in vacuum. The wave number within the crystal is larger compared to vacuum, owing to the electrostatic potential field (Section 3.1.1). The energy of the incident electron is $E = (h k_{vac})^2/(2m)$, and is conserved during elastic scattering within the crystal. For a perfect crystal the potential $V(\mathbf{r})$ can be expressed as a Fourier series over reciprocal vectors $\mathbf{g}$:

$$V(\mathbf{r}) = \sum_{\mathbf{g}} V_{\mathbf{g}} \exp(2\pi i \mathbf{g} \cdot \mathbf{r}) = \frac{h^2}{2me} \sum_{\mathbf{g}} U_{\mathbf{g}} \exp(2\pi i \mathbf{g} \cdot \mathbf{r}) \qquad (4.2)$$

where $V_{\mathbf{g}}$ are Fourier coefficients (the scaled Fourier coefficients $U_{\mathbf{g}}$ are used for mathematical convenience). Since the potential is a real quantity, that is, $V(\mathbf{r}) = V(\mathbf{r})^*$, it follows that $V_{\mathbf{g}} = V_{-\mathbf{g}}^*$. Furthermore, if the crystal is centrosymmetric with origin at the inversion centre, then $V(\mathbf{r}) = V(-\mathbf{r})$, so that $V_{\mathbf{g}} = V_{-\mathbf{g}} = V_{\mathbf{g}}^*$ (i.e. the Fourier coefficients are real numbers). Given Eq. (4.2) an appropriate solution to Schrödinger's equation is

$$\psi(\mathbf{r}) = \sum_{\mathbf{g}} C_{\mathbf{g}} \exp\left[2\pi i(\mathbf{k} + \mathbf{g}) \cdot \mathbf{r}\right] \qquad (4.3)$$

where the $C_{\mathbf{g}}$ terms are constants and $\mathbf{k}$ is a vector to be determined. First, consider evaluating the $\nabla^2 \psi(\mathbf{r})$ term in Eq. (4.1). The simpler case of differentiation along a single spatial dimension x gives,

$$\partial \psi(\mathbf{r}) / \partial x = 2\pi i \sum_{\mathbf{g}} (\mathbf{k} + \mathbf{g})_x C_{\mathbf{g}} \exp\left[2\pi i(\mathbf{k} + \mathbf{g}) \cdot \mathbf{r}\right]$$

$$\partial^2 \psi(\mathbf{r}) / \partial x^2 = -4\pi^2 \sum_{\mathbf{g}} \left[(\mathbf{k} + \mathbf{g})_x\right]^2 C_{\mathbf{g}} \exp\left[2\pi i(\mathbf{k} + \mathbf{g}) \cdot \mathbf{r}\right] \qquad (4.4)$$

where $(\mathbf{k} + \mathbf{g})_x$ is the x-component of the vector $(\mathbf{k} + \mathbf{g})$. Therefore,

$$\nabla^2 \psi(\mathbf{r}) = \frac{\partial^2 \psi}{\partial x^2} + \frac{\partial^2 \psi}{\partial y^2} + \frac{\partial^2 \psi}{\partial z^2} = -4\pi^2 \sum_{\mathbf{g}} |\mathbf{k} + \mathbf{g}|^2 C_{\mathbf{g}} \exp\left[2\pi i(\mathbf{k} + \mathbf{g}) \cdot \mathbf{r}\right]$$

$$(4.5)$$

Substituting Eqs. (4.2) and (4.5) in Eq. (4.1) gives

$$\sum_{\mathbf{g}} \left\{ \left(k_{mean}^2 - |\mathbf{k} + \mathbf{g}|^2\right) C_{\mathbf{g}} + \sum_{\mathbf{h} \neq \mathbf{0}} U_{\mathbf{h}} C_{\mathbf{g}-\mathbf{h}} \right\} = 0 \qquad (4.6)$$

$$k_{mean}^2 = k_{vac}^2 + U_0 \qquad (4.7)$$

Physically k_{mean} represents the modified wave number of the incident electron due to the long range, mean inner potential of the crystal. This follows from the approximate relation $k^2_{mean} = [2me(E + V_0)]/h^2$. In order to satisfy Eq. (4.6), the terms within the curly brackets must equal zero for each reciprocal lattice vector g:

$$(k_{mean}^2 - |\mathbf{k} + \mathbf{g}|^2)C_g + \sum_{h \neq 0} U_h C_{g-h} = 0 \tag{4.8}$$

Equation (4.8) represents N equations, where N is the number of reciprocal vectors g used in the Fourier expansion of the potential $V(\mathbf{r})$ (Eq. (4.2)) and wavefunction $\psi(\mathbf{r})$ (Eq. (4.3)). The unknown quantities are the Bloch state wave vector $\mathbf{k}$ and coefficients C_g. The simplest way of extracting these is to transform Eq. (4.8) into an eigen problem, with $|\mathbf{k}|$ being determined by the eigenvalue and the corresponding eigenvector yielding the C_g coefficients (expressed as an $N \times 1$ column vector). There are N eigenvector–eigenvalue solutions, corresponding to N different Bloch states. To see how Eq. (4.8) can be transformed into an eigen problem first consider the wave vector $\mathbf{k}$. The total electron wavefunction within the crystal is a linear superposition of all N Bloch states, that is,

$$\psi(\mathbf{r}) = \sum_{j=1}^{N} \varepsilon^{(j)} \left\{ \sum_g C_g^{(j)} \exp[2\pi i (\mathbf{k}^{(j)} + \mathbf{g}) \cdot \mathbf{r}] \right\} \tag{4.9}$$

where the jth Bloch state has wave vector $\mathbf{k}^{(j)}$, coefficients $C_g^{(j)}$ and excitation $\varepsilon^{(j)}$. The boundary conditions require that the wavefunction within the crystal (Eq. (4.9)) is identical to the wavefunction $\exp(2\pi i \mathbf{k}_{vac} \cdot \mathbf{r})$ in vacuum at *all points* at the specimen entrance surface. This is only possible if $\mathbf{k}^{(j)}$ and $\mathbf{k}_{vac}$ have identical transverse components parallel to the entrance surface plane (at the entrance surface plane the contribution of the longitudinal wave vector component to the terms $\mathbf{k}^{(j)} \cdot \mathbf{r}$ and $\mathbf{k}_{vac} \cdot \mathbf{r}$ is independent of position).[1] Since a similar reasoning must apply to $\mathbf{k}_{mean}$ and $\mathbf{k}_{vac}$, $\mathbf{k}^{(j)}$ can be rewritten as

$$\boxed{\mathbf{k}^{(j)} = \mathbf{k}_{mean} - \gamma^{(j)} \mathbf{n}} \tag{4.10}$$

where $\mathbf{n}$ is a unit reciprocal vector pointing outwards from and normal to the specimen entrance surface, while $\gamma^{(j)}$ represents the change in longitudinal wave vector component of the jth Bloch state as a result of

---

[1] Apart from a constant transverse wave vector component, there is also a second boundary condition which states that $\partial \psi / \partial n$ must be constant at the entrance surface, where differentiation is along the surface normal $\mathbf{n}$. This condition ensures that the electron flux is conserved. For high energy electron diffraction however, reflection at the entrance surface is negligible. The condition therefore simplifies to the requirement that the electron intensity within the crystal is equal to the incident intensity.

interaction with the periodic crystal potential. The $(k_{mean}^2 - |\mathbf{k}^{(j)} + \mathbf{g}|^2)$ term in Eq. (4.8) can now be simplified as

$$\left(k_{mean}^2 - |\mathbf{k}^{(j)} + \mathbf{g}|^2\right) = \left(k_{mean}^2 - |\mathbf{k}_{mean} + \mathbf{g}|^2\right) + 2(\mathbf{k}_{mean} + \mathbf{g}) \cdot \gamma^{(j)}\mathbf{n} - \left[\gamma^{(j)}\right]^2$$

$$= 2\mathbf{k}_{mean} \cdot \mathbf{s_g} + 2(\mathbf{k}_{mean} + \mathbf{g}) \cdot \gamma^{(j)}\mathbf{n} - [\gamma^{(j)}]^2 \qquad (4.11)$$

where the deviation parameter s_g for the **g** diffracted beam has the following relationship:

$$2\mathbf{k}_{mean} \cdot \mathbf{s_g} = \left(k_{mean}^2 - |\mathbf{k}_{mean} + \mathbf{g}|^2\right) \qquad (4.12)$$

This readily follows from the fact that if $\mathbf{k}'$ is the wave vector within the crystal following diffraction of $\mathbf{k}_{mean}$, then $\mathbf{k}' = (\mathbf{k}_{mean} + \mathbf{g} + \mathbf{s_g})$. Squaring both sides and noting that $|\mathbf{k}'|^2 = |\mathbf{k}_{mean}|^2$ leads to Eq. (4.12), with the following assumptions: $\mathbf{g} \cdot \mathbf{s_g} = 0$ (valid for zero order Laue zone (ZOLZ) reflections in an untilted thin foil) and $|\mathbf{s_g}|^2$ is small compared to $\mathbf{k}_{mean} \cdot \mathbf{s_g}$.

The $\gamma^{(j)}$ term is large for backscattering ($\gamma^{(j)} > |\mathbf{k}_{mean}|$; Eq. (4.10)), but for high energy electron diffraction scattering is predominantly in the forward direction and $\gamma^{(j)}$ is then sufficiently small to ignore the $[\gamma^{(j)}]^2$ term in Eq. (4.11). Consider the case of normal incidence in an untilted specimen, so that **n** is anti-parallel to $\mathbf{k}_{mean}$. This means that $|\mathbf{k}_{mean} \cdot \mathbf{n}| \gg |\mathbf{g} \cdot \mathbf{n}|$ for all reflections, including higher order Laue zone (HOLZ) reflections. With these simplifications to Eq. (4.11), Eq. (4.8) can be rearranged as

$$2k_{mean}s_g C_g^{(j)} + \sum_{h \neq 0} U_h C_{g-h}^{(j)} = 2k_{mean}\gamma^{(j)}C_g^{(j)} \qquad (4.13)$$

or

$$\boxed{\mathbf{A}\mathbf{C}^{(j)} = \gamma^{(j)}\mathbf{C}^{(j)}} \qquad (4.14)$$

where the eigenvector $\mathbf{C}^{(j)}$ is an $(N \times 1)$ column vector whose elements are the $C_g^{(j)}$ coefficients and **A** is a so-called $(N \times N)$ 'structure' matrix, where the diagonal elements A_{gg} are equal to s_g and the off-diagonal elements A_{gh} are equal to $U_{g-h}/(2k_{mean})$, or equivalently $1/(2\xi_{g-h})$, where ξ_{g-h} is the 'two-beam' extinction distance for the $(\mathbf{g}-\mathbf{h})$ reflection (Section 4.1.2). The fact that $U_{g-h} = U_{h-g}^*$ (real crystal potential), or equivalently $A_{gh} = A_{hg}^*$, means that **A** is a Hermitian matrix, and in such cases the $\gamma^{(j)}$ eigenvalues are real.[2] These are obtained from the secular equation:

$$\det(\mathbf{A} - \gamma^{(j)}\mathbf{I}) = 0 \qquad (4.15)$$

---

[2] Assume that λ and w are an eigenvalue–eigenvector pair of the Hermitian matrix A. Since $Aw = \lambda w$, taking the transpose followed by the complex conjugate gives $w^\dagger A = \lambda^* w^\dagger$, where

where $\mathbf{I}$ is an $(N \times N)$ identity matrix and 'det' denotes the determinant. Equation (4.15) is an Nth order polynomial in $\gamma^{(j)}$. Individual solutions of $\gamma^{(j)}$ can be substituted into Eq. (4.14) to obtain the eigenvector $\mathbf{C}^{(j)}$ and hence $C_g^{(j)}$ coefficients. Furthermore, a separate $(N \times N)$ matrix $\mathbf{C}$ can be constructed where individual columns correspond to the eigenvectors $\mathbf{C}^{(j)}$. For the Hermitian structure matrix $\mathbf{A}$ it can be shown that $\mathbf{C}$ is unitary,[3] that is, the inverse of $\mathbf{C}$ ($\mathbf{C}^{-1}$) is the complex conjugate of $\mathbf{C}^T$, where 'T' denotes the matrix transpose. From the two identities, $\mathbf{CC}^{-1} = \mathbf{I}$ and $\mathbf{C}^{-1}\mathbf{C} = \mathbf{I}$, it then follows that

$$\sum_i C_g^{(i)} C_h^{(i)*} = \delta_{gh} \qquad (4.16)$$

$$\sum_g C_g^{(i)} C_g^{(j)*} = \delta_{ij} \qquad (4.17)$$

where the δ terms represent Kronecker delta functions. Equation (4.17) indicates that the Bloch states form an orthonormal set.

At this stage, it is useful to revisit the expression for the electron wavefunction within the crystal (Eq. (4.9)). For convenience assume an untilted foil, so that the two-dimensional spatial coordinate $\mathbf{R}$ is defined in the specimen plane and 'z' is the depth within the specimen (the positive z-axis is in the direction of increasing depth from the entrance surface). The so-called *projection approximation* assumes that at high energies the fast moving electron is largely insensitive to the potential variation along the z-direction, so that the important reciprocal vectors are confined to the ZOLZ. Equation (4.9) then simplifies to

$$\psi(\mathbf{R}, z) = \sum_{j=1}^{N} \varepsilon^{(j)} \left[\left\{ \sum_{g \in \text{ZOLZ}} C_g^{(j)} \exp\left[2\pi i \left((\mathbf{k}_{\text{mean}})_t + \mathbf{g}\right) \cdot \mathbf{R}\right] \right\} \exp(2\pi i \gamma^{(j)} z) \right]$$

$$= \sum_{j=1}^{N} \varepsilon^{(j)} b^{(j)}(\mathbf{R}, z) \qquad (4.18)$$

where $(\mathbf{k}_{\text{mean}})_t$ is the transverse component of $\mathbf{k}_{\text{mean}}$ and $b^{(j)}(\mathbf{R}, z)$ is the jth Bloch state. For simplicity the constant phase factor due to the longitudinal component of $\mathbf{k}_{\text{mean}}$ has been ignored. The electron wavefunction

---

the † sign represents the Hermitian operation. Multiplying the former equation on the left by $w^\dagger$ and the latter equation on the right by w and subtracting gives $(\lambda - \lambda^*)w^\dagger w = 0$. Since $w^\dagger w \neq 0$ (w is not a null vector) it follows that $\lambda = \lambda^*$, that is, the eigenvalue is real (Dass and Sharma, 1998).

[3] The generalisation of Eq. (4.14) is $\mathbf{AC} = \mathbf{C}\{\gamma^{(j)}\}$, where { } denotes a diagonal matrix in which only the diagonal elements are non-zero. Performing the Hermitian operation and noting that the $\gamma^{(j)}$ values are real gives $(\mathbf{C}^T)^*\mathbf{A} = \{\gamma^{(j)}\}(\mathbf{C}^T)^*$. Right multiplying by $\mathbf{C}$ results in $[\mathbf{C}(\mathbf{C}^T)^*]\mathbf{A} = \mathbf{C}\{\gamma^{(j)}\}(\mathbf{C}^T)^* = \mathbf{A}[\mathbf{C}(\mathbf{C}^T)^*]$, which means $\mathbf{C}^{-1} = (\mathbf{C}^T)^*$.

can also be expressed in terms of diffracted beams:

$$\psi(\mathbf{r}) = \sum_{g} \phi'_g(z) \exp\left[2\pi i(\mathbf{k}_{mean} + \mathbf{g} + \mathbf{s}_g) \cdot \mathbf{r}\right] \tag{4.19}$$

where ϕ'_g is the diffracted beam amplitude for the $\mathbf{g}$ reciprocal vector. If the important diffracted beams are confined to the ZOLZ, comparing Eqs. (4.18) and (4.19) gives

$$\phi_g(z) = \phi'_g(z)\exp(2\pi i s_g z) = \sum_{j=1}^{N} \varepsilon^{(j)} C_g^{(j)} \exp\left(2\pi i \gamma^{(j)} z\right) \tag{4.20}$$

Note that since the foil is untilted $\mathbf{s}_g \cdot \mathbf{R} = 0$. ϕ_g is a modified amplitude that contains the additional phase factor $\exp(2\pi i s_g z)$, which nevertheless does not affect the intensity of the diffracted beam. Equation (4.20) can be expressed in matrix form as

$$\boldsymbol{\varphi} = \mathbf{C}\left\{\exp\left(2\pi i \gamma^{(j)} z\right)\right\} \boldsymbol{\varepsilon} \tag{4.21}$$

where $\boldsymbol{\varphi}$ and $\boldsymbol{\varepsilon}$ are $(N \times 1)$ column vectors whose elements are the (modified) diffracted beam amplitudes and Bloch wave excitations respectively. The curly brackets in $\{\exp(2\pi i \gamma^{(j)} z)\}$ indicate an $(N \times N)$ diagonal matrix, where the diagonal elements are $\exp(2\pi i \gamma^{(j)} z)$ and the off-diagonal elements are zero. Equation (4.21) can be used to calculate the excitation $\varepsilon^{(j)}$ of a given Bloch wave via boundary conditions. At the specimen entrance surface ($z = 0$) the (normalised) amplitude of the transmitted beam is unity and zero for all diffracted beams. Using the fact that $\mathbf{C}^{-1} = (\mathbf{C}^T)^*$ then gives $\varepsilon^{(j)} = [C_0^{(j)}]^*$. Thus, the electron wavefunction at any given depth can be calculated using Bloch waves (Eq. (4.18)) and related to the diffracted beam intensities via Eq. (4.21). The advantage of this method is that the computation time is governed by calculation of the Bloch wave parameters and is independent of the specimen depth z. Compare this to multislice calculations where simulation of thick specimens can become time consuming.

4.1.2 Application to Two-Beam Theory

In this section, Bloch wave solutions are derived for 'two-beam' conditions, where only the transmitted beam (0) and a single diffracted beam (g) have appreciable intensity. Only two Fourier coefficients (V_0 and V_g)

are required to reproduce the scattering potential, so that the problem consists of only two Bloch states. Despite its simplicity, results from the two-beam theory provide much insight into the nature of Bloch waves and are an important starting point for understanding dynamic diffraction.

For a centrosymmetric crystal under two-beam conditions, Eq. (4.14) simplifies to

$$
\begin{pmatrix} -\gamma^{(j)} & \dfrac{1}{2\xi_g} \\[2mm] \dfrac{1}{2\xi_g} & s_g - \gamma^{(j)} \end{pmatrix} \begin{pmatrix} C_0^{(j)} \\[2mm] C_g^{(j)} \end{pmatrix} = \begin{pmatrix} 0 \\[2mm] 0 \end{pmatrix}
\tag{4.22}
$$

For non-trivial solutions of $C_0^{(j)}$ and $C_g^{(j)}$ the determinant of the square matrix on the left-hand side of Eq. (4.22) must equal zero (this is equivalent to Eq. (4.15)), so that

$$
\gamma^{(1)} = \frac{w + \sqrt{1 + w^2}}{2\xi_g} \quad \text{and} \quad \gamma^{(2)} = \frac{w - \sqrt{1 + w^2}}{2\xi_g}
\tag{4.23}
$$

where $w = s_g \xi_g$ is a dimensionless deviation parameter. Furthermore,

$$
C_g^{(j)} / C_0^{(j)} = 2\gamma^{(j)} \xi_g = \begin{cases} w + \sqrt{1 + w^2} & (j = 1) \\ w - \sqrt{1 + w^2} & (j = 2) \end{cases}
\tag{4.24}
$$

In the Bragg orientation ($w = 0$), $C_g^{(1)} = C_0^{(1)}$ and $C_g^{(2)} = -C_0^{(2)}$. From the orthonormal property of Bloch states, $|C_0^{(j)}|^2 + |C_g^{(j)}|^2 = 1$ (Eq. (4.17)), it follows that $|C_0^{(j)}| = |C_g^{(j)}| = 1/\sqrt{2}$. Since the origin of the crystal is at its centre of symmetry, the Fourier coefficients of the crystal potential and therefore the ξ_g and $C_g^{(j)}$ values are real. The Bloch states $b^{(j)}(\mathbf{R}, z)$ can be derived from Eq. (4.18) and are given by

$$
\begin{aligned}
b^{(1)}(\mathbf{R}, z) &= \frac{1}{\sqrt{2}} \left[1 + \exp(2\pi i \mathbf{g} \cdot \mathbf{R}) \right] \exp\left[2\pi i (\mathbf{k}_{\text{mean}})_t \cdot \mathbf{R} \right] \exp\left(\frac{\pi i}{\xi_g} z \right) \\
&= \sqrt{2} \cos(\pi \mathbf{g} \cdot \mathbf{R}) \exp\left[2\pi i \left((\mathbf{k}_{\text{mean}})_t + \frac{\mathbf{g}}{2} \right) \cdot \mathbf{R} \right] \exp\left(\frac{\pi i}{\xi_g} z \right) \\
b^{(2)}(\mathbf{R}, z) &= \frac{1}{\sqrt{2}} \left[1 - \exp(2\pi i \mathbf{g} \cdot \mathbf{R}) \right] \exp\left[2\pi i (\mathbf{k}_{\text{mean}})_t \cdot \mathbf{R} \right] \exp\left(-\frac{\pi i}{\xi_g} z \right) \\
&= -i \sqrt{2} \sin(\pi \mathbf{g} \cdot \mathbf{R}) \exp\left[2\pi i \left((\mathbf{k}_{\text{mean}})_t + \frac{\mathbf{g}}{2} \right) \cdot \mathbf{R} \right] \exp\left(-\frac{\pi i}{\xi_g} z \right)
\end{aligned}
\tag{4.25}
$$

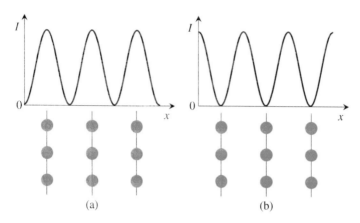

Figure 4.1 Intensity (I) of Bloch states (a) 1 and (b) 2 as a function of position x for two-beam conditions. In the former the electrons channel along the diffracting crystal planes, while for the latter the channelling is between the planes.

If **R** is varied normal to the **g** diffracting planes, then Bloch state 1 is found to have intensity maxima at the atomic planes (due to the $\cos(\pi \mathbf{g} \cdot \mathbf{R})$ term) while Bloch state 2 has intensity minima ($\sin(\pi \mathbf{g} \cdot \mathbf{R})$ term). This is schematically illustrated in Figure 4.1. Since Bloch state 1 is concentrated along a region of low potential it will acquire a greater kinetic energy, although the total energy is conserved, as required for elastic scattering. The increase in kinetic energy is the origin of a positive γ-value and hence larger wave number for Bloch state 1. The opposite scenario is valid for Bloch state 2.

A dispersion surface plots the variation in electron wave vector within the crystal as a function of deviation parameter. The dispersion surface for the two-beam case is shown in Figure 4.2. $\mathbf{k}_{\text{mean}}$ and its Ewald sphere are superimposed on the reciprocal lattice with origin **O** and reciprocal lattice point **G**. Bragg condition is satisfied if the start point of $\mathbf{k}_{\text{mean}}$ lies on the perpendicular bisector of the reciprocal lattice vector **g** (i.e. Brillouin zone boundary). If $\mathbf{k}_{\text{mean}}$ is tilted further towards **G** the deviation parameter is positive (i.e. **G** lies inside the Ewald sphere), while if $\mathbf{k}_{\text{mean}}$ is tilted in the opposite direction the deviation parameter is negative (i.e. **G** lies outside the Ewald sphere). The wave vectors within the crystal are constructed by shifting the start point of $\mathbf{k}_{\text{mean}}$ along the vector **n** by a distance equal to $\gamma^{(i)}$ (Eq. (4.23)). The transverse wave vector component remains unchanged as required by the boundary conditions (Section 4.1.1). The upper and lower solid lines in Figure 4.2 are the wave vectors for Bloch states 1 and 2 respectively, with a gap of $(1/\xi_g)$ at the Brillouin zone boundary (Eq. (4.23)). The dashed lines in the figure are circles of radii k_{mean} centred on

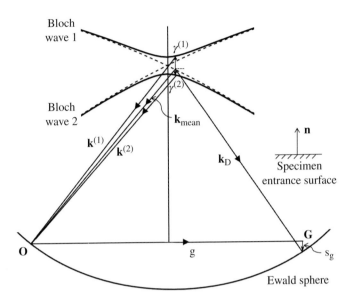

Figure 4.2 Dispersion surface diagram for two-beam diffracting conditions (not drawn to scale). $\mathbf{k}_{mean}$ is the incident beam wave vector after correcting for the mean inner potential. $\mathbf{k}^{(1)}$ and $\mathbf{k}^{(2)}$ are the wave vectors for the two Bloch waves; note that the wave vector component along the specimen surface normal $\mathbf{n}$ is altered by $\gamma^{(1)}$ and $\gamma^{(2)}$ respectively. The Bloch wave dispersion surfaces are highlighted by the solid lines, while the two circles of radius $\mathbf{k}_{mean}$ centred about the origin O and reciprocal lattice point G are indicated by the dashed lines. The diffracted beam wave vector is along $\mathbf{k}_D$ and has deviation parameter s_g.

O and G. If the electrons only interacted with the mean inner potential of the crystal the start point of $\mathbf{k}_{mean}$ would lie on the circle centred about O. From Figure 4.2 it is clear that the Bloch state wave vectors are asymptotic to these circles at large deviation parameters. For example, at positive deviation parameters Bloch state 2 is asymptotic to the circle centred about O, while Bloch state 1 is asymptotic to the circle centred about G. From Eq. (4.24) and the condition $|C_0^{(j)}|^2 + |C_g^{(j)}|^2 = 1$ it follows that $|\varepsilon^{(j)}| = |C_0^{(j)*}| = 1/\sqrt{[1 + (2\gamma^{(j)}\xi_g)^2]}$. Therefore, at large positive deviation parameters only Bloch state 2 will be excited by the electron beam and furthermore $C_g^{(2)} \to 0$, so that Bloch state 2 has the form of a plane wave. This is consistent with the physical picture that at large transverse wave vector components the incident electrons can no longer couple efficiently with the projected potential of the atom planes and are therefore only affected by the long range, mean inner potential of the crystal. Similar reasoning applies to large negative deviation parameters, where only Bloch state 1 is excited.

The intensity of the diffracted (I_g) and transmitted (I_0) beams can be calculated from Eq. (4.20), with the aid of Eqs. (4.23) and (4.24):

$$I_g = |\phi_g|^2 = \frac{\sin^2 \left[\pi(\sqrt{1 + w^2}/\xi_g)z \right]}{1 + w^2} = \frac{\sin^2 [\pi(\Delta\gamma)z]}{1 + w^2} \qquad (4.26)$$

$$I_0 = 1 - I_g \qquad (4.27)$$

where $\Delta\gamma$ is the difference in γ-values for the two Bloch states or equivalently the separation between the two branches of the dispersion surface at a given deviation parameter. The difference in Bloch state wave vector gives rise to 'beating', such that I_g and I_0 oscillate with depth z. At Bragg orientation the periodicity of oscillation is ξ_g, while off-Bragg it is shortened to $\xi_g/\sqrt{(1 + w^2)}$. This phenomenon is readily observed in wedge-shaped thin foils in the form of thickness fringes. An example is shown in Figure 4.3a and b, which are bright-field and dark-field images from the same area of a thin silicon foil in the $\mathbf{g} = \pm 220$ two-beam orientation. Changes in the deviation parameter, such as, across a bend contour, also give rise to oscillations in I_g and I_0 (Eqs. (4.26) and (4.27)). Figure 4.3c shows the transmitted and diffracted convergent beam discs for a $\sim$90 nm thick region of silicon in the $\mathbf{g} = 220$ two-beam orientation, where the above mentioned intensity oscillations are apparent. The equivalent result for a thicker ($\sim$215 nm) region of the foil is shown in Figure 4.3d. There are now many more oscillations due to the fact that for the larger z only a relatively small change in deviation parameter w is required to produce a large intensity variation.

Although Eqs. (4.26) and (4.27) describe the overall features of the experimental results in Figure 4.3, there are nevertheless subtle differences. For example, the contrast of the thickness fringes in Figure 4.3a and b are gradually damped as the foil thickness increases, such that in the thicker regions they are no longer observed. Furthermore, for thicker crystals the convergent transmitted beam disc is asymmetrical, with the *average* intensity being higher for positive deviation parameters compared to negative deviation parameters (Figure 4.3d). The diffracted beam disc is however symmetrical, as predicted by Eq. (4.26). These discrepancies are due to neglecting TDS in the thicker crystals. As discussed in Section 3.1.1, TDS can be phenomenologically modelled by treating the crystal potential as a complex quantity, with the imaginary term accounting for 'absorption' effects, that is, depletion of the elastic wave due to scattering into the thermal diffuse intensity background. In the next section, Bloch wave calculations incorporating TDS scattering are discussed.

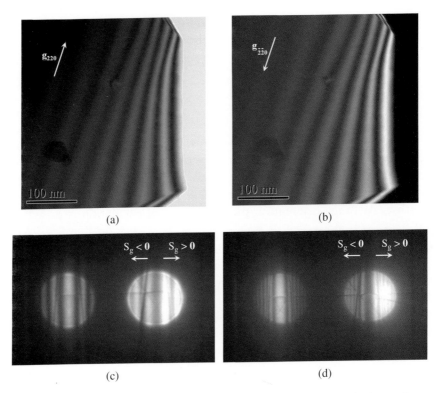

(a)   (b)

(c)   (d)

Figure 4.3 $g = \pm 220$ (a) bright-field and (b) dark-field images of silicon showing thickness pendellösung fringes. (c) and (d) are $g = 220$ two-beam, convergent beam electron diffraction patterns from approximately 90 and 215 nm thick (estimated using electron energy loss spectroscopy) regions of silicon respectively. Positive and negative deviation parameters (s_g) are indicated. The electron beam energy is 200 keV.

4.1.3 Phenomenological Modelling of Thermal Diffuse Scattering

TDS can be modelled phenomenologically via a complex crystal potential (so-called *optical* potential model), that is, Eq. (4.2) is modified to

$$V(\mathbf{r}) = \sum_g (V_g + iV_g') \exp(2\pi i \mathbf{g} \cdot \mathbf{r}) = \frac{h^2}{2me} \sum_g (U_g + iU_g') \exp(2\pi i \mathbf{g} \cdot \mathbf{r}) \quad (4.28)$$

Later it will be demonstrated how V_g', or equivalently U_g', may be calculated. With this new potential Eq. (4.13) becomes

$$\left(2k_{\text{mean}} s_g + iU_0'\right) C_g^{(j)} + \sum_{h \neq 0} \left(U_h + iU_h'\right) C_{g-h}^{(j)} = 2k_{\text{mean}} \gamma^{(j)} C_g^{(j)} \quad (4.29)$$

which can be expressed as an eigen problem similar to Eq. (4.14), that is, $\mathbf{AC}^{(j)} = \gamma^{(j)}\mathbf{C}^{(j)}$, but with the diagonal elements of $\mathbf{A}$ now equal to $[s_g + (i/2\xi'_0)]$ and the off-diagonal elements equal to $[(1/2\xi_{g-h}) + (i/2\xi'_{g-h})]$, where $\xi'_0 = k_{mean}/U'_0$ and $\xi'_{g-h} = k_{mean}/U'_{g-h}$. Since the structure matrix $\mathbf{A}$ is no longer Hermitian the $\gamma^{(j)}$ eigenvalues are complex quantities. For the two-beam case, it follows that

$$
\begin{pmatrix}
-\gamma^{(j)} + \dfrac{i}{2\xi'_0} & \dfrac{1}{2\xi_g} + \dfrac{i}{2\xi'_g} \\[2ex]
\dfrac{1}{2\xi_g} + \dfrac{i}{2\xi'_g} & s_g - \gamma^{(j)} + \dfrac{i}{2\xi'_0}
\end{pmatrix}
\begin{pmatrix}
C_0^{(j)} \\[2ex]
C_g^{(j)}
\end{pmatrix}
=
\begin{pmatrix}
0 \\[2ex]
0
\end{pmatrix}
\tag{4.30}
$$

This should be compared with Eq. (4.22). For non-zero eigenvector solutions the determinant of the square matrix in Eq. (4.30) must equal zero. Writing $\gamma^{(j)}$ as $(\gamma_0^{(j)} + iq^{(j)})$ and equating real and imaginary parts gives

$$
\left(\gamma_0^{(j)}\right)^2 - \gamma_0^{(j)} s_g - \left[(q^{(j)})^2 - \frac{q^{(j)}}{\xi'_0} + \frac{1}{4\xi_g'^2}\right] = 0 \tag{4.31}
$$

$$
(2\gamma_0^{(j)} - s_g)q^{(j)} - \frac{(2\gamma_0^{(j)} - s_g)}{2\xi'_0} - \frac{1}{2\xi_g\xi'_g} = 0 \tag{4.32}
$$

where terms involving $1/(\xi'_0)^2$ and $1/(\xi'_g)^2$ have been ignored, since it is assumed that the imaginary potentials U'_0 and U'_g are small. For the same reason, $q^{(j)}$ will also be small compared to $\gamma_0^{(j)}$. Indeed, ignoring terms involving $q^{(j)}$ in Eq. (4.31) gives a quadratic equation in $\gamma_0^{(j)}$ with solutions identical to Eq. (4.23). $q^{(j)}$ is determined by substituting these $\gamma_0^{(j)}$ values in Eq. (4.32):

$$
q^{(j)} =
\begin{cases}
\dfrac{1}{2\xi'_0} + \dfrac{1}{2\xi'_g\sqrt{1 + w^2}} & (j = 1) \\[3ex]
\dfrac{1}{2\xi'_0} - \dfrac{1}{2\xi'_g\sqrt{1 + w^2}} & (j = 2)
\end{cases}
\tag{4.33}
$$

The z-dependence of a Bloch state appears in the phase factor $\exp(2\pi i\gamma^{(j)}z)$, so that when $\gamma^{(j)}$ is imaginary the Bloch state is damped by the factor $\exp(-2\pi q^{(j)}z)$. Equation (4.18) therefore predicts a depletion of the total electron intensity with increasing z. In fact, no electrons are lost; the depletion is due to the missing electrons being scattered from the elastic wavefield (which Eq. (4.18) represents) into the TDS background. The $1/(2\xi'_0)$ term in $q^{(j)}$ (Eq. (4.33)) represents an average, background TDS scattering contribution from the crystal as a whole. The additional terms are due to the channelling characteristics of individual Bloch states. Thus Bloch state 1, which has intensity maxima along the atom planes (Figure 4.1), has a larger q-value and undergoes more TDS scattering than the average. The opposite is true for Bloch state 2, which channels between the atomic planes. A complex $\gamma^{(j)}$ also means that the eigenvector $\mathbf{C}^{(j)}$ is different to that of a 'non-absorbing' crystal (i.e. a crystal with no imaginary potential terms). The Bloch state excitations can be determined from the boundary conditions via Eq. (4.21); the relationship $\varepsilon^{(j)} = \left[C_0^{(j)}\right]^*$ is not valid for an absorbing crystal, since the matrix $\mathbf{C}$ is no longer unitary. Nevertheless, provided the imaginary part of the crystal potential is small compared to the real part, the excitation and eigenvectors of Bloch states are similar to those for a non-absorbing crystal. The only modification is the $\exp(-2\pi q^{(j)} z)$ damping term. Under these assumptions the diffracted and transmitted beam intensities for two-beam conditions can be shown to be

$$
I_g = \frac{\exp\left(\frac{-2\pi z}{\xi'_0}\right)}{1 + w^2}
$$

$$
\times \left[\cos^2(\pi\Delta\gamma z)\sinh^2\left(\frac{\pi z}{\xi'_g\sqrt{1+w^2}}\right) + \sin^2(\pi\Delta\gamma z)\cosh^2\left(\frac{\pi z}{\xi'_g\sqrt{1+w^2}}\right)\right] \quad (4.34)
$$

$$
I_0 = \frac{\exp\left(\frac{-2\pi z}{\xi'_0}\right)}{1 + w^2}
$$

$$
\times \left[\cos^2(\pi\Delta\gamma z)\left(\sqrt{1+w^2}\cosh\left(\frac{\pi z}{\xi'_g\sqrt{1+w^2}}\right) + w\sinh\left(\frac{\pi z}{\xi'_g\sqrt{1+w^2}}\right)\right)\right]^2 + \cdots
$$

$$
\sin^2(\pi\Delta\gamma z)\left(\sqrt{1+w^2}\sinh\left(\frac{\pi z}{\xi'_g\sqrt{1+w^2}}\right) + w\cosh\left(\frac{\pi z}{\xi'_g\sqrt{1+w^2}}\right)\right)^2\right] \quad (4.35)
$$

where 'sinh' and 'cosh' are hyperbolic sine and cosine respectively. $\Delta\gamma$ is the difference in $\gamma_0^{(j)}$ between the two Bloch states. In the limit of a weakly absorbing crystal ξ'_0 and ξ'_g become increasingly large and the above equations converge to the expressions for a non-absorbing crystal (Eqs. (4.26) and (4.27)).

Equations (4.34) and (4.35) can be used to resolve the discrepancies noted previously in Section 4.1.2, namely the damping of thickness fringes and asymmetry in the transmitted intensity as a function of deviation parameter. The intensity oscillations in z are still governed by the separation $\Delta\gamma$ of the branches of the dispersion surface, indicating a beating effect between Bloch waves with different wave vectors. However, as z increases the oscillations are damped out, as is evident from the asymptotic form of Eqs. (4.34) and (4.35) at large z:

$$I_g = \frac{1}{4(1+w^2)} \exp\left[2\pi z\left(\frac{1}{\xi_g'\sqrt{1+w^2}} - \frac{1}{\xi_0'}\right)\right]$$

$$I_0 = \frac{1}{4}\left(1 + \frac{w}{\sqrt{1+w^2}}\right)^2 \exp\left[2\pi z\left(\frac{1}{\xi_g'\sqrt{1+w^2}} - \frac{1}{\xi_0'}\right)\right] \qquad (4.36)$$

This is easily explained by the fact that Bloch state 1 is preferentially TDS scattered by the atomic planes, so that only Bloch state 2 remains at large z, thus eliminating any beating effect. Eventually, Bloch state 2 will also fully undergo TDS scattering, so that the elastic intensity in I_0 and I_g gradually decays to zero. The total *elastic* intensity $(I_0 + I_g)$ for an absorbing crystal is therefore not conserved. As for rocking beam patterns, that is, intensity as a function of deviation parameter, it is clear that the diffracted beam intensity is symmetrical with respect to deviation parameter w even for an absorbing crystal (Eq. (4.34)). However, this is not the case for the transmitted beam (Eq. (4.35)), with I_0, and hence the total elastic intensity $(I_0 + I_g)$, being larger for positive values of w. This is because for positive deviation parameters Bloch state 2 is preferentially excited over Bloch state 1 (Section 4.1.2), so that less intensity is lost via TDS. Figure 4.4 shows the rocking beam curves for the transmitted and $\mathbf{g} = 220$ diffracted beam in a 200 nm thick silicon sample, where the above-mentioned trends are clearly evident.

Thus far no mention has been made on how to calculate ξ_0' and ξ_g' in Eqs. (4.34) and (4.35). Several methods have been put forward in the literature, including Yoshioka's theory of elastic and inelastic scattering (Yoshioka, 1957; Yoshioka and Kainuma, 1962; see also Chapter 5) and a Green's function approach (Peng, 2005), but here we present the semi-classical, frozen phonon model of Hall and Hirsch (1965). The physical concepts underpinning the frozen phonon method were presented in detail in Sections 3.2.2 and 3.3.2; to summarise, it is assumed that the interaction time of a fast electron with the specimen is much shorter than the phonon oscillation period, so that the electron effectively sees a 'snapshot' of the vibrating atoms frozen in time. TDS is then modelled as a quasi-elastic scattering process of the incident electron by

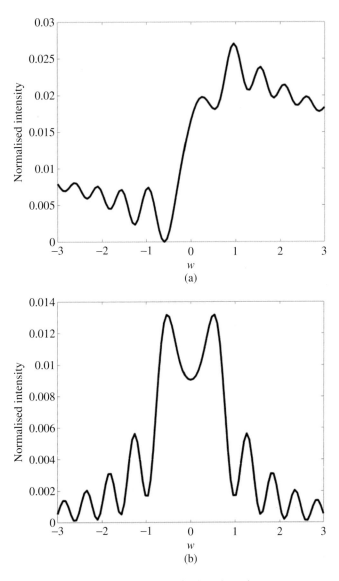

Figure 4.4 $g = 220$, bright-field and dark-field rocking beam patterns in 200 nm thick silicon, calculated using Eqs. (4.34) and (4.35). A $(\xi_g/\xi_g') = 0.1$ ratio was assumed. In (a) the intensity of the transmitted beam (I_0) is plotted as a function of the dimensionless deviation parameter w ($=s_g\xi_g$), while in (b) the intensity is that of the diffracted beam (I_g). The intensities are normalised, so that the total intensity incident on the sample is unity. The electron beam energy is 200 keV.

the displaced atoms in the solid. Many frozen phonon configurations are averaged in order to reproduce the experimental TDS intensity.

For Bragg orientation and two-beam conditions the Bloch waves have the form

$$b^{(j)}(\mathbf{r}) = \frac{1}{\sqrt{2}} \left[\exp\{2\pi i \mathbf{k}^{(j)} \cdot \mathbf{r}\} + (-1)^{j+1} \exp\{2\pi i (\mathbf{k}^{(j)} + \mathbf{g}) \cdot \mathbf{r}\} \right] \quad (4.37)$$

This equation is simply a generalisation of the Bloch waves in Eq. (4.25). The first term represents a plane wave within the crystal propagating along the incident beam direction, while the second term represents the Bragg diffracted plane wave. In the far-field the total intensity $I(\mathbf{q})$ scattered from a given Bloch wave by atoms at positions $\mathbf{r}'_m$ is given by

$$I(\mathbf{q}) = \frac{1}{2} \left| \sum_m (f_q \exp\{-2\pi i \mathbf{q} \cdot \mathbf{r}'_m\} + (-1)^{j+1} f_{q-g} \exp\{-2\pi i (\mathbf{q} - \mathbf{g}) \cdot \mathbf{r}'_m\}) \right|^2$$

$$= \frac{1}{2} \left[f_q^2 \sum_{m,n} \exp\{2\pi i \mathbf{q} \cdot (\mathbf{r}'_n - \mathbf{r}'_m)\} + f_{q-g}^2 \sum_{m,n} \exp\{2\pi i (\mathbf{q} - \mathbf{g}) \cdot (\mathbf{r}'_n - \mathbf{r}'_m)\} \right.$$

$$\left. + 2(-1)^{j+1} f_q f_{q-g} \sum_{m,n} \cos\{2\pi (\mathbf{q} \cdot (\mathbf{r}'_n - \mathbf{r}'_m) + \mathbf{g} \cdot \mathbf{r}'_m)\} \right] \quad (4.38)$$

where $\mathbf{q}$ is the scattering vector for the 'incident' plane wave and $(\mathbf{q} - \mathbf{g})$ is the scattering vector for the diffracted plane wave (see Figure 4.5a).

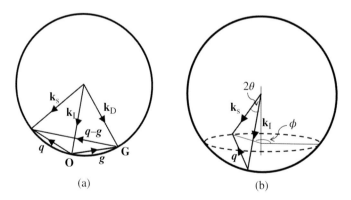

(a)   (b)

Figure 4.5 (a) Scattering geometry used for calculating the TDS intensity distribution. $\mathbf{k}_I$ and $\mathbf{k}_D$ are the incident and Bragg diffracted beam wave vectors superimposed on the Ewald sphere. The TDS wave with wave vector $\mathbf{k}_S$ is due to TDS scattering of the incident beam along the vector q. The diffracted beam can also contribute via scattering along $(q - g)$. (b) shows the scattering geometry used for calculating the total TDS intensity. Integration is carried out over all scattering vectors q in three-dimensional reciprocal space.

f_q is the atom scattering factor for scattering vector $\mathbf{q}$ and similarly for f_{q-g}. The exponential terms following f_q and f_{q-g} are the relative phase shifts due to scattering from an atom at position $\mathbf{r}'_m$ (compare with the expression for the structure factor F_g; Eq. (3.12)). $I(\mathbf{q})$ includes elastic scattering as well as TDS, although in a perfect crystal the net contribution for the former is zero for all $\mathbf{q}$ which are not a reciprocal lattice vector. The atom position $\mathbf{r}'_m$ can be written as $(\mathbf{r}_m + \mathbf{u}_m)$, where $\mathbf{r}_m$ is the equilibrium lattice position and $\mathbf{u}_m$ is the thermal vibration displacement. Averaging Eq. (4.38) over frozen phonon configurations gives

$$I(\mathbf{q}) = \frac{1}{2}\left[f_q^{\,2}\left(N + \sum_{m\neq n} \exp\{2\pi i \mathbf{q} \cdot (\mathbf{r}_n - \mathbf{r}_m)\}\overline{\exp\{2\pi i \mathbf{q} \cdot (\mathbf{u}_n - \mathbf{u}_m)\}}\right) + \cdots \right.$$

$$f_{q-g}^{\,2}\left(N + \sum_{m\neq n} \exp\{2\pi i (\mathbf{q}-\mathbf{g}) \cdot (\mathbf{r}_n - \mathbf{r}_m)\}\overline{\exp\{2\pi i (\mathbf{q}-\mathbf{g}) \cdot (\mathbf{u}_n - \mathbf{u}_m)\}}\right) + \cdots$$

$$\left. 2(-1)^{j+1} f_q f_{q-g}\left(\sum_n \overline{\cos(2\pi \mathbf{g} \cdot \mathbf{r}'_n)} + \sum_{m\neq n}\overline{\cos\{2\pi \mathbf{q} \cdot (\mathbf{r}'_n - \mathbf{r}'_m) + 2\pi \mathbf{g} \cdot \mathbf{r}'_m\}}\right)\right]$$

$$(4.39)$$

where the vinculum represents the term to be averaged and N is the number of atoms. Consider first the average of an exponential term, say, $\overline{\exp\{2\pi i \mathbf{q} \cdot (\mathbf{u}_n - \mathbf{u}_m)\}}$. Expanding up to quadratic powers only

$$\overline{\exp\{2\pi i \mathbf{q} \cdot (\mathbf{u}_n - \mathbf{u}_m)\}} = 1 - 2\pi^2\overline{(\mathbf{q} \cdot (\mathbf{u}_n - \mathbf{u}_m))^2}$$

$$= 1 - 4\pi^2 q^2\overline{u^2} \approx \exp(-2Bq^2) \qquad (4.40)$$

where B is the Debye–Waller factor (Eq. (3.16)) and $\overline{u^2}$ is the mean square displacement along a given direction. The Einstein approximation has been assumed, so that $\overline{(\mathbf{q} \cdot \mathbf{u}_m)(\mathbf{q} \cdot \mathbf{u}_n)} = 0$, that is, atom motions are uncorrelated. Furthermore,

$$\sum_n \overline{\cos(2\pi \mathbf{g} \cdot \mathbf{r}'_n)} = \sum_n \cos(2\pi \mathbf{g} \cdot \mathbf{r}_n)\overline{\cos(2\pi \mathbf{g} \cdot \mathbf{u}_n)} \quad (\because\ \overline{\sin(2\pi \mathbf{g} \cdot \mathbf{u}_n)} = 0)$$

$$= \sum_n \cos(2\pi \mathbf{g} \cdot \mathbf{r}_n)\left[1 - 2\pi^2 g^2\overline{u^2}\right] \approx N\exp(-Bg^2)$$

$$(4.41)$$

and

$$\sum_{m\neq n} \overline{\cos\{2\pi \mathbf{q} \cdot (\mathbf{r}'_n - \mathbf{r}'_m) + 2\pi \mathbf{g} \cdot \mathbf{r}'_m\}}$$

$$= \sum_{m \neq n} \cos\{2\pi \mathbf{q} \cdot (\mathbf{r}_n - \mathbf{r}_m) + 2\pi \mathbf{g} \cdot \mathbf{r}_m\} \overline{\cos\{2\pi \mathbf{q} \cdot (\mathbf{u}_n - \mathbf{u}_m) + 2\pi \mathbf{g} \cdot \mathbf{u}_m\}}$$

$$\approx \exp\{-(Bq^2 + B(q-g)^2)\} \sum_{m \neq n} \cos\{2\pi \mathbf{q} \cdot (\mathbf{r}_n - \mathbf{r}_m) + 2\pi \mathbf{g} \cdot \mathbf{r}_m\} \quad (4.42)$$

In Eq. (4.41) a primitive cell is assumed, so that $\cos(2\pi \mathbf{g} \cdot \mathbf{r}_n) = 1$ (examples include face-centred cubic and body-centred cubic crystals). The intensity $I(\mathbf{q})$ is therefore

$$I(\mathbf{q}) = \frac{1}{2} \left[f_q^2 \left(N + \exp\{-2Bq^2\} \sum_{m \neq n} \exp\{2\pi i q \cdot (\mathbf{r}_n - \mathbf{r}_m)\} \right) + \cdots \right.$$

$$f_{q-g}^2 \left(N + \exp\{-2B(q-g)^2\} \sum_{m \neq n} \exp\{2\pi i (\mathbf{q}-\mathbf{g}) \cdot (\mathbf{r}_n - \mathbf{r}_m)\} \right) + \cdots$$

$$2(-1)^{j+1} f_q f_{q-g} \left(N \exp\{-Bg^2\} + \exp\{-(Bq^2 + B(q-g)^2)\} \right.$$

$$\left. \left. \times \sum_{m \neq n} \cos\{2\pi \mathbf{q} \cdot (\mathbf{r}_n - \mathbf{r}_m) + 2\pi \mathbf{g} \cdot \mathbf{r}_m\} \right) \right] \quad (4.43)$$

As mentioned previously, Eq. (4.43) is the total scattered intensity. The elastic contribution can be determined from Eq. (4.38) by replacing $\mathbf{r}'_m$ by $\mathbf{r}_m$ and f_q by $f_q \exp(-Bq^2)$, that is, the atom position and atom scattering factor are time averaged. Hence,

$$I_{\text{elastic}}(\mathbf{q})$$

$$= \frac{1}{2} \left[f_q^2 \exp\{-2Bq^2\} \sum_{m,n} \exp\{2\pi i q \cdot (\mathbf{r}_n - \mathbf{r}_m)\} + f_{q-g}^2 \exp\{-2B(q-g)^2\} \right.$$

$$\times \sum_{m,n} \exp\{2\pi i (\mathbf{q}-\mathbf{g}) \cdot (\mathbf{r}_n - \mathbf{r}_m)\} + \cdots$$

$$\left. 2(-1)^{j+1} f_q f_{q-g} \exp\{-B(q^2 + (q-g)^2)\} \sum_{m,n} \cos\{2\pi(\mathbf{q} \cdot (\mathbf{r}_n - \mathbf{r}_m) + \mathbf{g} \cdot \mathbf{r}_m)\} \right] \quad (4.44)$$

Subtracting Eq. (4.44) from (4.43) gives the TDS intensity:

$$I_{\text{TDS}}(\mathbf{q}) = \frac{1}{2} N \left[f_q^2 (1 - \exp\{-2Bq^2\}) + f_{q-g}^2 (1 - \exp\{-2B(q-g)^2\}) + \cdots \right.$$

$$\left. 2(-1)^{j+1} f_q f_{q-g} (\exp\{-Bg^2\} - \exp\{-B(q^2 + (q-g)^2)\}) \right] \quad (4.45)$$

TDS causes depletion of the elastic wavefield by the damping factor $\exp(-2\pi q^{(j)} z)$. From Eq. (4.33) at Bragg orientation

$$2\pi q^{(j)} = 2\pi \left(\frac{1}{2\xi'_0} + (-1)^{j+1} \frac{1}{2\xi'_g} \right) = \mu_0 + (-1)^{j+1} \Delta \mu_g \quad (4.46)$$

The rate of loss of elastic intensity per unit area is

$$\frac{dI_{elastic}}{dz} = -\frac{I_{elastic}}{\Omega N} \iint_{\theta,\phi} [I_{TDS}(\mathbf{q}) \sin 2\theta] d(2\theta) d\phi$$

$$= -\frac{\lambda^2 I_{elastic}}{\Omega N} \iint_{q,\phi} I_{TDS}(\mathbf{q}) q(dq \ d\phi)$$

$$= -\left(\mu_0 + (-1)^{j+1} \Delta\mu_g\right) I_{elastic} \tag{4.47}$$

where Ω is the atomic volume and here N is specifically the number of atoms within a (unit area $\times$ dz) volume element. Integration is carried out over the entire solid angle of a sphere containing all possible scattered wave vectors (see Figure 4.5b). The polar angle 2θ is related to q by $q = (2\sin \theta)/\lambda$ and ϕ is the azimuthal angle. Integration of the first two terms for I_{TDS} in Eq. (4.45) yields results independent of $\mathbf{g}$, so that

$$\mu_0 = \frac{\pi}{\xi_0'} = \left(\frac{2\pi\lambda^2}{\Omega}\right) \int f_q^2 \left(1 - \exp\{-2Bq^2\}\right) q(dq) \tag{4.48}$$

while integration of the third term in Eq. (4.45) gives

$$\Delta\mu_g = \frac{\pi}{\xi_g'} = \left(\frac{\lambda^2}{\Omega}\right) \iint_{q,\phi} f_q f_{q-g} \left(\exp\{-Bg^2\} - \exp\{-B(q^2 + (q-g)^2)\}\right)$$

$$\times q(dq \ d\phi) \tag{4.49}$$

Since $\xi_g' = k_{mean}/U_g'$ it is possible to determine the imaginary part of the complex crystal potential via Eq. (4.49). A method for evaluating the integral in Eq. (4.49) can be found in Bird and King (1990). As a general rule, $\xi_g = 0.1\xi_g'$ is a good approximation for metals (Hirsch et al., 1965). The TDS intensity distribution, as given by Eq. (4.45) however, is not strictly correct, since the model did not take into account re-scattering by the crystal, so that Kikuchi lines are not reproduced. Nevertheless, the *total* TDS intensity is sufficiently accurate for estimating $\xi_{g'}$ (see Section 3.3.2 for validity of the frozen phonon model). Rossouw and Bursill (1985) have provided an equivalent, frozen phonon expression for the TDS intensity in many-beam orientations.

4.1.4 Bloch States in Zone-Axis Orientations

In this section, Bloch states for a beam incident along a zone-axis orientation will be examined, since this is the condition for lattice imaging.

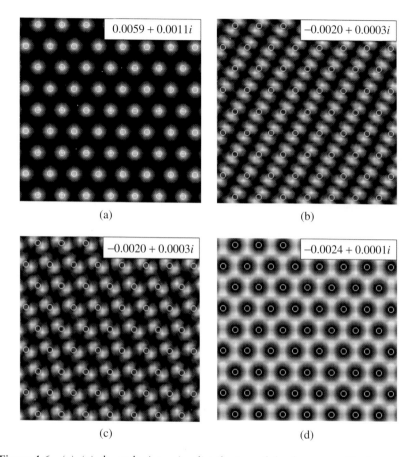

(a) 0.0059 + 0.0011*i*

(b) −0.0020 + 0.0003*i*

(c) −0.0020 + 0.0003*i*

(d) −0.0024 + 0.0001*i*

Figure 4.6 (a)–(g) show the intensity distribution of the first seven Bloch states in [111]-Mo at 300 kV normal incidence. The atom column positions are indicated by the open circles. The inset in each figure lists the real and imaginary parts of γ for the Bloch state. From Mendis and Hemker (2008). Reproduced with permission; copyright Elsevier science. (h) Shows the dispersion surfaces for the seven Bloch states as a function of beam tilt along the **g** = 110 reciprocal vector. The transverse wave vector is in units of **g**, while the longitudinal wave vector is in inverse Angstroms. The numbering of the dispersion surfaces corresponds to the order of the Bloch states as they appear in (a) to (g). For visual clarity dispersion surfaces for degenerate Bloch states at normal incidence are shown as dotted lines.

At high energies, the swift incident electron is largely insensitive to the potential variation along the atom columns, so that the point group of the projected crystal structure imposes certain symmetry restrictions on the Bloch waves (Hirsch *et al.*, 1965; Kogiso and Takahashi, 1977). As an example of many-beam Bloch wave solutions, Figure 4.6a–g shows the first seven Bloch states, arranged in descending order of the real part of γ,

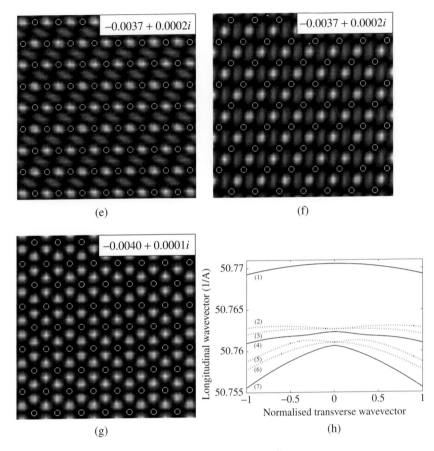

Figure 4.6 *(Continued)*

in [111]-oriented Mo at 300 kV and normal beam incidence (Mendis and Hemker, 2008). In each figure, the Bloch wave intensity is superimposed on the projected atom columns that are outlined by the open circles. The real and imaginary parts of γ for a given Bloch state are indicated in the top right-hand corner of each figure. Only Bloch state 1, which is localised at the atom columns, has a positive real value of γ, indicating a gain in kinetic energy as the electrons channel along the low potential energy sites within the crystal. For the other Bloch states the intensity is largely concentrated between the atom columns and hence the real value of γ is negative. Furthermore, TDS scattering is greater for Bloch state 1, as is evident from the larger value for the imaginary part of γ compared to other Bloch states. It should also be pointed out that Bloch states 2 and 3 are degenerate (i.e. have identical γ-values) at normal incidence; this is also the case for Bloch states 5 and 6. This is to be expected since these Bloch waves have the same symmetry and therefore identical channelling characteristics. Figure 4.6h plots the dispersion surfaces for the Bloch states along the 110 reciprocal vector, that is, variation of the longitudinal wave vector component of the Bloch state as the beam is tilted about g_{110}. Upper Bloch states, such as Bloch state 1, show little variation with respect to beam tilt and are therefore called *non-dispersive*. For the beam tilt angles considered in the figure the transverse momentum of the incident electrons is too small to significantly perturb the Bloch state, which remains tightly bound to the atom column 'strings'. This is, however, not the case for the lower lying 'dispersive' Bloch states (e.g. Bloch states 5–7), for which the electrons are only weakly coupled with the atom columns. As will be seen later, non-dispersive Bloch states play a crucial role in HREM and HAADF lattice imaging.

Examination of the Bloch states in Figure 4.6a–g reveals similarities with the electronic states for a cylindrically symmetric potential (Buxton *et al.*, 1978). This is because these Bloch states are relatively strongly bound to the atomic columns (i.e. large γ_0). For high energy electron diffraction the potential variation along an atom column can be ignored and since (in this case) there is little interaction between the relatively widely spaced atomic columns the effective potential for a bound Bloch state is cylindrically symmetric. As γ_0 decreases the Bloch states are less strongly bound to the atomic columns and the similarity breaks down. Schrödinger's equation for a cylindrical potential has been analysed by Berry and de Almeida (1973). If (R, θ) denote polar coordinates, the angular dependence of the wavefunction has the form $\cos(l\theta)$, and hence there are l radial nodal lines (fixed θ, variable R), where l is the angular momentum quantum number. The number of nodal circles (fixed R, variable θ) surrounding the potential well is $n - (l + 1)$, where n is the principal quantum number. Based on these rules (Buxton *et al.*, 1978) Bloch state 1 (Figure 4.6a) is found to have $l = 0$ (s-state), $n = 1$ or '1s' symmetry. There are, however, subtle differences between the 1s state for a cylindrical potential and that for a spherical atom. For the latter, the electron density at the atomic nucleus is zero due to the $l(l + 1)$ centrifugal barrier term in the radial Schrödinger equation. For a cylindrical potential the equivalent term is however $(l^2 - 1/4)$, so that for $l = 0$ the centripetal force gives rise to electron accumulation at the origin (Berry and de Almeida, 1973). The 1s Bloch state is therefore strongly coupled to the atom column. From the above rules the symmetry of Bloch states 2 and 3 are '2p' (Figure 4.6b and c), Bloch state 4 is '2s' (Figure 4.6d), Bloch states 5 and 6 are '3d' (Figure 4.6e and f) and Bloch state 7 is '4f' symmetry (Figure 4.6g).

The analogy between bound Bloch states and atom column orbitals can be further extended to the eigenvalues (i.e. energies) as well. To see this, a given Bloch state $b(\mathbf{R}, z)$ at normal incidence can be rewritten in the general form (Anstis *et al.*, 2003; compare with Eq. (4.18)):

$$b(\mathbf{R}, z) = \Phi(\mathbf{R}) \exp\left[2\pi i \frac{\sqrt{2me(E_o - E_t)}}{h} z \right] \qquad (4.50)$$

where eE_o is the incident electron energy ($= (hk_{vac})^2/2m$) and eE_t is the so-called *Bloch state transverse energy*. Substitution in Schrödinger's equation (Eq. (4.1)) gives

$$\frac{\hbar^2}{2m} \nabla_{\mathbf{R}}^2 \Phi(\mathbf{R}) + e[E_t + V(\mathbf{R}, z)]\Phi(\mathbf{R}) = 0 \qquad (4.51)$$

where $\nabla_{\mathbf{R}}^2$ is the Laplacian over $\mathbf{R}$ coordinates. In high energy electron diffraction $V(\mathbf{R}, z)$ is approximately independent of z, and is therefore replaced with an equivalent projected potential (i.e. projection approximation; Section 4.1.1). From Eq. (4.51) it is clear that eE_t is the eigenvalue energy of $\Phi(\mathbf{R})$. Furthermore, if E_t is small compared to E_o, the exponential term in Eq. (4.50) simplifies to

$$\exp\left[2\pi i \frac{\sqrt{2me(E_o - E_t)}}{h} z \right] = \exp(2\pi i k_{vac} z) \exp\left(-\pi i \frac{E_t}{E_o} k_{vac} z \right) \qquad (4.52)$$

The second exponential on the right-hand side of Eq. (4.52) represents the change in wave vector for the Bloch wave within the crystal (cf. Eq. (4.25)), so that E_t is directly related to γ_0. E_t is negative for tightly bound Bloch states with positive values of γ_0. Anstis *et al.* (2003) present a numerical method for estimating E_t, and hence γ_0, for the 1s eigenstate of an isolated atom column.

The discussion of bound Bloch states assumed that the atom columns are relatively widely spaced, so that there is little interaction between neighbouring columns. This is, however, not the case for (say) specimens with a diamond or sphalerite crystal structure in the [110] orientation, where 'dumbbells' of closely spaced atom columns are observed[4] (Section

---

[4]For diamond and sphalerite crystal structures dumbbells are found for [11n]-type orientations, where n is an even integer, including zero. The spacings of the dumbbell atom columns decrease monotonically with n.

3.2.1). Here, the atomic column orbitals will interact with one another to form 'molecular' orbitals; the wavefunction and energies of these molecular orbitals can be calculated using methods such as linear combination of atomic orbitals (Buxton *et al.*, 1978; Anstis *et al.*, 2003; Levine, 2006). Interaction between two atomic orbitals gives rise to bonding and anti-bonding molecular orbitals, as illustrated in Figure 4.7g (Levine, 2006). For the former the electron density is accumulated between the atom columns, so that the molecular orbital energy is lower than that of the individual atomic column orbitals. This is similar to the formation of a stable covalent bond between two nuclei. For the anti-bonding molecular orbital the electron density is peaked on the atom columns and hence the energy is larger. Figure 4.7a–f shows the first six Bloch

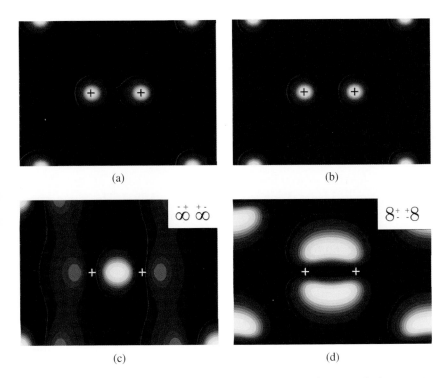

(a)

(b)

(c)

(d)

Figure 4.7 (a)–(f) show the intensity distribution of the first six Bloch states in [110]-Si at 300 kV normal incidence. The crosses in each figure represent the atom column positions of the central dumbbell. The symmetries of the Bloch states are 1s, 1s*, $2p_x$, $2p_y$, 2s and $2p_y$* respectively. The insets in (c), (d) and (f) indicate the orientation and polarity of the overlapping 2p atom column orbitals. (g) is a schematic of the energy (*E*) levels of two atomic orbitals (AO1 and AO2) and the molecular orbitals (MO) created by their overlap.

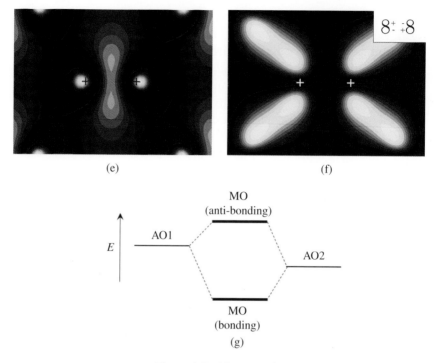

Figure 4.7 (*Continued*)

states, arranged in descending order of γ_0, for [110]-Si at 300 kV and normal beam incidence (Pennycook and Jesson, 1991). The symmetry of the interacting atomic column orbitals determines the molecular orbital symmetry and hence corresponding Bloch waves. Thus, the 1s bonding and 1s* anti-bonding Bloch states (Figure 4.7a and b, respectively) are formed by the 1s atom column orbitals; note the accumulation (1s) and depletion (1s*) of electron intensity between the atom columns for the two Bloch states. The 1s Bloch state has even polarity (i.e. the sign of the Bloch wavefunction is identical at the two dumbbell atom columns) while the 1s* state has odd (i.e. opposite signs) polarity. Furthermore, these two Bloch states are also non-dispersive (Pennycook and Jesson, 1991).

The 300 kV, non-dispersive Bloch states for GaAs, a sphalerite crystal, are shown in Figure 4.8a and b for the [110]-orientation and in Figure 4.8c and d for the [112]-orientation. The dumbbell atom columns are now heteronuclear (i.e. one column is Ga and the other is As). Consequently, for [110]-GaAs the bonding molecular orbital corresponds

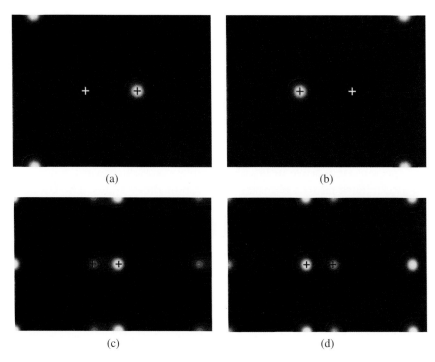

(a)   (b)

(c)   (d)

Figure 4.8 (a) and (b) show the As 1s and Ga 1s Bloch states for [110]-GaAs at 300 kV normal incidence. The corresponding Bloch states for [112]-GaAs are shown in (c) and (d) respectively. The atom column positions in the central dumbbell are indicated by the crosses in each figure (As is the atom column to the right).

largely to As 1s (Figure 4.8a), the inter-mixing with the Ga 1s atom column orbital in this case being negligible. Similarly the anti-bonding molecular orbital is largely Ga 1s (Figure 4.8b). Note that the lower energy bonding molecular orbital is associated with the higher atomic number As atom column and vice versa for the anti-bonding molecular orbital. This is also evident from the γ_0 values for the two Bloch states (0.0068 vs 0.0060 Å^{-1} respectively). In the [112]-orientation the dumbbell atom column spacing is only ~0.8 Å, compared to ~1.4 Å along [110], so that the stronger interaction leads to more inter-columnar mixing. This is reflected in the [112] bonding (Figure 4.8c) and anti-bonding (Figure 4.8d) molecular orbitals, which show both Ga 1s and As 1s character. The γ_0 values (0.0026 and 0.0014 Å^{-1} respectively) are smaller than for the [110] case. A possible explanation for the lower γ_0, or equivalently larger transverse energy, is the role of nuclear repulsion at small inter-columnar spacings (Levine, 2006).

4.2 APPLICATIONS OF BLOCH WAVE THEORY

4.2.1 HREM Imaging

In Section 4.1.2, it was shown that two-beam diffraction contrast features, such as thickness fringes and rocking beam patterns, were due to 'beating' of two Bloch states. When a crystal is tilted to a zone-axis for HREM the number of strongly excited Bloch states is still quite small, although there are many diffracted beams. This makes interpretation of the specimen exit wave in HREM relatively straightforward, since it is an interference pattern between the different Bloch states. The electron wavefunction, $\psi(\mathbf{R}, z)$, within the crystal (Eq. (4.18)) for normal incidence can be rewritten as

$$\psi(\mathbf{R}, z) = \sum_j \varepsilon^{(j)} \Phi^{(j)}(\mathbf{R}) \exp(2\pi i \gamma^{(j)} z) \qquad (4.53)$$

where the constant phase term $\exp(2\pi i k_{mean} z)$ has been ignored. The boundary condition at the specimen entrance surface $z = 0$ requires that $\sum_j \varepsilon^{(j)} \Phi^{(j)}(\mathbf{R}) = 1$. Van Dyck and Op de Beeck (1996) have considered the case where only one bound Bloch state, the 1s state, with positive γ_0 or equivalently negative transverse energy is excited. All other excited Bloch states are assumed to have γ_0 and transverse energies close to zero (i.e. they are only weakly bound). For a non-absorbing crystal (i.e. real γ) Eq. (4.53) then simplifies to

$$\psi(\mathbf{R}, z) = \left(\sum_j \varepsilon^{(j)} \Phi^{(j)}(\mathbf{R}) \right) + \varepsilon^{1s} \Phi^{1s}(\mathbf{R}) \left[\exp(2\pi i \gamma^{1s} z) - 1 \right]$$

$$= 1 + \varepsilon^{1s} \Phi^{1s}(\mathbf{R}) \left[\exp\left(2\pi i \gamma^{1s} z\right) - 1 \right] \qquad (4.54)$$

$$= 1 - 2\varepsilon^{1s} \Phi^{1s}(\mathbf{R}) \sin(\pi \gamma^{1s} z) \exp\left[-i \left(\frac{\pi}{2} - \pi \gamma^{1s} z \right) \right]$$

The spatial (i.e. $\mathbf{R}$) variation of the electron wavefunction is therefore governed by the 1s Bloch state. Van Dyck and Op de Beeck (1996) suggest that the close resemblance of many HREM images to the projected crystal structure under ideal imaging conditions (e.g. Scherzer imaging) is due to the 1s state being the only strongly bound state that is excited. The wavefunction oscillates with thickness with a periodicity of $(2/\gamma^{1s})$, due to interference of the 1s state with the other non-1s states, which are plane wave like. There are several factors that could lead to multiple bound Bloch states (not just 1s) being excited. These include an increase in incident electron energy, higher atomic number and closely spaced atom columns. The relativistic electron mass increases with incident energy, so that the relative contribution of the potential term in the Schrödinger's equation (Eq. (4.51)) is magnified. An alternative physical description of this effect is that at higher energies Lorentz contraction leads to an increase in the projected potential, as seen by the incident electron (Hovden et al., 2012). An increase in projected potential means that Bloch states will be more bound to the atom columns; a similar effect is observed with atom columns of larger atomic number. On the other hand, when the inter-columnar spacing decreases interaction between the atom column orbitals gives rise to many more bound Bloch states. This is the case for [110]-Si, where the 1s, $2p_x$ and 2s Bloch states are excited under normal beam incidence (Figure 4.7a, c and e respectively). Kambe (1982) has analysed the exit wave due to three strongly excited Bloch states, arbitrarily labelled u, v and w. Assume at a certain depth D the first two Bloch states are in phase. Equation (4.53) then gives

$$\psi(\mathbf{R}, D) = \left[\varepsilon^{(u)}\Phi^{(u)}(\mathbf{R}) + \varepsilon^{(v)}\Phi^{(v)}(\mathbf{R})\right]\exp(2\pi i\gamma^{(u)}D)$$

$$+ \varepsilon^{(w)}\Phi^{(w)}(\mathbf{R})\exp(2\pi i\gamma^{(w)}D)$$

$$= \left[1 - \varepsilon^{(w)}\Phi^{(w)}(\mathbf{R})\right]\exp(2\pi i\gamma^{(u)}D) + \varepsilon^{(w)}\Phi^{(w)}(\mathbf{R})\exp(2\pi i\gamma^{(w)}D)$$

$$= \exp(2\pi i\gamma^{(u)}D)\{1 + \varepsilon^{(w)}\Phi^{(w)}(\mathbf{R})[\exp(2\pi i(\gamma^{(w)} - \gamma^{(u)})D) - 1]\}$$

$$\tag{4.55}$$

The spatial variation (as a function of $\mathbf{R}$) of the electron wavefunction therefore depends on Bloch wave w. Furthermore, if it is assumed that the phase of Bloch wave w at depth D differs from that of u and v by only a small amount θ_o (modulo 2π), then $\exp[2\pi i(\gamma^w - \gamma^u)D] = \cos\theta_o + i\sin\theta_o = 1 + i\theta_o$, and Eq. (4.55) becomes

$$\psi(\mathbf{R}, D) = \exp(2\pi i\gamma^{(u)}D)[1 + i\theta_o\varepsilon^{(w)}\Phi^{(w)}(\mathbf{R})] \tag{4.56}$$

Bloch wave w now acts as a weak phase object (Section 3.1.1); for an ideal lens a negative image of the Bloch wave is formed when θ_o is positive (i.e. phase of w is ahead of u and v) and vice versa for when θ_o is negative (Kambe, 1982). If w is the $2p_x$ Bloch state in [110]-Si then, even under ideal imaging conditions, the HREM image will not display the dumbbell structure if the specimen thickness D satisfies the conditions for Eq. (4.56). Hence for accurate structure determination the HREM images must be compared with image simulation or else imaging must be carried out with an aberration-free lens and a sufficiently thin specimen to avoid dynamic diffraction artefacts. Van Dyck and Op de Beeck (1996) have shown that, despite many Bloch states potentially being excited, there will always be a one-to-one correspondence with the projected crystal structure provided the specimen is thin. To see this, Eq. (4.53) can be written in the alternative form via transverse energies, that is,

$$\psi(\mathbf{R}, z) = \sum_{j} \varepsilon^{(j)} \Phi^{(j)}(\mathbf{R}) \exp \left(-\pi i \frac{E_t^{(j)}}{E_o} k_{vac} z \right) \qquad (4.57)$$

For small z, using the boundary condition $\Sigma_j \varepsilon^{(j)} \Phi^{(j)}(\mathbf{R}) = 1$ and the fact that from Eq. (4.51), $\Sigma_j \varepsilon^{(j)} E_t^{(j)} \Phi^{(j)}(\mathbf{R}) = -V(\mathbf{R}, z)$, we have

$$\psi(\mathbf{R}, z) = \sum_{j} \varepsilon^{(j)} \Phi^{(j)}(\mathbf{R}) - \left(\frac{\pi i}{E_o} k_{vac} z \right) \sum_{j} \varepsilon^{(j)} E_t^{(j)} \Phi^{(j)}(\mathbf{R})$$

$$= 1 + i \left(\frac{\pi k_{vac}}{E_o} \right) [V(\mathbf{R}, z)z] \qquad (4.58)$$

This is the weak phase object approximation (Section 3.1.1) with $(\pi k_{vac}/E_o)$ being the interaction constant σ and $[V(\mathbf{R}, z)z]$ the projected potential. The convergence of the Bloch wave result with the multislice formula is not surprising, since both methods are ultimately derived from Schrödinger's equation.

4.2.2 HAADF Imaging

Incoherent HAADF images are formed when the inner angle of the annular dark-field detector is relatively large (i.e. several Bragg angles). The image intensity is then a convolution of the probe intensity profile and an object function (e.g. specimen potential). Unlike HREM, HAADF images do not exhibit contrast reversals as the defocus or specimen thickness is varied (Pennycook and Jesson, 1991; Klenov and Stemmer, 2006). At large scattering angles, the thermal diffuse scattered intensity can exceed coherent elastic scattering (Section 3.1.1). TDS involves a small energy loss and is therefore incoherent; this has led some authors to attribute the incoherence of a HAADF image to the predominance of TDS over coherent elastic scattering. However, TDS is not a pre-requisite for incoherence; even if coherent elastic scattering only is present the integration of this intensity over the annular detector suppresses any interference effects, provided the detector inner angle is large (Pennycook and Jesson, 1990, 1991; Nellist and Pennycook, 1999).

It is possible to define a 'coherence volume' for HAADF imaging, such that scattering by any two atoms within this volume can interfere with one another (Treacy and Gibson, 1993). The coherence volume is most easily calculated for the optically equivalent case of hollow cone dark-field imaging in the TEM, which is related to HAADF imaging in the STEM by the principle of reciprocity.[5] A schematic for hollow cone illumination is shown in Figure 4.9. An inclined plane wave of wave vector $\mathbf{k}$ is coherently scattered by the atom j along $\mathbf{k}'$, which passes within the objective aperture. The far-field scattered wave is $f(\Delta \mathbf{k}) \exp(-2\pi i \Delta \mathbf{k} \cdot \mathbf{r}_j)$, where $f(\Delta \mathbf{k})$ is the atom scattering factor for $\Delta \mathbf{k} = \mathbf{k}' - \mathbf{k}$ and $\mathbf{r}_j$ is the position vector of the jth atom. The real space wavefunction $\psi(\mathbf{r}, \mathbf{k})$ due to a single incident plane wave of wave vector $\mathbf{k}$ is obtained by inverse Fourier transforming the waves scattered by all the atoms:

$$\psi(\mathbf{r}, \mathbf{k}) = \sum_j \int_{\text{obj.apt.}} \left[f(\Delta \mathbf{k}) \exp(-2\pi i \Delta \mathbf{k} \cdot \mathbf{r}_j) \right]$$

$$\times \exp[-i\chi(\mathbf{u})] \exp(2\pi i \Delta \mathbf{k} \cdot \mathbf{r}) d\mathbf{k}' \qquad (4.59)$$

The second exponential is the phase shift due to the objective lens aberration function $\chi(\mathbf{u})$, where $\mathbf{u}$ is the transverse vector component of

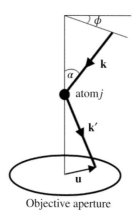

Objective aperture

Figure 4.9 Scattering geometry used for calculating the hollow cone dark-field intensity and coherence volume. The incident wave vector $\mathbf{k}$, with polar angle α and azimuthal angle ϕ, is scattered by an atom along the wave vector $\mathbf{k}'$. The scattering lies within the objective aperture. See text for further details.

---

[5]This states that the direction of the electron rays can be reversed, so that the annular detector in STEM is equivalent to the hollow cone illumination source and the STEM probe angle is equivalent to the detection angle defined by the TEM objective aperture.

$\mathbf{k}'$ measured with respect to the optic axis (Figure 4.9). The domain of integration is such that $\mathbf{k}'$ passes through the objective aperture. Writing $\Delta\mathbf{k} = (\mathbf{k}' - \mathbf{k}_o) + (\mathbf{k}_o - \mathbf{k}) = \mathbf{u} + \boldsymbol{\kappa}$, where $\mathbf{k}_o$ is a wave vector parallel to the optic axis with magnitude $|\mathbf{k}|$, Eq. (4.59) can be expressed in the alternative, but simpler, form:

$$\psi(\mathbf{r}, \mathbf{k}) = \sum_j a_j(\mathbf{r}) \exp[2\pi i \boldsymbol{\kappa} \cdot (\mathbf{r} - \mathbf{r}_j)]$$

$$a_j(\mathbf{r}) = \int_{\text{obj.apt.}} f(\Delta\mathbf{k}) \exp[-i\chi(\mathbf{u})] \exp[2\pi i \mathbf{u} \cdot (\mathbf{r} - \mathbf{r}_j)] d\mathbf{k}' \qquad (4.60)$$

It is reasonably assumed that the objective aperture is sufficiently small so that $(\mathbf{k}' - \mathbf{k}_o)$ has a negligible component along the optic axis, and is therefore approximately equal to $\mathbf{u}$ (Figure 4.9). Furthermore, if the k-vector has a large transverse component (as, for example, in the optically equivalent case of HAADF imaging), $|\Delta\mathbf{k}|$ is large and $f(\Delta\mathbf{k})$ is therefore approximately constant over the objective aperture (Treacy and Gibson, 1993; Section 3.1.1). For a radially symmetric lens aberration function the a_j term is therefore real and independent of $\mathbf{k}$.

The intensity $I(\mathbf{r}, \mathbf{k})$ is given by the square modulus of $\psi(\mathbf{r}, \mathbf{k})$:

$$I(\mathbf{r}, \mathbf{k}) = \sum_j a_j(\mathbf{r})^2 + 2 \sum_{i \neq j} a_i(\mathbf{r}) a_j(\mathbf{r}) \cos(2\pi\boldsymbol{\kappa} \cdot \mathbf{r}_{ij}) \qquad (4.61)$$

where $\mathbf{r}_{ij} = \mathbf{r}_i - \mathbf{r}_j$. The first term is the incoherent contribution, while the coherent contribution is due to interference between the scattered intensity from atoms i and j. The hollow cone intensity is obtained by integrating Eq. (4.61) over all wave vectors $\mathbf{k}$ with constant polar angle α and variable azimuthal angle ϕ (Figure 4.9). Since a_i and a_j are independent of $\mathbf{k}$, the integrand is effectively $\cos(2\pi\boldsymbol{\kappa} \cdot \mathbf{r}_{ij})$. If $\mathbf{r}_{ij}$ has components z_{ij} and R_{ij} parallel and perpendicular to the optic axis, then

$$\cos(2\pi\boldsymbol{\kappa} \cdot \mathbf{r}_{ij}) = \cos[2\pi(\mathbf{k}_o - \mathbf{k}) \cdot \mathbf{r}_{ij}]$$

$$= \cos\left[\frac{2\pi}{\lambda}(z_{ij} - z_{ij} \cos\alpha - R_{ij} \sin\alpha \sin\phi)\right]$$

$$= \cos\left[\frac{4\pi}{\lambda} z_{ij} \sin^2\left(\frac{\alpha}{2}\right) - \frac{2\pi}{\lambda} R_{ij} \sin\alpha \sin\phi\right] \qquad (4.62)$$

where λ is the electron wavelength. For simplicity, ϕ is defined such that R_{ij} has zero azimuthal angle. Therefore,

$$\int_0^{2\pi} \cos(2\pi\boldsymbol{\kappa} \cdot \mathbf{r}_{ij}) d\phi$$

$$= \cos\left[\frac{4\pi}{\lambda}z_{ij}\sin^2\left(\frac{\alpha}{2}\right)\right]\int_0^{2\pi}\cos\left[\frac{2\pi}{\lambda}R_{ij}\sin\alpha\sin\phi\right]d\phi$$

$$= 2\pi\cos\left[\frac{4\pi}{\lambda}z_{ij}\sin^2\left(\frac{\alpha}{2}\right)\right]J_0\left(\frac{2\pi}{\lambda}R_{ij}\sin\alpha\right) \qquad (4.63)$$

where J_0 is a zero order Bessel function of the first kind. Equation (4.63) is the coherence volume for hollow cone illumination; for the coherence volume in HAADF imaging a second integration with respect to the polar angle α is carried out spanning the inner and outer collection angles of the annular detector (Treacy and Gibson, 1993). Figure 4.10 shows plots of the coherence volume for inner detector angles of 50 and 100 mrad, with and without thermal vibration (the detector outer angle is fixed at 200 mrad). Thermal vibration is taken into account by introducing a Debye–Waller factor to $f(\Delta k)$. Bright and dark regions in Figure 4.10 correspond to regions of constructive and destructive interference respectively, while grey regions represent no interference (i.e. fully incoherent). Also superimposed in Figure 4.10a are the atom positions of [001]-oriented Si, with the (vertical) optic axis along [110]. The coherence volume is cigar shaped. The Bessel function in Eq. (4.63) is highly effective in damping any interference between neighbouring atom columns (i.e. transverse coherence along R_{ij}), although interference is still possible between atoms along a given column (i.e. longitudinal coherence along z_{ij}) due to the cosine term. This explains why HAADF images faithfully resolve the projected crystal structure, provided the probe is sufficiently small. From Figure 4.10 the longitudinal coherence can be suppressed by increasing the detector inner angle and via thermal vibrations, although the effect of the latter is more pronounced at larger detector inner angles. In the limit of zero transverse and longitudinal coherence, only the first term in Eq. (4.61) remains, and the HAADF image is truly incoherent.

It is of interest to establish which Bloch states contribute most to the HAADF signal. The electron wavefunction within the crystal for a STEM probe incident at position vector $\mathbf{R}_p$ is expressed as

$$\Psi(\mathbf{R},z) = \int_{\text{obj.apt.}}\left\{\sum_j \varepsilon^{(j)}(\mathbf{k}_t)\Phi^{(j)}\left(\mathbf{R},\mathbf{k}_t\right)\exp(2\pi i\gamma^{(j)}(\mathbf{k}_t)z)\right\}$$

$$\times\, e^{-i\chi(\mathbf{k}_t)}e^{2\pi i\mathbf{k}_t\cdot(\mathbf{R}-\mathbf{R}_p)}e^{2\pi ik_z z}d\mathbf{k}_t$$

$$\Phi^{(j)}(\mathbf{R},\mathbf{k}_t) = \sum_{\mathbf{g}}C_{\mathbf{g}}^{(j)}(\mathbf{k}_t)\exp(2\pi i\mathbf{g}\cdot\mathbf{R}) \qquad (4.64)$$

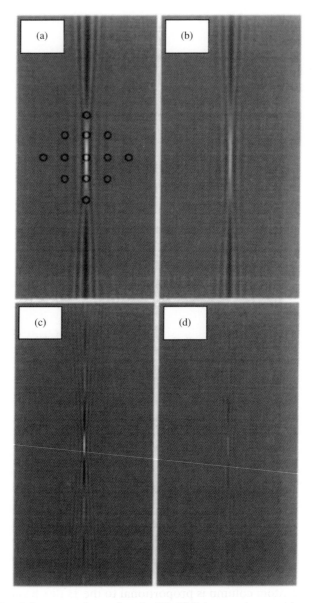

Figure 4.10 Coherence volume plots for HAADF detector inner angles of 50 mrad (a and b) and 100 mrad (c and d) in silicon. The detector outer angle is fixed at 200 mrad. Thermal vibrations are included in (b) and (d), but not (a) and (c). In (a) the atomic structure of silicon along [001], with [110] parallel to the (vertical) optic axis, is superimposed as a guide to the size of the coherence volume. The electron beam energy is 100 keV. From Treacy and Gibson (1993). Reproduced with permission; copyright Elsevier science.

where the summation is over all Bloch states j and the integration is over all transverse wave vectors $\mathbf{k}_t$ within the objective aperture (k_z is the longitudinal wave vector component). The phase terms in the integrand outside the curly brackets represent the incident partial plane wave in vacuum after correcting for lens aberrations (see also Eq. (3.44) in Section 3.2.2). The terms within the curly brackets are the crystal Bloch states excited by the partial plane wave. It is assumed that the important reciprocal vectors $\mathbf{g}$ for high energy electron diffraction lie within the ZOLZ. Furthermore, the $\exp(2\pi i k_z z)$ term does not vary significantly over a typical objective aperture and is therefore approximately constant at a given depth. The depth (z) variation of the electron wavefunction is then due to the $\exp[2\pi i \gamma^{(j)}(\mathbf{k}_t)z]$ phase term. As pointed out by Pennycook and Jesson (1990, 1991) integration of this term over the objective aperture will yield only a relatively small value for dispersive Bloch states, where $\gamma^{(j)}$ varies significantly with incident wave vector. The cycling of the phase term is especially enhanced as the depth z increases. On the other hand, $\gamma^{(j)}$ and hence the phase term, is approximately constant for non-dispersive Bloch states, such as the 1s state, and as a result a relatively large value is obtained upon integration. Furthermore, the 1s Bloch state intensity channels along the atom columns, so that there is a high probability for high angle scattering towards the annular detector. A more rigorous analysis by Nellist and Pennycook (1999) has shown the annular detector to be an efficient 1s Bloch state filter, provided the detector inner angle is sufficiently large (see also Rafferty et al., 2001). Assuming the 1s Bloch state to be dominant, the HAADF intensity can be expressed as

$$
I(\mathbf{R}_p) = \sum_i \sigma_i \int_0^t \left| \int_{\text{obj.apt.}} \left\{ \varepsilon^{1s}(\mathbf{k}_t) \Phi^{1s}(\mathbf{R}_i, \mathbf{k}_t) \exp[2\pi i \gamma^{1s}(\mathbf{k}_t)z] \right\} \right.
$$

$$
\left. \times \exp[-i\chi(\mathbf{k}_t)] \exp[2\pi i \mathbf{k}_t \cdot (\mathbf{R}_i - \mathbf{R}_p)] d\mathbf{k}_t \right|^2 dz \qquad (4.65)
$$

Here, σ_i is a high angle scattering cross-section per unit length of the ith column. Equation (4.65) simply states that the high angle scattering from a given atom column is proportional to the 1s Bloch state intensity along that column, integrated through the specimen thickness t. The total HAADF intensity is the sum of high angle scattering from all columns. For non-dispersive 1s Bloch states, Φ^{1s} and γ^{1s} are approximately independent of $\mathbf{k}_t$, so that the above equation can be rearranged as

$$
I(\mathbf{R}_p) = \left\{ \sum_i \sigma_i \int_0^t \left| \varepsilon^{1s}(\mathbf{0}) \Phi^{1s}(\mathbf{R}_i, \mathbf{0}) \exp[2\pi i \gamma^{1s}(\mathbf{0})z] \right|^2 dz \right\}
$$

$$\times \left\{ \left| \frac{1}{\varepsilon^{1s}(0)} \int_{\text{obj.apt.}} \varepsilon^{1s}(\mathbf{k}_t) \exp[-i\chi(\mathbf{k}_t)] \exp[2\pi i \mathbf{k}_t \cdot (\mathbf{R}_i - \mathbf{R}_p)] d\mathbf{k}_t \right|^2 \right\}$$
$$(4.66)$$

where $\varepsilon^{1s}(0)$ denotes excitation of the 1s Bloch state for normal incidence (i.e. null $\mathbf{k}_t$ vector) and similarly for $\Phi^{1s}(\mathbf{R}_i, 0)$ and $\gamma^{1s}(0)$. The first curly bracket in Eq. (4.66) is purely an object function (denoted by $O(\mathbf{R}_i)$), while the second curly bracket, apart from the $\varepsilon^{1s}(\mathbf{k}_t)$ term, has a similar form to the STEM probe wavefunction in vacuum (cf. Eq. (3.44) in Section 3.2.2). Replacing $\mathbf{k}_t$ in the second curly bracket by $-\mathbf{k}_t$ and noting that ε^{1s} and (typically) χ are even functions of $\mathbf{k}_t$, it is seen that Eq. (4.66) is a convolution of $O(\mathbf{R}_i)$ with the intensity of an *effective* probe wavefunction $P^{\text{eff}}(\mathbf{R}_i)$, given by (Pennycook and Jesson, 1990,1991):

$$P^{\text{eff}}(\mathbf{R}_i) = \frac{1}{\varepsilon^{1s}(0)} \int_{\text{obj.apt.}} \varepsilon^{1s}(\mathbf{k}_t) \exp[-i\chi(\mathbf{k}_t)] \exp(2\pi i \mathbf{k}_t \cdot \mathbf{R}_i) d\mathbf{k}_t \quad (4.67)$$

The HAADF intensity (Eq. (4.66)) is therefore $I(\mathbf{R}_p) = \Sigma_i[O(\mathbf{R}_i) \otimes |P^{\text{eff}}(\mathbf{R}_i)|^2]$, where $\otimes$ denotes convolution. This shows that HAADF images are incoherent even under dynamic diffraction conditions (note that Eqs. (4.61) and (4.63) are only valid for kinematic scattering). However, this relies on the detector inner angle being sufficiently large and/or the specimen sufficiently thick, so that the dominant contribution is due to the 1s Bloch state.

Interference can nevertheless still take place between Bloch states excited at the same position vector $\mathbf{R}$ as they propagate through the crystal. As an example, consider imaging [110]-oriented Si with a 300 kV, aberration-free STEM probe. The objective aperture 'radius' is either 10 or 30 mrad, and is sufficiently large to resolve the dumbbell structure. Figure 4.11a and b shows the intensity variation as a function of depth (i.e. pendellösung) along an atom column for 10 and 30 mrad STEM probes that are focused at the entrance surface of the same column. The pendellösung intensity is plotted with and without TDS, which was modelled phenomenologically by using a complex crystal potential (Section 4.1.3). The wavefunction due to a single Bloch state j is evaluated from Eq. (4.64), with the summation restricted to only the Bloch state of interest. From this, it follows that for a STEM probe focused on an atom column the 1s and 1s* Bloch states (Figure 4.7a and b) are strongly excited, along with relatively weak excitation of *individual* non-1s states (note however that the collective effect of non-1s states can be quite large). Strong excitation of 1s and 1s* Bloch states by an atomic-scale STEM probe is not surprising, since they are highly localised at the atom columns and therefore readily satisfy the boundary conditions, that is, at the specimen entrance surface the crystal and probe wavefunctions must be identical. The wavefunction due to 1s and 1s* Bloch states are in phase at the column in which the probe is incident, but the odd polarity of the latter means that at the neighbouring dumbbell atom column the two wavefunctions are out of phase and therefore interfere destructively. This is consistent with a weak probe intensity at that atom column.

The non-dispersive 1s and 1s* states have similar γ-values (0.0024 and 0.0021 Å^{-1} respectively at 300 kV for the real part with a beating periodicity of >3000 Å) and therefore propagate through the thin specimen with approximately equal phase. The intensity oscillations in Figure 4.11a and b are therefore due to interference between the strongly excited 1s/1s* Bloch states and non-1s states. First consider the pendel-lösung plots with no TDS. The intensity oscillations are damped at large depths and furthermore their periodicity decreases with increasing objective aperture angle (Peng et $al.$, 2004). As noted previously the depth dependence of the STEM wavefunction is due to the $\exp[2\pi i\gamma^{(j)}(\mathbf{k}_t)z]$ phase term in Eq. (4.64). The intensity $I = \Psi(\mathbf{R}, z) \cdot \Psi(\mathbf{R}, z)^*$ therefore contains two distinct terms (Nellist and Pennycook, 1999): (i) 'cross-terms', where two Bloch states i and j ($i \neq j$) interfere with one another, so that the depth dependence is $\exp[2\pi i(\gamma^{(i)}(\mathbf{k}_t) - \gamma^{(j)}(\mathbf{k}_t))z]$ and (ii) 'self-terms', where a given Bloch state interferes with itself and is therefore independent of depth. As z increases the cross-terms are gradually damped, owing to the $\exp[2\pi i(\gamma^{(i)}(\mathbf{k}_t) - \gamma^{(j)}(\mathbf{k}_t))z]$ phase term varying over many 2π radian cycles for different $\mathbf{k}_t$, provided that at least one of i or j is dispersive. This leaves only the depth independent self-terms at large z, as well as the 1s–1s* cross-term, which is nevertheless only weakly z-dependent, due to similar γ-values. The intensity oscillations are therefore damped. TDS dampens the pendellösung further through depletion of the elastic wavefield, and is strongest for the 1s/1s* Bloch states, which channel along the atom columns. The periodicity of the intensity oscillations is determined by the difference in γ-values between the 1s/1s* Bloch states and non-1s states. Figure 4.11c and d shows plots of the excitation $\varepsilon^{(j)}(\mathbf{k}_t)$ as a function of $\mathbf{k}_t$ for the non-dispersive 1s state and an arbitrary dispersive Bloch state (i.e. Bloch state number 50, if the Bloch waves are arranged in descending order of the real part of γ). The 1s state excitation peaks at normal incidence, whereas dispersive states, which are weakly bound to the atom columns and have large transverse energies, are only excited at inclined illumination angles. As

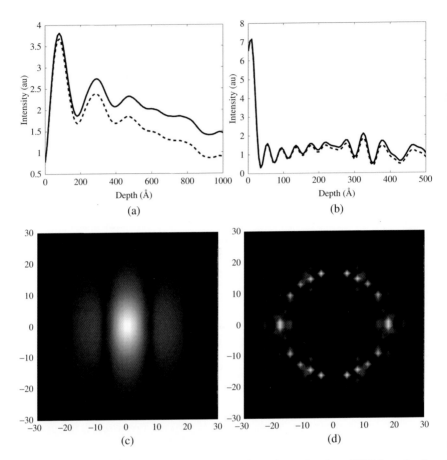

Figure 4.11 Pendellösung plots for a 300 kV, aberration-free STEM probe in [110]-Si with (a) 10 mrad and (b) 30 mrad probe semi-convergence angle. The probe is incident on a dumbbell atom column and the intensity is calculated along the same column as a function of depth. The dashed and solid lines represent results with and without TDS. Note the difference in scale (intensity and depth) for the two figures. (c) and (d) respectively show the excitation at 300 kV of the 1s Bloch state and Bloch state 50 in [110]-Si as a function of the incident plane wave transverse wave vector component (expressed as an angle in mrad units). The horizontal and vertical axes are along the 002 and $1\bar{1}0$ reciprocal vectors respectively. (e) and (f) are two-dimensional cross-sections of STEM probe propagation in [110]-Si. The contrast is inverted for visual clarity, so that black represents regions of high electron intensity. The horizontal axis measures distance along [001], with the dumbbell atom columns at ±0.7 Å, while the vertical depth axis is along [110]. In (e) the probe is incident on the dumbbell atom column at −0.7 Å, while in (f) it is incident at the dumbbell centre (0 Å). The probe parameters are 300 kV, 10 mrad semi-convergence angle and zero electron optic aberrations.

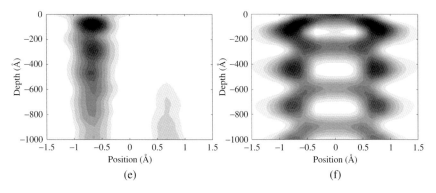

Figure 4.11   (*Continued*)

the objective aperture size is increased more dispersive Bloch states are excited, such that the mean γ-value for the non-1s states decreases. The oscillation period therefore decreases with objective aperture radius, as observed in the simulation results.

Figure 4.11e is a specimen cross-section through the 10 mrad probe intensity as a function of depth, calculated using Bloch waves with TDS included; the horizontal and vertical axes are along [001] and [110] respectively, with the dumbbell atom columns at positions ±0.7 Å. The probe is focused at the entrance surface, in line with the atom column at −0.7 Å. There is a gradual build-up of intensity in the neighbouring atom column (at 0.7 Å) as the probe propagates through the specimen. De-channelling or beam spreading has important implications for atomic resolution chemical analysis, as well as interpreting atomic number Z-contrast from HAADF images. Figure 4.11f is similar to (e), but now the probe is positioned at the mid-point of the dumbbell (i.e. at 0 Å). The probe tails weakly excite the 1s Bloch state, as well as non-1s states, on the dumbbell atom columns; the 1s* Bloch state is not excited, because of its odd polarity, which would otherwise lead to a symmetry violation. Although the electron intensity on the atom columns is weak at the entrance surface, Figure 4.11f shows that as the probe propagates the Bloch states can phase align and interfere constructively to generate appreciable intensity. This effect is also evident in Figure 4.11a and b, where the intensity of the first pendellösung peak is larger than the probe intensity at the specimen entrance surface. The atomic columns, with their net positive potential, act as 'atomic lenses' for focusing and channelling the electron intensity (Fertig and Rose, 1981). The channelling is however imperfect, since beating between the Bloch states gives rise to intensity oscillations. The fact that the 1s Bloch state is excited in Figure 4.11f means that some HAADF intensity is detected at the centre of the dumbbell, thereby reducing the visibility of the two atom columns. 1s Bloch state excitation is, however, suppressed for high resolution probes, owing to reduced overlap between the atom columns and narrower probe tails. The dumbbell is therefore better resolved as the probe size decreases, as required for incoherent imaging.

Before concluding this section, it is worth briefly mentioning very large objective apertures (e.g. 50 mrad), which can be used in conjunction with modern aberration correctors (see, e.g. Krivanek *et al.*, 2008). A large objective aperture is used for ultra-high resolution imaging (e.g. [112]-Si dumbbells) or for improving the counting statistics in atomic resolution chemical analysis. Peng *et al.* (2004) have shown that there is a critical objective aperture size that maximises excitation of the 1s Bloch state. At this critical aperture size the incident probe wavefunction matches that of the 1s Bloch state as closely as possible. For a 300 kV, aberration-free probe incident on an As column in [110]-GaAs the critical objective aperture 'radius' for As 1s Bloch state excitation is ~20 mrad (Peng *et al.*, 2004). If the objective aperture size is increased significantly beyond this the cumulative excitation of the non-1s states can be a considerable fraction of the 1s Bloch state. The highly dispersive, non-1s states excited at large illumination angles (e.g. Figure 4.11d) behave approximately as plane waves travelling through the mean inner potential of the crystal, and exhibit very little channelling, since the large transverse momentum of these states effectively decouples them from the atom columns. This raises the possibility of optical depth sectioning using very large apertures (Benthem *et al.*, 2005; Allen *et al.*, 2008). Here the probe has a reduced depth of field, so that different depths of the sample can be imaged by simply changing the probe defocus. The probe intensity must be highly concentrated at the required depth and rapidly diverge away from the focal point. This behaviour is evident on comparing Figure 4.11a and b; the first pendellösung peak is sharper and closer to the probe focus (i.e. specimen entrance surface) in the 30 mrad probe than the 10 mrad probe. Cosgriff and Nellist (2007) have given a full Bloch wave analysis of this phenomenon, including its limitations when applied to zone-axis oriented crystals.

4.2.3 Bloch Wave Scattering By Elastic Strain Fields

In a perfect crystal the Bloch wave excitations are constant throughout the specimen (Section 4.1.1). This is however not the case for a deformed crystal. So long as the strain is small and slowly varying the electron wavefunction can be represented as a sum of distorted (with respect to the perfect crystal) Bloch waves, whose excitations are a function of position $\mathbf{r}$. The change in Bloch wave excitation is due to scattering by the strain field, and can be determined within a small region of the deformed crystal via the column approximation. This states that for high energy electrons scattering is confined to only a narrow column in the direction of the diffracted beam (see Figure 3.8). In practice the column is assumed to be parallel to the z-axis (i.e. along the optic axis), rather than the diffracted beam, owing to the small Bragg angle (this will be discussed in more detail later on). The column can be further divided along its length into thin slices of thickness δz, such that the deformation vector $\tau(\mathbf{r})$ within a slice is approximately constant. If the Bloch wave excitations or diffracted beam amplitudes are known at the entrance surface of a given slice, it is possible to calculate the equivalent values at the exit surface of the slice for a given $\tau(\mathbf{r})$. The electron wavefunction at any given depth can be calculated in this manner, starting from the entrance surface of the specimen. This is similar to the multislice method, except that the lateral dimensions of the slices are confined to a narrow column, so that the process must be repeated for different values of the two-dimensional position vector $\mathbf{R}$.

In a perfect crystal the diffracted beam amplitudes ($\boldsymbol{\varphi}$) are related to the Bloch wave excitations ($\boldsymbol{\varepsilon}$) via the matrix equation (4.21) (Section 4.1.1). At the entrance surface of a given slice, $\boldsymbol{\varepsilon} = \{\exp(-2\pi i \gamma^{(j)} z)\} \mathbf{C}^{-1} \boldsymbol{\varphi}$, where $\boldsymbol{\varphi}$ is now the diffracted beam amplitude at the entrance surface (recall that the excitations are constant for a perfect crystal). If the diffracted beam amplitudes at the bottom of the slice of thickness δz are denoted by $\boldsymbol{\varphi}'$, Eq. (4.21) is modified to

$$\boldsymbol{\varphi}' = \mathbf{P}\boldsymbol{\varphi}; \quad \mathbf{P} = \mathbf{C}\left\{\exp(2\pi i \gamma^{(j)} \delta z)\right\} \mathbf{C}^{-1} \tag{4.68}$$

where $\mathbf{P}$ is the so-called scattering matrix for a perfect crystal slice. Assume now that the slice is part of a deformed crystal with local deformation vector $\tau(\mathbf{r})$. In the 'deformable-ion' approximation, the slice potential is transformed from $V(\mathbf{r})$ to $V(\mathbf{r} - \tau(\mathbf{r}))$ as a result of the deformation. Equation (4.2) then becomes

$$V(\mathbf{r}) = \sum_{\mathbf{g}} V_{\mathbf{g}} \exp[2\pi i \mathbf{g} \cdot (\mathbf{r} - \tau(\mathbf{r}))]$$

$$= \frac{h^2}{2me} \sum_{\mathbf{g}} [U_{\mathbf{g}} \exp(-2\pi i \mathbf{g} \cdot \tau(\mathbf{r}))] \exp(2\pi i \mathbf{g} \cdot \mathbf{r}) \tag{4.69}$$

The Fourier coefficients for the potential are no longer constants, but vary with position. The off-diagonal element A_{gh} of the structure matrix $\mathbf{A}$ is therefore modified from $U_{g-h}/(2k_{mean})$ to $[U_{g-h}\exp\{2\pi i(\mathbf{h} - \mathbf{g})\cdot\tau(\mathbf{r})\}]/(2k_{mean})$. By inspection, it follows that the structure matrix for the deformed crystal is $\mathbf{Q}^{-1}\mathbf{A}\mathbf{Q}$, where the diagonal matrix $\mathbf{Q} = \{\exp(2\pi i\beta_g)\} = \{\exp(2\pi i\mathbf{g}\cdot\tau(\mathbf{r}))\}$. From Eq. (4.14) $\mathbf{C}$ is therefore modified to $\mathbf{Q}^{-1}\mathbf{C}$, so that the Bloch wave coefficient C_g is $C_g\exp(-2\pi i\mathbf{g}\cdot\tau(\mathbf{r}))$ in the deformed crystal. The additional phase factor ensures that the Bloch wave is in 'registry' with the displacement field. For example, the 1s Bloch state will continue to channel along the atom columns despite the distortion (Mendis and Hemker, 2008). Equation (4.68) for a deformed crystal becomes

$$\boldsymbol{\varphi}' = \mathbf{P}^d\boldsymbol{\varphi}; \quad \mathbf{P}^d = \mathbf{Q}^{-1}\mathbf{C}\{\exp(2\pi i\gamma^{(j)}\delta z)\}\mathbf{C}^{-1}\mathbf{Q} = \mathbf{Q}^{-1}\mathbf{P}\mathbf{Q} \qquad (4.70)$$

where $\mathbf{P}^d$ is the corresponding scattering matrix. For a thin slice $\{\exp(2\pi i\gamma^{(j)}\delta z)\} = \mathbf{I} + 2\pi i\delta z\{\gamma^{(j)}\}$, where $\mathbf{I}$ is the identity matrix, so that

$$\delta\boldsymbol{\varphi} = \boldsymbol{\varphi}' - \boldsymbol{\varphi} = 2\pi i\mathbf{Q}^{-1}\mathbf{C}\{\gamma^{(j)}\}\mathbf{C}^{-1}\mathbf{Q}\boldsymbol{\varphi}\delta z$$

$$\text{or } \frac{d\boldsymbol{\varphi}}{dz} = 2\pi i(\mathbf{Q}^{-1}\mathbf{A}\mathbf{Q})\boldsymbol{\varphi} \qquad (4.71)$$

where $\mathbf{A} = \mathbf{C}\{\gamma^{(j)}\}\mathbf{C}^{-1}$, derived from Eq. (4.14), has been used to simplify the expression. For an absorbing crystal under two-beam conditions the above equation gives

$$\frac{d\phi_0}{dz} = \pi i\left(\frac{1}{\xi_0} + \frac{i}{\xi_0'}\right)\phi_0 + \pi i\left(\frac{1}{\xi_g} + \frac{i}{\xi_g'}\right)\phi_g\exp[2\pi i(s_g z + \mathbf{g}\cdot\tau(\mathbf{r})]$$

$$\frac{d\phi_g}{dz} = \pi i\left(\frac{1}{\xi_g} + \frac{i}{\xi_g'}\right)\phi_0\exp[-2\pi i(s_g z + \mathbf{g}\cdot\tau(\mathbf{r}))] + \pi i\left(\frac{1}{\xi_0} + \frac{i}{\xi_0'}\right)\phi_g$$

$$(4.72)$$

where the transmitted (ϕ_0) and diffracted (ϕ_g) beam amplitudes have been substituted by $\phi_0\exp(-\pi iz/\xi_0)$ and $\phi_g\exp(2\pi is_g z - \pi iz/\xi_0)$ respectively. The substitutions modify the phase of the beams but not their intensities, which is the experimentally measured quantity. These are the Howie–Whelan equations for two-beam scattering in a deformed crystal. They consist of a series of coupled ordinary differential equations, where the ϕ_g amplitude is governed by scattering from both ϕ_0 and ϕ_g, and similarly for the ϕ_0 amplitude. This is simply a result of multiple scattering (i.e. dynamic diffraction).

A second expression for $d\varphi/dz$ is obtained from Eq. (4.21) as applied to the deformed crystal (i.e. $\mathbf{C}$ is replaced by $\mathbf{Q}^{-1}\mathbf{C}$). Combining that expression with Eq. (4.71) gives

$$
\frac{d\boldsymbol{\varepsilon}}{dz} = 2\pi i \{\exp(-2\pi i\gamma^{(j)}z)\}\mathbf{C}^{-1}\{\beta'_g\}\mathbf{C}\{\exp(2\pi j\gamma^{(j)}z)\}\boldsymbol{\varepsilon};
$$

$$
\{\beta'_g\} = \left\{ \frac{\partial}{\partial z}\mathbf{g}\cdot\tau(\mathbf{r}) \right\} \tag{4.73}
$$

Equation (4.73) describes the evolution of Bloch wave excitations within a deformed crystal. For the two-beam condition the resulting equations are

$$
\frac{d\varepsilon^{(1)}}{dz} = 2\pi i \left\{ \beta'_g(C_g^{(1)})^2\varepsilon^{(1)} + \beta'_g(C_g^{(1)}C_g^{(2)})\varepsilon^{(2)}\exp[2\pi i(\gamma^{(2)} - \gamma^{(1)})z] \right\}
$$

$$
\frac{d\varepsilon^{(2)}}{dz} = 2\pi i \left\{ \beta'_g(C_g^{(2)})^2\varepsilon^{(2)} + \beta'_g(C_g^{(1)}C_g^{(2)})\varepsilon^{(1)}\exp[2\pi i(\gamma^{(1)} - \gamma^{(2)})z] \right\}
$$

$$
\tag{4.74}
$$

where it has been assumed that the undeformed crystal is centrosymmetric and absorption can be neglected, so that $\mathbf{C}$ is a unitary matrix with real elements (Section 4.1.1). The first term on the right-hand side of each equation represents scattering within the same Bloch state and is known as *intra*band scattering. The second term is due to scattering between different Bloch states and is called *inter*band scattering. Since this involves a transition between different branches of the dispersion surface there is also a phase change of $\exp[\pm2\pi i(\gamma^{(2)} - \gamma^{(1)})z]$. Equations (4.72) and (4.74) can be used to simulate the diffraction contrast of defects, such as stacking faults and dislocations (see Hirsch *et al.*, 1965 for a comprehensive discussion). Note that if $\mathbf{g}\cdot\tau(\mathbf{r}) = 0$ there is no change in Bloch wave excitations (Eq. (4.74)) or diffracted beam intensities (Eq. (4.72)) from that of a perfect crystal. This is because $\tau(\mathbf{r})$ has zero component along $\mathbf{g}$

and therefore does not change the inter-planar spacing of the diffracting planes. The defect is therefore invisible during bright-field or dark-field imaging; for a dislocation, this feature enables its Burgers vector and character to be determined. Nellist *et al.* (2008) have also shown that for 'symmetry' zone-axes (i.e. zone-axes such as [100] and [111] in cubic systems where all the main diffracted beams are symmetry related) intraband transitions are forbidden, irrespective of the nature of the strain field, while interband transitions are only allowed if they satisfy a dipole selection-type rule, that is, the change in the angular momentum quantum number of the Bloch states (Δl) must be equal to ± 1.

Strain can also affect the HAADF image intensity, despite it being an incoherent imaging technique sensitive to atomic number contrast. There are many examples where dislocations, grain boundaries and heavily doped layers in a semiconductor show higher HAADF intensity compared to the perfect crystal (Perovic *et al.*, 1993). Strain contrast in HAADF is primarily due to two reasons. First, interband scattering can modify the 1s Bloch state excitation and thereby the high angle scattered intensity. This is particularly true if the atom columns are not straight, but are continuously displaced laterally through the thickness of the foil. The STEM probe may then be unable to channel along the 'column', so that strong interband scattering from the 1s to non-1s Bloch states will take place, leading to probe spreading and lower HAADF intensities (Plamann and Hÿtch, 1999; Cosgriff *et al.*, 2010). The second effect is elastic diffuse scattering, due to static displacements of the atoms (the so-called *Huang scattering* in the case of point defects). This is similar to TDS, in the sense that some of the Bragg intensity is redistributed over a larger region of reciprocal space. Unlike TDS, however, it is not temperature sensitive. In Section 4.1.3 the TDS intensity was obtained by averaging the thermal vibrations (Eq. (4.45)); similarly the elastic diffuse scattered intensity is obtained by averaging static displacements, which may vary between different atom positions. A second Debye–Waller type-factor can therefore be used to model the effect of elastic diffuse scattering on the atom scattering factor (for a derivation see Wang, 1995). Indeed Yu *et al.* (2004) used such an approach to successfully simulate, using multislice, the strain contrast at a [110]-Si/amorphous Si interface.

Before concluding this section, it is worth briefly discussing some of the limitations of the column approximation. First, there is the requirement that the strain must be slowly varying, at least within the width of the column. This is valid for (say) the long-range elastic strain field of a dislocation, but breaks down at the dislocation core itself. The diffracted beam amplitudes are a function of z, and are expressed through coupled ordinary differential equations (Eq. (4.72)). This is because the columns are assumed to be along the z-axis. In reality, the column is parallel to the diffracted beam (Figure 3.8) and an ordinary differential equation is only obtained if the derivative is taken along this direction (Takagi, 1962; Howie and Basinski, 1968; Hirsch *et al.*, 1965). Taking derivatives along z results in a partial differential equation, with additional terms such as $\partial\phi_g/\partial x$ and $\partial\phi_g/\partial y$. Nevertheless, since high energy electron diffraction is primarily in the forward z-direction, the additional terms can be safely ignored. For two-beam conditions, Takagi (1962) has shown that the wavefunction at the exit surface point P in Figure 4.12a is governed by the transmitted and diffracted beams within the triangle APB, which has sides parallel to the wave vectors of the two beams. Thus the 'column' is strictly speaking a triangle, although due to the very small Bragg angle in electron diffraction this hardly makes a difference. Finally, it should also be pointed out that in the column approximation scattering is restricted to Bloch states with transverse wave vector identical to that of the incident plane wave (e.g. points A and B in the dispersion surface of Figure 4.12c). This has limitations when applied to a deformed crystal. As an example, Figure 4.12b shows an inclined stacking fault with surface normal $\mathbf{n}_2$ that is at an angle to the specimen surface normal $\mathbf{n}_1$. The crystal is effectively divided into two halves by the stacking fault. Boundary conditions at the stacking fault require that the Bloch waves at points A and B in the upper half of the crystal excite Bloch waves C and D respectively in the bottom half through interband scattering (Figure 4.12c; Whelan and Hirsch, 1957). In the column approximation however, this change in the transverse wave vector is not taken into account (Howie and Basinski, 1968; Howie and Sworn, 1970). Nevertheless, despite its limitations, it is still a highly successful theory for quantitative analysis of defect contrast (Hirsch *et al.*, 1965).

4.3 FURTHER TOPICS IN BLOCH WAVES

4.3.1 *Dopant Atom Imaging in STEM*

One of the main attractions of HAADF imaging is that it exhibits atomic number contrast. For large detector inner angles the HAADF intensity is due to high angle scattering of electrons that pass close to the atomic nucleus. The scattering cross-section is asymptotic to the unscreened Rutherford cross-section, which varies as Z^2 (Eq. (2.8)). This has

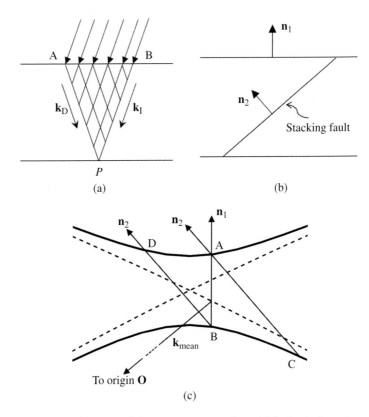

Figure 4.12 (a) Schematic of the true geometry for which the 'column' approximation holds under two-beam conditions. The bright-field, dark-field intensity at the specimen exit surface point P is governed by scattering within the triangle APB, which has sides parallel to the incident (k_I) and diffracted (k_D) beam wave vectors. (b) Schematic of an inclined stacking fault within a crystal. n_1 and n_2 are the surface normals for the specimen and stacking fault respectively. (c) Illustration of Bloch wave scattering by the stacking fault in (b) under two-beam conditions. Solid lines represent the Bloch wave dispersion surfaces, while the dashed lines are circles of radii k_{mean} (compare with Figure 4.2). The Bloch waves excited at the specimen entrance surface are given by the points A and B. At the stacking fault the two Bloch waves are scattered to points C and D respectively.

enabled imaging single atoms of a heavy element present either on the surface of a low atomic number substrate (Nellist and Pennycook, 1996) or as dopant atoms within a crystal (Voyles *et al.*, 2002). The recent technique of STEM annular bright-field (ABF) imaging has also proved successful in detecting light atom columns (Okunishi *et al.*, 2009), even down to hydrogen (Ishikawa *et al.*, 2011). In this section, a Bloch wave theory will be developed for these imaging methods.

Consider first HAADF imaging of a single dopant atom buried within a crystal. For weak scattering, the Bloch wavefunction profiles within the doped crystal are identical to that of a perfect crystal. However, their excitations will change due to dopant atom scattering, which is assumed to originate from the different atomic number of the dopant atom, rather than any strain that might be induced within the host lattice. Perturbation theory is used to estimate the change in excitation within a narrow column containing the dopant atom (i.e. the column approximation; Mendis, 2008). It is convenient to start with the relatively simple case of plane wave illumination, for a wave vector $\mathbf{k}_{vac} = (k_x, k_y, k_z)$. The jth Bloch wavefunction in a perfect crystal is $b^{(j)}(\mathbf{R}, z) \exp(2\pi i k_z z)$, where $b^{(j)}(\mathbf{R}, z) = \Phi^{(j)}(\mathbf{R}) \exp(2\pi i \gamma^{(j)} z)$. It is assumed that only ZOLZ reciprocal vectors are important in the Bloch wave expansion. Furthermore, the small change in z-component of the wave vector due to the mean inner potential has been ignored for convenience, but does not affect the final result even if included. Substituting the wavefunction of a single Bloch wave in Schrödinger's equation (Eq. (4.1)) results in

$$\frac{\partial b^{(j)}}{\partial z} = i\left[\left(\frac{\nabla_{xy}^2}{4\pi k_z} + \frac{\pi}{k_z}(k_x^2 + k_y^2)\right) + \frac{2\pi m e}{h^2 k_z} V(\mathbf{r})\right] b^{(j)} \qquad (4.75)$$

where $\nabla^2_{xy} = \partial^2/\partial x^2 + \partial^2/\partial y^2$. For high energy electron diffraction the $\partial^2 b^{(j)}/\partial z^2$ term is relatively small (Section 3.3.1) and has been ignored in Eq. (4.75). With a dopant atom the potential $V(\mathbf{r})$ is modified to $[V(\mathbf{r}) + v(\mathbf{r})]$, where $v(\mathbf{r})$ is the 'excess' potential due to the dopant. For a substitutional atom $v(\mathbf{r})$ is the difference in potential between the dopant and the host atom, while for an interstitial it is equal to the full dopant atom potential. Atom scattering factors can be used to calculate $v(\mathbf{r})$ (Eq. (3.10)). The total wavefunction within the column containing the dopant atom is $[\Sigma_j \varepsilon^{(j)}(z) b^{(j)}(\mathbf{R}, z)] \exp(2\pi i k_z z)$, where the Bloch state excitations $\varepsilon^{(j)}(z)$ can vary with depth due to scattering. Replacing $b^{(j)}$ in Eq. (4.75) with $\Sigma_j \varepsilon^{(j)}(z) b^{(j)}$ the equivalent result for the total wavefunction is

$$\sum_j \left(\frac{d\varepsilon^{(j)}}{dz} b^{(j)} + \varepsilon^{(j)} \frac{\partial b^{(j)}}{\partial z}\right)$$

$$= \sum_j i \left[\frac{\nabla_{xy}^2}{4\pi k_z} + \frac{\pi}{k_z}(k_x^2 + k_y^2) + \frac{2\pi m e}{h^2 k_z}(V(\mathbf{r}) + v(\mathbf{r}))\right] \varepsilon^{(j)} b^{(j)}$$

$$\text{or} \quad \sum_j \frac{d\varepsilon^{(j)}}{dz} b^{(j)} = \frac{2\pi i m e}{h^2 k_z} \sum_j v(\mathbf{r}) \varepsilon^{(j)} b^{(j)} \qquad (4.76)$$

Multiplying both sides of the equation by the complex conjugate of $b^{(i)}$ and integrating over $\mathbf{R}$ gives, after making use of the orthonormal property of Bloch states (Mendis, 2008),

$$\frac{d\varepsilon^{(i)}(z)}{dz} = \frac{2\pi ime}{h^2 k_z} \sum_j \varepsilon^{(j)}(z) \exp\left[2\pi i(\gamma^{(j)} - \gamma^{(i)})z\right] \int \Phi^{(j)}(\mathbf{R})\upsilon(\mathbf{r})\Phi^{(i)}(\mathbf{R})^* d\mathbf{R}$$

(4.77)

The $i \neq j$ terms represent interband scattering with the additional phase factor $\exp[2\pi i(\gamma^{(j)} - \gamma^{(i)})z]$, while the $i = j$ terms are due to intraband scattering. For scattering to occur the two Bloch states i and j must overlap with the excess potential $\upsilon(\mathbf{r})$, which is peaked at the dopant atom site. 1s intraband scattering is therefore particularly dominant for a substitutional atom. On the other hand, for an interstitial, both intra- and interband scattering are mainly due to the non-1s states, provided the dopant atom does not overlap with an atom column, when viewed along the incident beam direction.

For STEM calculations Eq. (4.77) is applied to each partial plane wave within the incident probe. TDS is modelled phenomenologically by adding an imaginary part to the atomic potential (Bird and King, 1990), so that $\upsilon(\mathbf{r})$ is a complex quantity. As an example, Figure 4.13 shows STEM Bloch wave results for a W substitutional atom at depths of 18, 46 and 66 Å respectively in a 100 Å thick, [111]-Fe specimen. The individual depths coincide with minima and maxima of the STEM probe pendellösung in the perfect crystal. The STEM probe is focussed at the entrance surface of the atom column containing the W atom. Simulations are carried out with and without the dopant atom, and from this the change in STEM probe intensity at the atom column (due to dopant atom scattering) is plotted in Figure 4.13. For each curve there are two important features, labelled A and B respectively. Feature A is at the dopant atom depth and represents an increase in scattering of the intensity away from the atom column. A breakdown of the probe intensity into individual Bloch state contributions shows that feature A is due to the 1s state, that is, the 1s state is preferentially scattered by the heavy W atom, thereby giving bright atom contrast in the HAADF image. On the other hand, feature B in Figure 4.13 shows a 'sharpening' of the probe profile (i.e. the intensity is more localised on the atom column) at a depth slightly beyond the dopant atom. This is due to Coulomb attraction of the non-1s states by the W atom and is similar to the focusing effect of atom columns discussed in Section 4.2.2, although here the stronger focusing is due to the higher atomic number of W compared to Fe, and therefore occurs over a single atom. Bloch wave calculations

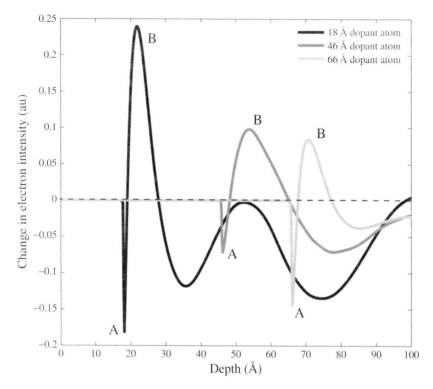

Figure 4.13   Change in STEM probe intensity (with respect to a perfect crystal) as a function of depth for [111]-Fe containing a single substitutional atom of W. Curves for the W atom at depths of 18, 46 and 66 Å are shown. The STEM probe is incident on the atom column containing the W atom, and the intensity is calculated along the same column. Simulations are for a 300 kV aberration-free probe with 20 mrad semi-convergence angle. From Mendis (2010). Reproduced with permission; copyright International Union of Crystallography.

have also been carried out for interstitial atoms, where similar effects are observed, although the scattering mechanisms are different, that is, the non-1s states play a more important role (Mendis, 2010).

HAADF is highly successful for imaging heavy atoms, while ABF is ideal for light atom columns. The annular detector in ABF has inner and outer collection angles within the STEM bright-field disc (Figure 4.14a). At optimum probe defocus and detector angles atom columns appear dark against a white background with high contrast; the image is also relatively insensitive to changes in specimen thickness. ABF therefore shows some characteristics of incoherent imaging, although, strictly speaking, coherence effects are present even for optimum imaging

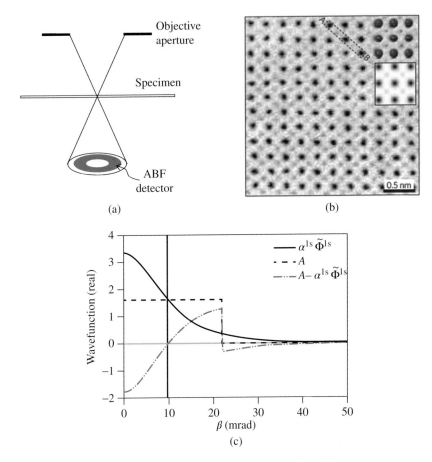

Figure 4.14 (a) Schematic of the detection geometry used in ABF imaging. An annular detector is placed within the bright-field disc in STEM mode. (b) An ABF image of [010]-oriented YH_2. The insets show the projected crystal structure (purple and green atoms represent yttrium and hydrogen respectively) as well as the multislice simulated ABF image. The image was acquired using an aberration corrected STEM operating at 200 kV with 22 mrad probe semi-convergence angle. The ABF detector inner and outer angles are 11 and 22 mrad respectively. From Ishikawa *et al.* (2011). Reproduced with permission; copyright Nature Publishing Group. (c) Shows the individual contributions to the ABF wavefunction, plotted as a function of the scattering angle (β). See text for further details. From Findlay *et al.* (2009). Reproduced with permission; copyright Elsevier.

conditions (Findlay *et al.*, 2010). An example of imaging hydrogen atom columns in an [010]-oriented, YH_2 crystal is shown in Figure 4.14b (Ishikawa *et al.*, 2011). Findlay *et al.* (2009, 2010) have developed a Bloch wave theory for ABF imaging, assuming interference between the

1s Bloch state and other dispersive Bloch states, which are collectively modelled as an unscattered wave propagating through the crystal. The STEM electron wavefunction in reciprocal space $\mathbf{u}$ can then be expressed as (Findlay *et al.*, 2009, 2010):

$$\tilde{\Psi}(\mathbf{u}, z) = \alpha^{1s}\tilde{\Phi}^{1s}(\mathbf{u}) \exp\left(-\pi i \frac{E_t^{1s}}{E_o} k_{vac} z \right)$$
$$+ \left[A(\mathbf{u}) - \alpha^{1s}\tilde{\Phi}^{1s}(\mathbf{u}) \right] \exp(-\pi i \lambda u^2 z) \qquad (4.78)$$

where the first term is the Fourier transform of the 1s Bloch state (Eq. (4.57)), and α^{1s} is its excitation by the STEM probe (note that $\alpha^{1s} \neq \varepsilon^{1s}$, since ε^{1s} is the excitation by a single plane wave). $A(\mathbf{u})$ is the Fourier transform of the STEM probe at the specimen entrance surface and is equal to the aperture function for an aberration-free probe. At $z = 0$ the wavefunction equals $A(\mathbf{u})$ as required, so that $(A(\mathbf{u}) - \alpha^{1s}\tilde{\Phi}^{1s}(\mathbf{u}))$ is the unscattered wave due to non-1s Bloch states. The phase term $\exp(-\pi i \lambda u^2 z)$ represents propagation of the unscattered wave through the crystal (Section 3.1.2). Figure 4.14c plots the (real quantity) terms $\alpha^{1s}\tilde{\Phi}^{1s}$, A and $(A-\alpha^{1s}\tilde{\Phi}^{1s})$ as a function of the scattering angle for an aberration-free probe. Interference between the first and third terms gives the second term at $z = 0$. Furthermore, at $z = 0$, $\alpha^{1s}\tilde{\Phi}^{1s}$ and $(A-\alpha^{1s}\tilde{\Phi}^{1s})$ are out of phase by π-radians for scattering angles smaller than the first zero crossing of the $(A-\alpha^{1s}\tilde{\Phi}^{1s})$ term (at ~10 mrad in Figure 4.14c). Any interference at $z \neq 0$ between these two terms[6] can therefore only be less destructive within this angular range, so that the resultant intensity is greater than the probe intensity (i.e. square modulus of $A(\mathbf{u})$). Similarly, it follows that interference at $z \neq 0$ between $\alpha^{1s}\tilde{\Phi}^{1s}$ and $(A-\alpha^{1s}\tilde{\Phi}^{1s})$ will give an intensity less than the probe intensity for scattering angles between the first zero crossing of $(A-\alpha^{1s}\tilde{\Phi}^{1s})$ and the objective aperture 'radius'. If the ABF detector angles are chosen such that it spans this particular angular range, the image will consist of dark atoms on a white background, irrespective of specimen thickness (i.e. contrast reversals are not possible). The atom column contrast depends on its chemical composition, since the phase of the first term in Eq. (4.78) is a function of the 1s Bloch state transverse energy. However, there is not always a direct correlation between column contrast and chemical composition. Indeed, multislice simulations have shown that

---

[6] Propagation along z changes the phase of $\alpha^{1s}\tilde{\Phi}^{1s}$ by a constant amount, while for the $(A-\alpha^{1s}\tilde{\Phi}^{1s})$ term the phase change is dependent on the scattering angle (Eq. (4.78)).

light atom columns can appear darker in the ABF image compared to heavier atom columns within certain specimen thickness ranges (Findlay *et al.*, 2010). TDS will also transfer some of the intensity within the bright-field disc out to larger scattering angles, thereby preferentially enhancing the ABF contrast of heavy elements (Findlay *et al.*, 2010).

4.3.2 Electron Channelling and Its Uses

Channelling of the electron beam is essential for revealing the crystal structure in HREM and HAADF imaging; it also has an important effect on other 'localised' signals that are generated close to the atom positions, such as X-ray and Auger electron emission from ionised atoms as well as backscattered electrons. Consider characteristic X-ray emission under two-beam diffraction conditions. For negative deviation parameters Bloch wave 1, which channels along the atomic planes, is preferentially excited. The X-ray intensity is therefore enhanced for this crystal orientation, while the opposite is true for positive deviation parameters, where Bloch wave 2 is preferentially excited. While channelling does not affect composition analysis of a disordered crystal (apart from the X-ray emission rate), it does however have an important effect on ordered crystals, where the constituent elements are distributed on individual sub-lattices. An example is the 200 reflection in sphalerite structures, such as GaAs. Positive and negative deviation parameters result in channelling along the Ga and As sub-lattices respectively. For accurate chemical analysis the crystal must therefore be tilted away from strong Bragg diffracting conditions. In the ALCHEMI (*atom location by channeling enhanced microanalysis*) technique channelling is used to determine the site occupancy of a substitutional impurity in an otherwise ordered crystal. The crystal is tilted so that the electron beam channels along the atomic planes of a given sub-lattice; by comparing the X-ray spectrum with that obtained under non-channelling conditions the site occupancy of the impurity can be quantified. The technique has been applied successfully using X-rays (Spence and Taftø, 1983; Jones, 2002) as well as electron energy loss spectroscopy (Taftø and Krivanek, 1982).

Coates (1967) first reported observation of Kikuchi-like patterns in SEM backscattered images of flat, single crystal samples recorded over a large (e.g. several millimetres) field of view. These so-called *electron channelling patterns* (ECP) are formed due to the beam being slightly

tilted by the scan coils as it is rastered over a large area.[7] The origin of
ECP patterns is similar to what has been described previously, that is,
Bloch waves that channel along (between) the atom columns give rise to
a larger (smaller) backscatter yield. This dependence of the backscatter
yield on crystal orientation is also the basis of electron channelling contrast imaging (ECCI). An example is shown in Figure 4.15a, which is a
backscattered image of a polycrystalline CdTe specimen. Despite the lack
of any atomic number contrast, individual grains, as well as twin boundaries, are visible due to the channelling effect. A similar contrast effect
should also apply to crystal defects such as dislocations. Although the
orientation within the undistorted regions of the crystal is fixed, close to
the dislocation core the atomic planes can be bent either towards or away
from the Bragg angle, thereby resulting in a larger or smaller backscatter
yield compared to the background, which renders the dislocation visible. Figure 4.15b, for example, shows dislocations in a deformed sample
of the mineral calcite. Application of ECCI to deformed microstructures
has been reviewed by Joy *et al.* (1982) and Wilkinson and Hirsch (1997).

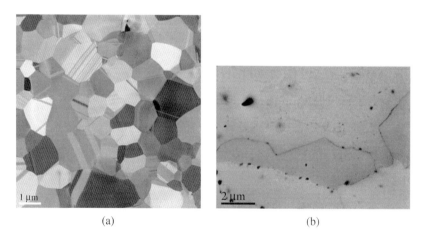

(a)   (b)

Figure 4.15   SEM backscattered images of (a) a polycrystalline CdTe thin film.
Micrograph courtesy of Dr Aidan Taylor. (b) Individual dislocations in deformed
calcite. Sample courtesy of Dr Nicola De Paola.

---

[7] In modern SEMs, ECPs are recorded by rocking the electron beam about a 'point' on the specimen, rather than scanning over a large area (Wilkinson and Hirsch, 1997). Electron backscattered diffraction (EBSD) patterns, which are widely used for orientation imaging, are related to
ECP through the principle of reciprocity (Wells, 1999).

Accurate simulation of ECCI is complicated by the fact that both elastic and inelastic scattering must be taken into account. Spencer *et al.* (1972) have developed a simplified Bloch wave model that can also simulate defects within the column approximation and is summarised here for completeness (note that more advanced methods are also available, such as Dudarev *et al.* (1995), but are beyond the scope of this book). The jth Bloch wave current $I^{(j)}$ decays exponentially (Eq. (4.47)) due to TDS scattering, that is, $I^{(j)}(z) = I^{(j)}(0)\exp(-\mu^{(j)}z)$. The TDS intensity is broken down into a forward current ($I_F^{(j)}$) and a backscattered current ($I_B^{(j)}$), which are assumed to be plane waves for simplicity. From Figure 4.16a the change in backscattered intensity within a thin slice dz is

$$dI_B^{(j)}(z) = \left[p^{(0)}I_B^{(j)}(z) - p^{(j)}I^{(j)}(z) - p^{(0)}I_F^{(j)}(z) \right] dz \qquad (4.79)$$

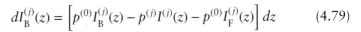

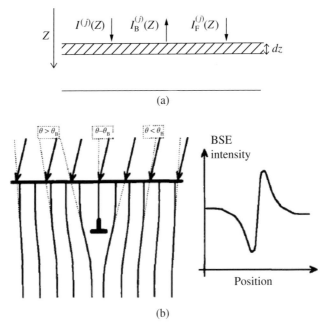

(b)

Figure 4.16 (a) Schematic of the scattering geometry assumed for calculating the ECCI signal intensity. See text for further details. (b) illustrates the backscattered electron (BSE) intensity expected under two-beam conditions for an edge dislocation lying parallel, and in close proximity, to the specimen entrance surface. The electron beam is incident at the Bragg angle within the undistorted regions of the crystal. From Wilkinson and Hirsch (1997). Reproduced with permission; copyright Elsevier.

where $p^{(0)}$ and $p^{(j)}$ are the backscattering probability per unit cross-section area and unit depth for a plane wave and jth Bloch wave respectively ($p^{(j)}$ is due to TDS scattering of the Bloch wave). Note that the direction of $I_B^{(j)}$ is opposite to that of $I^{(j)}$ and $I_F^{(j)}$ (Figure 4.16a) and this is reflected in Eq. (4.79). For example, considering the backscattered current only, then $I_B^{(j)}(z) = (1 - p^{(0)}dz)I_B^{(j)}(z + dz)$, where $(1 - p^{(0)}dz)$ represents the forward scattering probability over the slice thickness dz. Hence, $dI_B^{(j)} = p^{(0)}I_B^{(j)}(z + dz)dz \approx p^{(0)}I_B^{(j)}(z)dz$. This is the first term in Eq. (4.79); similar arguments can be used to explain the remaining terms. Clearly with increasing depth both $I^{(j)}$ and $I_B^{(j)}$ must decrease (i.e. $dI^{(j)}, dI_B^{(j)} < 0$), while $I_F^{(j)}$ increases (i.e. $dI_F^{(j)} > 0$). Furthermore,

$$dI^{(j)}(z) = dL^{(j)}(z) - \mu^{(j)}I^{(j)}(z)dz$$
$$= dL^{(j)}(z) + [dI_B^{(j)} - dI_F^{(j)}] \qquad (4.80)$$

where $dL^{(j)}$ represents any change to the jth Bloch wave intensity due to scattering by a defect (Eq. (4.73)). The $-\mu^{(j)}I^{(j)}(z)dz$ term is the approximate (to first order) intensity lost from the jth Bloch wave through TDS scattering into $I_F^{(j)}$ and $I_B^{(j)}$. From Eqs. (4.79) and (4.80) it then follows that

$$\frac{d^2 I_B^{(j)}}{dz^2} = (p^{(0)} - p^{(j)})\frac{dI^{(j)}}{dz} - p^{(0)}\frac{dL^{(j)}}{dz} \qquad (4.81)$$

Integrating twice, while substituting Eq. (4.79) for $[dI_B^{(j)}/dz]_{z=0}$ and the boundary conditions $L^{(j)}(0) = 0, I_F^{(j)}(0) = 0, I_B^{(j)}(t) = 0$, gives

$$I_B^{(j)}(0) = \frac{1}{1 + p^{(0)}t}\left[p^{(0)}tI^{(j)}(0) + (p^{(j)} - p^{(0)})\int_0^t I^{(j)}(z)dz + p^{(0)}\int_0^t L^{(j)}(z)dz \right] \qquad (4.82)$$

Here, t represents the maximum depth beyond which no backscattered electrons are generated. The total backscattered intensity $I_B(0)$ is obtained by summing Eq. (4.82) over all Bloch states j:

$$I_B(0) = \frac{1}{1 + p^{(0)}t}\left[p^{(0)}t + \sum_j (p^{(j)} - p^{(0)})\int_0^t I^{(j)}(z)dz \right] \qquad (4.83)$$

where the additional constraints $\Sigma_j I^{(j)}(0) = 1$ and $\Sigma_j L^{(j)}(z) = 0$ have been used to simplify the expression (the latter constraint is due to the fact that the electron intensity must be conserved during defect scattering).

$p^{(j)}$, and similarly $p^{(0)}$, are determined by integrating the frozen phonon derived TDS intensity distribution (Eq. (4.45), Section 4.1.3) over all scattering vectors that correspond to backscattering events. For more details, see Spencer *et al.* (1972), who provide expressions for backscattering probabilities that are valid for many-beam conditions and any crystal system (not just primitive cells). The value of t to be used in Eq. (4.83) is somewhat arbitrary, since the model does not take into account energy loss of the incident electrons. In practice t is made a fraction ($\sim$0.4; Spencer *et al.*, 1972) of the electron beam penetration range, which can be determined independently from (say) Monte Carlo simulations. Another weakness of the model is that it assumes backscattering to be a single scattering event. The cumulative effect of many forward scattered events that eventually lead to backscattering is not taken into account, so that the total backscatter yield $I_B(0)$ is underestimated.

Spencer *et al.* (1972) have applied Eq. (4.83) to stacking faults and dislocations and find good qualitative agreement with experiment. As an example, Figure 4.16b is a schematic of the contrast expected for an edge dislocation lying parallel to the surface and where the electron beam is incident at the Bragg angle within the undeformed parts of the crystal. The crystal planes to the right of the dislocation core are tilted such that Bloch wave 1 is preferentially excited and hence the backscattered yield is larger. The opposite is true to the left of the dislocation, so that the net result is black–white contrast for an edge dislocation. This has been confirmed by the simulations of Spencer *et al.* (1972). For edge dislocations lying at greater depths (i.e. several extinction distances ξ_g) from the surface, however, the simulations show only white contrast. This is because at these depths Bloch wave 1 is heavily absorbed, so that the contrast is due to interband scattering from Bloch wave 2 to Bloch wave 1 (Spencer *et al.*, 1972).

4.4 SUMMARY

Bloch waves are solutions to Schrödinger's equation for a high energy electron within a crystal. The total electron wavefunction is a linear combination of individual Bloch waves. In electron diffraction a large number of diffracted beams can have significant intensity (e.g. near a zone-axis orientation), but the number of strongly excited Bloch waves is relatively small. Interpretation of dynamic diffraction effects is therefore relatively straightforward using Bloch waves. For example, two-beam diffraction contrast effects, such as thickness fringes and bend contours, are due to interference between two Bloch waves, which channel along and between the atom planes respectively. For quantitative agreement, particularly at larger specimen thicknesses, TDS must be included phenomenologically by using a complex crystal potential. HREM and HAADF images are also readily understood using Bloch waves, and here the 1s Bloch state is found to play a crucial role.

The Bloch wave method is ideal for crystals with small unit cells, since the number of Fourier components required to accurately represent the Bloch wave is relatively small. As the unit cell size increases, however, the number of Fourier components increase rapidly, so that other simulation methods, such as multislice, are computationally less expensive. It should, however, be pointed out that for Bloch waves the calculation time is independent of specimen thickness, and this has obvious advantages over multislice for thick specimens. For the same reason, amorphous materials and deformed crystals are non-ideal specimens for Bloch wave simulations; the method therefore lacks versatility, compared to multislice. Nevertheless, the column approximation can be used to extend Bloch wave analysis to deformed crystals with slowly varying strain fields in a straightforward manner.

REFERENCES

Allen, J.E., Hemesath, E.R., Perea, D.E., Lensch-Falk, J.L., Li, Z.Y., Yin, F., Gass, M.H., Wang, P., Bleloch, A.L., Palmer, R.E. and Lauhon, L.J. (2008). *Nature Nanotech.* **3**, 168.

Anstis, G.R., Cai, D.Q. and Cockayne, D.J.H. (2003) *Ultramicroscopy* **94**, 309.

Benthem, K. van, Lupini, A.R., Kim, M., Baik, H.S., Doh, S.J., Lee, J.-H., Oxley, M.P., Findlay, S.D., Allen, L.J. and Pennycook, S.J. (2005) *Appl. Phys. Lett.* **87**, 034104.

Berry, M.V. and Ozorio de Almeida, A.M. (1973) *J. Phys.* **A6**, 1451.

Bird, D.M. and King, Q.A. (1990) *Acta Cryst.* **A46**, 202.

Buxton, B.F., Loveluck, J.E. and Steeds, J.W. (1978) *Phil. Mag.* **A38**, 259.

Coates, D.G. (1967) *Phil. Mag.* **16**, 1179.

Cosgriff, E.C. and Nellist, P.D. (2007) *Ultramicroscopy* **107**, 626.

Cosgriff, E.C., Nellist, P.D., Hirsch, P.B., Zhou, Z. and Cockayne, D.J.H. (2010) *Phil. Mag.* **90**, 4361.

Dass, T. and Sharma, S.K. (1998) *Mathematical Methods in Classical and Quantum Physics*, University Press (India) Ltd, Hyderabad, India.

Dudarev, S.L., Rez, P. and Whelan, M.J. (1995) *Phys. Rev. B* **51**, 3397.

Fertig, J. and Rose, H. (1981) *Optik* **59**, 407.

Findlay, S.D., Shibata, N., Sawada, H., Okunishi, E., Kondo, Y., Yamamoto, T. and Ikuhara, Y. (2009) *Appl. Phys. Lett.* **95**, 191913.

Findlay, S.D., Shibata, N., Sawada, H., Okunishi, E., Kondo, Y. and Ikuhara, Y. (2010) *Ultramicroscopy* **110**, 903.

Hall, C.R. and Hirsch, P.B. (1965) *Proc. Roy. Soc. A* **286**, 158.

Hirsch, P.B., Howie, A., Nicholson, R.B., Pashley, D.W. and Whelan, M.J. (1965) *Electron Microscopy of Thin Crystals*, Butterworths, Great Britain.

Hovden, R., Xin, H.L. and Muller, D.A. (2012) *Phys. Rev. B* **86**, 195415.

Howie, A. and Basinski, Z.S. (1968) *Phil. Mag.* **17**, 1039.

Howie, A. and Sworn, C.H. (1970) *Phil. Mag.* **22**, 861.

Ishikawa, R., Okunishi, E., Sawada, H., Kondo, Y., Hosokawa, F. and Abe, E. (2011) *Nature Mat.* **10**, 278.

Jones, I.P. (2002) *Adv. Imaging Electron Phys.* **125**, 63.

Joy, D.C., Newbury, D.E. and Davidson, D.L. (1982) *J. Appl. Phys.* **53**, R81.

Kambe, K. (1982) *Ultramicroscopy* **10**, 223.

Klenov, D.O. and Stemmer, S. (2006) *Ultramicroscopy* **106**, 889.

Kogiso, M. and Takahashi, H. (1977) *J. Phys. Soc. Japan* **42**, 223.

Krivanek, O.L., Corbin, G.J., Dellby, N., Elston, B.F., Keyse, R.J., Murfitt, M.F., Own, C.S., Szilagyi, Z.S. and Woodruff, J.W. (2008) *Ultramicroscopy* **108**, 179.

Levine, I.R. (2006) *Quantum Chemistry, 5th Edition*, Prentice-Hall of India, New Delhi.

Mendis, B.G. (2008) *Acta Cryst.* **A64**, 613.

Mendis, B.G. and Hemker, K.J. (2008) *Ultramicroscopy* **108**, 855.

Mendis, B.G. (2010) *Acta Cryst.* **A66**, 407.

Nellist, P.D. and Pennycook, S.J. (1996) *Science* **274**, 413.

Nellist, P.D. and Pennycook, S.J. (1999) *Ultramicroscopy* **78**, 111.

Nellist, P.D., Cosgriff, E.C., Hirsch, P.B. and Cockayne, D.J.H. (2008) *Phil. Mag.* **88**, 135.

Okunishi, E., Ishikawa, I., Sawada, H., Hosokawa, F., Hori, M. and Kondo, Y. (2009) *Microsc. Microanal.* **15**, 164.

Peng, L.M. (2005) *J. Electr. Microscopy* **54** 199.

Peng, Y., Nellist, P.D. and Pennycook, S.J. (2004) *J. Electron Microsc.* **53**, 257.

Pennycook, S.J. and Jesson, D.E. (1990) *Phys. Rev. Lett.* **64**, 938.

Pennycook, S.J. and Jesson, D.E. (1991) *Ultramicroscopy* **37**, 14.

Perovic, D.D., Rossouw, C.J. and Howie, A. (1993) *Ultramicroscopy* **52**, 353.

Plamann, T. and Hÿtch, M.J. (1999) *Ultramicroscopy* **78**, 153.

Rafferty, B., Nellist, P.D. and Pennycook, S.J. (2001) *J. Electron Microsc.* **50**, 227.

Rossouw, C.J. and Bursill, L.A. (1985) *Acta Cryst.* **A41**, 320.

Spence, J.C.H. and Taftø, J. (1983) *J. Microsc.* **130**, 147.

Spence, J.C.H. and Zuo, J.M. (1992) *Electron Microdiffraction*, Plenum Press, New York.

Spencer, J.P., Humphreys, C.J. and Hirsch, P.B. (1972) *Phil. Mag.* **26**, 193.

Taftø, J. and Krivanek, O.L. (1982) *Phys. Rev. Lett.* **48**, 560.

Takagi, S. (1962) *Acta Cryst.* **15**, 1311.

Treacy, M.M.J. and Gibson, J.M. (1993) *Ultramicroscopy* **52**, 31.

Van Dyck D. and Op de Beeck, M. (1996) *Ultramicroscopy* **64**, 99.

Voyles, P.M., Muller, D.A., Grazul, J.L., Citrin, P.H. and Gassmann, H.-J.L. (2002) *Nature* **416**, 826.

Wang, Z.L. (1995) *Elastic and Inelastic Scattering in Electron Diffraction and Imaging*, Plenum Press, New York.

Wells, O.C. (1999) *Scanning* **21**, 368.

Whelan, M.J. and Hirsch, P.B. (1957) *Phil. Mag.* **2**, 1121.
Wilkinson, A.J. and Hirsch, P.B. (1997) *Micron* **28**, 279.
Yoshioka, H. (1957) *J. Phys. Soc. Japan* **12**, 618.
Yoshioka, H. and Kainuma, Y. (1962) *J. Phys. Soc. Japan* **17**, Supplementary Bulletin B-II, 134.
Yu, Z., Muller, D.A. and Silcox, J. (2004) *J. Appl. Phys.* **95**, 3362.

5

Single Electron Inelastic Scattering

Recent developments in both spherical (Haider *et al.*, 1998; Krivanek *et al.*, 2008) and chromatic (Kabius *et al.*, 2009) aberration corrected electron microscopy have made chemical analysis at the atomic scale now almost routine. Examples include scanning transmission electron microscopy (STEM) mapping using electron energy loss spectroscopy (EELS; Bosman *et al.*, 2007; Muller *et al.*, 2008) and energy dispersive X-ray (EDX) signals (Chu *et al.*, 2010; D'Alfonso *et al.*, 2010), as well as energy-filtered TEM (EFTEM) imaging (Lugg *et al.*, 2010; Urban *et al.*, 2013). The inelastic signal is due to excitation of a single core electron to an unoccupied conduction band state by the incident electron. In the previous chapters it was shown that accurate modelling of the elastic wavefield within a crystal required incorporating inelastic scattering as well, especially thermal diffuse scattering; similarly for measurements using an inelastic signal the effects of elastic scattering must also be considered. A more holistic theory that takes into account the generation, propagation and interaction of elastic and inelastic waves is therefore necessary. The theory is due to Yoshioka (1957) and most high energy, inelastic scattering simulations are derived from this. Before proceeding to Yoshioka's theory, however, the excitation of a single atom by a high energy electron plane wave will first be considered. This is the simplest case of inelastic scattering and is a good starting point to introduce several key concepts, such as the dynamic form factor and selection rules. This will then be generalised to inelastic scattering within a crystal consisting of

Electron Beam-Specimen Interactions and Simulation Methods in Microscopy,
First Edition. Budhika G. Mendis.
© 2018 John Wiley & Sons Ltd. Published 2018 by John Wiley & Sons Ltd.

many atoms, and where the incident electron may not necessarily be a plane wave, but rather a focused STEM probe formed by the coherent superposition of many plane waves.

In several cases the EELS edge fine structure, that is, the intensity oscillations superimposed on the gross shape of the energy loss signal, is sensitive to the bonding and coordination environment of the excited atom. This is apparent in the energy loss near edge structure (ELNES) and extended energy loss fine structure (EXELFS) respectively and can provide complementary information to conventional lattice imaging and electron diffraction, especially when the region of interest is spatially non-uniform, such as dopant atoms, grain boundaries and interfaces. Here, it is the shape of the inelastic loss signal, rather than its total intensity, that is of interest. This does not, however, mean that elastic scattering of electrons is not important. For example, if the crystal is thick the measured signal will be due to excitation of several atoms within the analytical volume, some of which may have different chemical and/or bonding environments. The final part of this chapter deals with the physical origin of EELS edge fine structure. The focus is on the scattering mechanisms of the excited 'photoelectron' that gives rise to ELNES and EXELFS features, rather than a survey of the vast literature on EELS edge shapes of different compounds and their interpretation. Furthermore, band structure and real space methods to simulate the fine structure, such as density functional theory and multiple scattering theory, are not described, but can be found elsewhere (e.g. Martin, 2004; Gonis and Butler, 2000). Application of these methods to EELS is discussed in Hébert (2007; WIEN density functional theory code) and Moreno et al. (2007; FEFF multiple scattering code).

5.1 FUNDAMENTALS OF INELASTIC SCATTERING

5.1.1 Electron Excitation in a Single Atom by a Plane Wave

Single electron excitation of a free atom by a plane wave will first be considered. The derivation is a summary of the treatment found in McDaniel (1989). Within the Born–Oppenheimer approximation[1] (Levine, 2006) the Schrödinger equation for a system consisting of an incident electron

---

[1]This assumes that, compared to electrons, the heavier nuclei are effectively frozen in space, so that for any given nuclear configuration the system energy is minimised via the electron density.

and target atom of atomic number Z is given by

$$\left[-\frac{\hbar^2}{2m} \nabla_\mathbf{r}^2 - \sum_\alpha \frac{\hbar^2}{2m} \nabla_{\mathbf{r}_\alpha}^2 - eV(\mathbf{r}, \mathbf{r}_\alpha) \right] \Psi_s(\mathbf{r}, \mathbf{r}_\alpha) = E\Psi_s(\mathbf{r}, \mathbf{r}_\alpha) \quad (5.1)$$

$$V(\mathbf{r}, \mathbf{r}_\alpha) = \frac{Ze}{4\pi\varepsilon_o |\mathbf{r}|} + \sum_\alpha \left[\frac{Ze}{4\pi\varepsilon_o |\mathbf{r}_\alpha|} - \frac{e}{4\pi\varepsilon_o |\mathbf{r} - \mathbf{r}_\alpha|} \right] \quad (5.2)$$

The first two terms in the left-hand side of Eq. (5.1) represent the kinetic energies of the incident and N atomic electrons (position coordinates $\mathbf{r}$ and $\mathbf{r}_\alpha$ respectively; m is the electron mass and $\hbar = h/2\pi$ is the reduced Planck's constant). The potential $V(\mathbf{r}, \mathbf{r}_\alpha)$ consists of three terms, representing Coulomb interaction between (i) the nucleus and incident electron, (ii) atomic electrons and nucleus and (iii) incident and atomic electrons respectively (Eq. (5.2)). It is assumed that the atom is at the origin and that the incident electron energy is sufficiently large so that exchange effects can be neglected (Inokuti, 1971; with exchange the incident electron is 'absorbed' by the atom, while an atomic electron is ejected as the 'scattered' particle). Furthermore, ε_o is the permittivity of free space. The energy E of the system is the sum of the incident electron kinetic energy (E_{kin}) and the target atom (E_0), initially assumed to be in the ground state. Both energies are evaluated when the atom and incident electron are far apart. The total energy of the *system* is conserved during inelastic scattering.

To solve Eq. (5.1) the system wavefunction $\Psi_s(\mathbf{r}, \mathbf{r}_\alpha)$ is expressed in terms of atomic wavefunctions, $u_n(\mathbf{r}_1, \dots, \mathbf{r}_N)$, and the incident electron wavefunction $\psi_n(\mathbf{r})$, that is,

$$\Psi_s(\mathbf{r}, \mathbf{r}_\alpha) = \sum_n u_n(\mathbf{r}_1, \dots, \mathbf{r}_N)\psi_n(\mathbf{r}) \quad (5.3)$$

The subscript n in $u_n(\mathbf{r}_1, \dots, \mathbf{r}_N)$ denotes the atomic state, with $n = 0$ being the ground state and $n > 0$ representing excited states of energy E_n. Similarly ψ_0 is the elastic wave, while $\psi_{n>0}$ are inelastic waves corresponding to the nth excited state of the target atom. $u_n(\mathbf{r}_1, \dots, \mathbf{r}_N)$ are solutions to the Schrödinger equation for the target atom:

$$\left[\sum_\alpha \left(-\frac{\hbar^2}{2m} \nabla_{\mathbf{r}_\alpha}^2 - \frac{Ze^2}{4\pi\varepsilon_o |\mathbf{r}_\alpha|} \right) \right] u_n(\mathbf{r}_1, \dots, \mathbf{r}_N) = E_n u_n(\mathbf{r}_1, \dots, \mathbf{r}_N) \quad (5.4)$$

Substituting Eqs. (5.3) and (5.4) in (5.1) it follows that for each value of n,

$$-\frac{\hbar^2}{2m}(\nabla_r^2\psi_n)u_n + \left[-\frac{Ze^2}{4\pi\varepsilon_o|r|} + \sum_\alpha \frac{e^2}{4\pi\varepsilon_o|r-r_\alpha|}\right]u_n\psi_n = (E-E_n)u_n\psi_n$$

(5.5)

Multiplying by $u_m^*(r_1, \ldots, r_N)$ and integrating over all atomic electron coordinates gives

$$\left(\nabla_r^2 + 4\pi^2 k_m^2 + \frac{2me}{\hbar^2}V_{mm}(r)\right)\psi_m(r) = -\frac{2me}{\hbar^2}\sum_{n\neq m}V_{mn}(r)\psi_n(r) \quad (5.6)$$

where the orthonormal property of the atomic wavefunctions u_n has been used to simplify the above equation. Furthermore,

$$V_{mn}(r) = \int u_m^*(r_1, \ldots, r_N)\left(\frac{Ze}{4\pi\varepsilon_o|r|} - \sum_\alpha \frac{e}{4\pi\varepsilon_o|r-r_\alpha|}\right)$$

$$u_n(r_1, \ldots, r_N)dr_1 \ldots dr_N$$

$$= \frac{Ze}{4\pi\varepsilon_o|r|}\delta_{mn} - \int u_m^*(r_1, \ldots, r_N)\left(\sum_\alpha \frac{e}{4\pi\varepsilon_o|r-r_\alpha|}\right)$$

$$u_n(r_1, \ldots, r_N)dr_1 \ldots dr_N \quad (5.7)$$

$$k_m^2 = \frac{2m[E_{kin} + (E_0 - E_m)]}{\hbar^2} \quad (5.8)$$

where δ_{mn} is the Kronecker delta, which is equal to unity for $m=n$ and zero otherwise. From Eq. (5.8) it is clear that k_m is the wave number of the inelastically scattered electron (energy loss $E_m - E_0$) at infinite distance from the atom.

Equation (5.6) is the fundamental equation for inelastic scattering and represents a set of coupled equations in $\psi_m(r)$. A solution can be found by invoking the Born approximation, which assumes weak scattering. Under these conditions $|\psi_0(r)| \gg |\psi_m(r)|$ and hence only the $n=0$ term remains in the right-hand side of Eq. (5.6). Furthermore, it is assumed that V_{mm} is negligible, which effectively means that elastic scattering of the inelastic wave is ignored. For an incident plane wave of the form $\psi_0(r) = \exp(2\pi i k_0 n_0 \cdot r)$ Eq. (5.6) therefore simplifies to

$$(\nabla_r^2 + 4\pi^2 k_m^2)\psi_m(r) = -\frac{2me}{\hbar^2}V_{m0}(r)\exp(2\pi i k_0 n_0 \cdot r) \quad (5.9)$$

where $\mathbf{n}_0$ is a unit vector parallel to the wave vector of the incident wave. V_{m0} is the potential for scattering of the elastic wave into the ψ_m inelastic channel. The inhomogeneous solution to Eq. (5.9) is given by (see Appendix A for an outline of the general method):

$$\psi_m(\mathbf{r}) = \frac{me}{2\pi\hbar^2} \int \frac{\exp(2\pi i k_m |\mathbf{r} - \mathbf{r}'|)}{|\mathbf{r} - \mathbf{r}'|} V_{m0}(\mathbf{r}') \exp(2\pi i k_0 \mathbf{n}_0 \cdot \mathbf{r}') d\mathbf{r}' \quad (5.10)$$

For large $|\mathbf{r}|$ the elastic and inelastic scattered waves have the following form:

$$\psi_0(\mathbf{r}) \sim \exp(2\pi i k_0 \mathbf{n}_0 \cdot \mathbf{r}) + f_0(\theta) \frac{e^{2\pi i k_0 r}}{r} \quad (5.11)$$

$$\psi_m(\mathbf{r}) \sim f_m(\theta) \frac{e^{2\pi i k_m r}}{r} \quad (5.12)$$

where $f_0(\theta)$, $f_m(\theta)$ are the atom scattering factors for elastic and inelastic scattering respectively. The first and second terms in Eq. (5.11) represent an unscattered plane wave and an elastic scattered (distorted) spherical wave respectively. From Section 2.3.1, the flux of inelastic scattered electrons within a solid angle $d\Omega$ is proportional to $k_m |f_m(\theta)|^2 d\Omega$, while the incident flux is proportional to k_0. The differential inelastic scattering cross-section is therefore given by $d\sigma_m = (k_m/k_0)|f_m(\theta)|^2 d\Omega$. By comparing the asymptotic form of Eq. (5.10) at large $|\mathbf{r}|$ with Eq. (5.12) it can be shown that (see Appendix A):

$$d\sigma_m = \frac{m^2 e^2}{4\pi^2 \hbar^4} \left(\frac{k_m}{k_0} \right) \left| \int \exp[2\pi i (k_0 \mathbf{n}_0 - k_m \mathbf{n}_m) \cdot \mathbf{r}'] V_{m0}(\mathbf{r}') d\mathbf{r}' \right|^2 d\Omega \quad (5.13)$$

with $\mathbf{n}_m$ being a unit vector along the direction of the inelastic scattering wave vector. The above can be simplified by using the general formula:

$$\int \frac{e^{i\mathbf{q}' \cdot \mathbf{r}'}}{|\mathbf{r} - \mathbf{r}'|} d\mathbf{r}' = \frac{4\pi}{(q')^2} e^{i\mathbf{q}' \cdot \mathbf{r}} \quad (5.14)$$

It then follows that

$$\int \exp[2\pi i (k_0 \mathbf{n}_0 - k_m \mathbf{n}_m) \cdot \mathbf{r}'] V_{m0}(\mathbf{r}') d\mathbf{r}'$$

$$= -\frac{e}{4\pi^2 \varepsilon_o q^2}$$

$$\times \left(\sum_\alpha \int u_m{}^*(\mathbf{r}_1, \ldots, \mathbf{r}_N) \exp(-2\pi i \mathbf{q} \cdot \mathbf{r}_\alpha) u_0(\mathbf{r}_1, \ldots, \mathbf{r}_N) d\mathbf{r}_1 \ldots d\mathbf{r}_N \right) \quad (5.15)$$

where $\mathbf{q} = k_m \mathbf{n}_m - k_0 \mathbf{n}_0$ is the inelastic scattering vector (the momentum transferred to the atom is $-\hbar\mathbf{q}$). Substituting in Eq. (5.13),

$$\frac{d\sigma_m}{d\Omega} = \left(\frac{1}{2\pi^2 a_o}\right)^2 \left(\frac{k_m}{k_0}\right) \frac{|\varepsilon_{m0}(\mathbf{q})|^2}{q^4} \qquad (5.16)$$

where $a_o = 4\pi\varepsilon_o \hbar^2/me^2$ is the Bohr radius and the transition matrix element $\varepsilon_{m0}(\mathbf{q})$ represents the bracketed term on the right-hand side of Eq. (5.15). It can be shown that the above equation has the same form as the Rutherford inelastic scattering cross-section (Eq. (2.23)), apart from the transition matrix element term (Inokuti, 1971). In the Rutherford model the atomic electrons are treated as free particles, whereas the inclusion of $\varepsilon_{m0}(\mathbf{q})$ in the cross-section takes into account the full quantum mechanical details of the atom. For a free atom, the final state (u_m) may either be a discrete energy level or within the continuum in the case of ionisation. Therefore, it is more appropriate to express Eq. (5.16) in the general form:

$$\frac{\partial^2 \sigma_m}{\partial(\Delta E)\partial\Omega} = \left(\frac{1}{2\pi^2 a_o}\right)^2 \left(\frac{k_m}{k_0}\right) \frac{S(\mathbf{q}, \Delta E)}{q^4} \qquad (5.17)$$

$$S(\mathbf{q}, \Delta E) = \sum_{i,f} p_i \left| \sum_\alpha \langle f | \exp(-2\pi i \mathbf{q} \cdot \mathbf{r}_\alpha) | i \rangle \right|^2 \delta(E_f - E_i - \Delta E) \qquad (5.18)$$

where $S(\mathbf{q}, \Delta E)$ is known as the dynamic form factor and $|i\rangle$, $|f\rangle$ are the wavefunctions of the initial occupied and final unoccupied electronic states with energies E_i and E_f respectively. Dirac's bra–ket notation is used to represent $\varepsilon_{m0}(\mathbf{q})$. The Dirac delta function (δ) ensures that the initial and final states are separated by the measured energy loss ΔE, while summation over i and f takes into account degenerate energy levels, with p_i being the probability that the excited electron is in the initial state i. The dynamic form factor is therefore averaged over all relevant initial states and summed over all possible final states (Kohl and Rose, 1985). The latter requires some explanation. The wavefunction of the system, as expressed by Eq. (5.3), depends on both the quantum state of the high energy electron as well as the target atom. In an EELS experiment, however, only the energy loss of the incident electrons is measured and there is no information about the final state of the target. Quantum mechanically the measurement corresponds to an incoherent sum over all possible transitions to (unobserved) final states of the atomic electron[2] (cf. Eq. (5.18)). Schattschneider *et al.*, (2000) have prescribed the use of the density matrix in such situations, although this approach will not be adopted here.

For a given energy loss the inelastic scattering vector can only have certain values. This is illustrated in Figure 5.1, which relates $\mathbf{q}$ to the incident $(k_0\mathbf{n}_0)$ and scattered $(k_m\mathbf{n}_m)$ wave vectors. If θ denotes the scattering angle the magnitude of $\mathbf{q}$ is given by

$$q^2 = k_0^2 + k_m^2 - 2k_0k_m \cos \theta \qquad (5.19)$$

The minimum (q_{min}) and maximum (q_{max}) values of q are for $\theta = 0$ and π respectively, that is, $\mathbf{n}_0$, $\mathbf{n}_m$ are parallel and anti-parallel respectively.

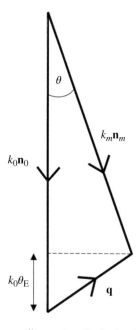

Figure 5.1 Schematic diagram illustrating the inelastic scattering vector $\mathbf{q}$ and its relation to the incident $(k_0\mathbf{n}_0)$ and scattered $(k_m\mathbf{n}_m)$ wave vectors. θ_E is the characteristic scattering angle. See text for further details.

---

[2] From Eq. (5.3) the measured intensity when no target information is available is given by $I(\mathbf{r}) = \int |\sum_n u_n (\mathbf{r}_1, \ldots, \mathbf{r}_N) \psi_n(\mathbf{r})|^2 d\mathbf{r}_1 \ldots d\mathbf{r}_N$, or equivalently $\sum_n |\psi_n(\mathbf{r})|^2$ after taking into account the orthonormal property of the atomic eigenfunctions u_n (Rae, 2008).

For small energy losses,

$$q_{min} = k_0 - k_m \approx \frac{k_0^2 - k_m^2}{2k_0} = k_0 \left(\frac{\Delta E}{2E_o} \right)$$

$$q_{max} = k_0 + k_m \approx 2k_0 \qquad\qquad (5.20)$$

where Eq. (5.8) has been used to simplify q_{min}. $\theta_E = (\Delta E/2E_o)$, where E_o is the incident electron energy, is known as the *characteristic scattering angle*. For small scattering angles $\cos\theta \approx (1 - \theta^2/2)$, so that from Eq. (5.19)

$$q^2 = (k_0\theta_E)^2 + (k_0\theta)^2 \qquad\qquad (5.21)$$

Referring back to Figure 5.1, it is clear that the second term in Eq. (5.21) is approximately the square transverse component of $\mathbf{q}$, which makes the first term the square longitudinal component.

The transition matrix element $\varepsilon_{m0}(\mathbf{q})$ will now be discussed in more detail. If the magnitude of the momentum change is small, then $\exp(-2\pi i \mathbf{q} \cdot \mathbf{r}_\alpha)$ in Eq. (5.15) can be approximated as $(1 - 2\pi i \mathbf{q} \cdot \mathbf{r}_\alpha)$; this is known as the *dipole approximation* and is valid when $(qr_c) \ll 1$, where r_c is the spatial extent of the relevant atomic electron wavefunction. If r_c is taken as the Bohr radius (≈ 0.5 Å) then for 200 kV electrons and 100 eV energy loss the scattering angle must be less than ~ 8 mrad for the dipole approximation to be satisfied. It follows that for small q, the term $(|\varepsilon_{m0}(\mathbf{q})|/q)^2$ in Eq. (5.16) is independent of the magnitude of q, so that, assuming single inelastic scattering and no elastic scattering from the specimen, the inelastic scattered electron intensity varies as q^{-2} (Eqs. (5.16) and (5.17) and Eq. (5.21)); that is, the intensity follows a Lorentzian distribution with respect to θ. Hence, the inelastic scattered intensity decreases to 50% of its peak value for $\theta = \theta_E$, which for 200 kV electrons undergoing an energy loss of 100 eV is only 0.25 mrad. The EELS spectrometer collection angle can therefore easily satisfy the dipole approximation whilst still maintaining good collection efficiency, provided the specimen is thin enough. The situation for EDX is however, different, since here it is the X-rays emitted that are detected. Characteristic X-ray emission is isotropic, so that for any given collection solid angle inelastic scattering events from all allowed values of q could potentially be detected.

The first term in $(1 - 2\pi i \mathbf{q} \cdot \mathbf{r}_\alpha)$ vanishes due to the orthonormal property of atomic eigenstates, so that only the second term remains in the transition matrix element. For a spherically symmetric atomic potential, the angular part of the atomic eigenfunction is given by a spherical harmonic (Rae, 2008):

$$Y_{lm}(\theta, \phi) = (-1)^m \sqrt{\frac{(2l+1)(l-|m|)!}{4\pi(l+|m|)!}} P_l^m(\cos\theta)e^{im\phi} \quad (5.22)$$

where l, m are the orbital and magnetic quantum numbers and P_l^m is the associated Legendre polynomial of order $|m|$. θ and ϕ are the polar and azimuthal angles (note that θ here is not to be confused with the scattering angle). First, consider the case where $\mathbf{q}$ is parallel to the (z-) axis of quantization of the atom, so that the transition matrix element is proportional to

$$\int [L_{n'l'}(r_\alpha)Y_{l'm'}(\theta,\phi)]^* z[L_{nl}(r_\alpha)Y_{lm}(\theta,\phi)]r_\alpha^2 \sin\theta \, dr_\alpha d\theta d\phi$$

$$\rightarrow \left(\int_0^\infty L_{n'l'}(r_\alpha)^* L_{nl}(r_\alpha)r_\alpha^3 dr_\alpha \right)$$

$$\cdot \left(\int_0^\pi [P_{l'}^{m'}(\cos\theta)P_l^m(\cos\theta)]\sin\theta \cos\theta d\theta \right)$$

$$\cdot \left(\int_0^{2\pi} e^{i(m-m')\phi} d\phi \right) \quad (5.23)$$

where the substitution $z = r_\alpha \cos\theta$ has been made. (n, l, m) and (n', l', m') are the quantum numbers for the initial and final states respectively (n, n' are principal quantum numbers) and $L_{nl}(r_\alpha)$ denotes the radial part of the wavefunction. The azimuthal angle integral is zero unless $m = m'$. To evaluate the polar angle integral the following recursive relation for associated Legendre polynomials is used (Gradshteyn and Ryzhik, 1980):

$$[(2l+1)\cos\theta]P_l^m = (l-m+1)P_{l+1}^m + (l+m)P_{l-1}^m \quad (5.24)$$

so that

$$\int_0^\pi (P_{l'}^{m'} P_l^m)\sin\theta \cos\theta \, d\theta$$

$$= \frac{l-m+1}{2l+1} \int_0^\pi P_{l'}^m P_{l+1}^m \sin\theta \, d\theta + \frac{l+m}{2l+1} \int_0^\pi P_{l'}^m P_{l-1}^m \sin\theta d\theta \quad (5.25)$$

where the substitution $m = m'$, due to the result from the azimuthal integration, has been made in the right-hand side. The integrals are simplified using the orthonormal property of associated Legendre polynomials (Gradshteyn and Ryzhik, 1980):

$$\int_0^\pi (P_{l'}^m P_l^m) \sin \theta \; d\theta = \delta_{ll'} \tag{5.26}$$

where $\delta_{ll'}$ is the Kronecker delta. From this, it follows that the polar angle integral is zero unless $l' = l \pm 1$. If $\mathbf{q}$ is parallel to $x = r_\alpha \sin \theta \cos \phi$ instead of z, the azimuthal angle integral becomes

$$\int_0^{2\pi} e^{i(m-m')\phi} \cos \phi \; d\phi$$

$$= \frac{1}{2} \left[\int_0^{2\pi} e^{i(m-m'+1)\phi} \; d\phi + \int_0^{2\pi} e^{i(m-m'-1)\phi} \; d\phi \right] \tag{5.27}$$

and hence the integral is zero unless $m' = m \pm 1$. To solve the polar angle integral the following recursive relations are used (Gradshteyn and Ryzhik, 1980):

$$(\sin \theta) P_l^m = \frac{1}{2l+1}[(l-m+1)(l-m+2)P_{l+1}^{m-1} - (l+m-1)(l+m)P_{l-1}^{m-1}]$$

$$(\sin \theta) P_l^m = \frac{1}{2l+1}[P_{l-1}^{m+1} - P_{l+1}^{m+1}] \tag{5.28}$$

Combining these with Eq. (5.26) gives the condition $l' = l \pm 1$. A similar result is obtained for $\mathbf{q}$ parallel to $y = r_\alpha \sin \theta \sin \phi$. Furthermore, arbitrary directions of $\mathbf{q}$ can be resolved into components along x, y and z-axes. For a dipole transition the following selection rules therefore apply:

$$\boxed{\Delta l = \pm 1; \quad \Delta m = 0, \; \pm 1} \tag{5.29}$$

where Δl, Δm represent changes in the orbital and magnetic quantum numbers during the electronic transition. The physical meaning of Eq. (5.29) can be described as follows. Before excitation the electron is in the atomic eigenstate $\psi_{a1} = u_1(\mathbf{r}_\alpha) \exp(-iE_1 t/\hbar)$, where E_1 is the energy and t is time. The electron density, given by $|\psi_{a1}|^2$, is time independent. During excitation to a higher state, $\psi_{a2} = u_2(\mathbf{r}_\alpha) \exp(-iE_2 t/\hbar)$, the time-dependent atomic wavefunction (Ψ_a) is a coherent superposition of the initial and final states:

$$\Psi_a = c_1\psi_{a1} + c_2\psi_{a2} \qquad (5.30)$$

Here c_1 and c_2 are coefficients that vary with time, but at a much slower rate compared to the bound electron oscillation periods h/E_1 and h/E_2, so that at any given moment they are effectively constant. The electron density of this coherent state is then

$$|\Psi_a|^2 = |c_1 u_1|^2 + |c_2 u_2|^2 + (c_1 u_1)(c_2 u_2)^* e^{i\omega t} + (c_1 u_1)^*(c_2 u_2)e^{-i\omega t} \quad (5.31)$$

where $\omega = (E_2 - E_1)/\hbar$. The atomic electron density is now time dependent. Consider first the case of a 1s to 2p ($m = 0$) transition, corresponding to the $\Delta l = 1$, $\Delta m = 0$ selection rule. If, for simplicity, c_1 and c_2 are assumed to be unity, the time dependence of the $(1s + 2p_0)$ coherent state electron density is as illustrated in Figure 5.2a. The electron density oscillates about the origin along the z-axis of quantisation, and is equivalent to an oscillating electric dipole. The atom will then radiate in a manner

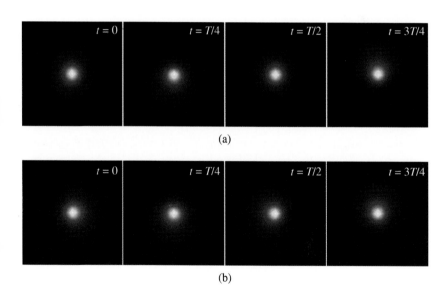

(a)

(b)

Figure 5.2   Two-dimensional electron density plots for the $(1s + 2p_0)$ and $(1s + 2p_1)$ coherent states at quarter period (T) time intervals are shown in (a) and (b) respectively. The axis of quantisation for the atom is parallel to the vertical axis in (a) and along the viewing direction in (b). The $(1s + 2p_0)$ state is an oscillating linear dipole, while $(1s + 2p_1)$ is a rotating dipole.

similar to a dipole antenna, with polarisation vector along the z-axis. On the other hand, consider the 1s to 2p ($m = \pm 1$) transition, corresponding to the $\Delta l = 1$, $\Delta m = \pm 1$ selection rule. The time-dependent electron density of the $(1s + 2p_1)$ coherent state is shown in Figure 5.2b and is now equivalent to a rotating electric dipole. The radiation emitted by the atom is more complicated, with photons emitted along the z-axis being circular polarised, while photons in the xy-plane are linearly polarised with polarisation vector along z (Fowles, 1989).

The transition matrix element $\varepsilon_{m0}(\mathbf{q})$ can be directly calculated for hydrogen (Inokuti, 1971), since the atomic wavefunctions have exact solutions, while approximate numerical methods are required for the other elements. In this section, the results for hydrogen are presented as an illustrative example. It is useful to calculate the so-called *generalised oscillator strength*, which is related to $\varepsilon_{m0}(\mathbf{q})$ by (Inokuti, 1971):

$$f_{m0}(\mathbf{q}) = \frac{2m(E_m - E_0)}{q^2 \hbar^2} |\varepsilon_{m0}(\mathbf{q})|^2 \tag{5.32}$$

At small values of q the generalised oscillator strength is equal to the absorption coefficient of photons of equivalent energy, so that optical absorption measurements can be directly compared with EELS. The similarity between electrons and photons is apparent from Fermi's golden rule, which predicts the photoexcitation cross-section to be proportional to $|\langle f|(\varepsilon \cdot \mathbf{r}_\alpha)|i\rangle|^2$, where ε is the electric field vector of the electromagnetic radiation (Rae, 2008). Comparing with Eq. (5.18) it follows that in the dipole limit the vector $\mathbf{q}$ for inelastic electron scattering is equivalent to ε.

Figure 5.3a shows the generalised oscillator strengths for promotion of a 1s electron in hydrogen to excited states with $n = 2$ principal quantum number (i.e. 2s, 2p sub-shells) plotted as a function of the momentum transfer. Also shown is the sum of the individual transitions as well as the value obtained from optical absorption; for small momentum transfers the total generalised oscillator strength is asymptotic to the optical absorption coefficient, as required. Furthermore, in this region the oscillator strength for the 1s to 2p transition is considerably greater than 1s to 2s, owing to the dipole selection rule (Eq. (5.29)). For larger momentum transfers, however, the 1s to 2p oscillator strength decreases rapidly, such that the dipole-forbidden 1s to 2s transition becomes comparatively significant. Non-dipole intensity is observable when there is a high density of unoccupied, dipole-forbidden states above the Fermi energy level, such as 3d to 4d transitions in second row transition metals ($M_{4,5}$ peaks) and 2p to higher energy p-states in $L_{2,3}$ edges (Auerhammer and Rez, 1989).

Note that the quadratic $(2\pi i q \cdot \mathbf{r}_\alpha)^2/2$ term in the Taylor expansion of $\exp(-2\pi i q \cdot \mathbf{r}_\alpha)$ can give rise to both monopole $(\Delta l = 0)$ and quadrupole $(\Delta l = \pm 2)$ transitions, although in most cases it is the former that is dominant (Auerhammer and Rez, 1989).

Figure 5.3b plots the oscillator strengths for ionisation of hydrogen, that is, the energy transfer is sufficiently large to remove the 1s electron from the atom. The energy transfer (E) is represented by the dimensionless parameter (E/R), where R is the ionisation energy. Furthermore, the oscillator strength is now expressed as df/dE, because of the fact that the 1s electron is excited into the continuum, rather than discrete, bound states (see Inokuti, 1971 for more details). As the (E/R) ratio increases the peak in the oscillator strength gradually shifts towards a

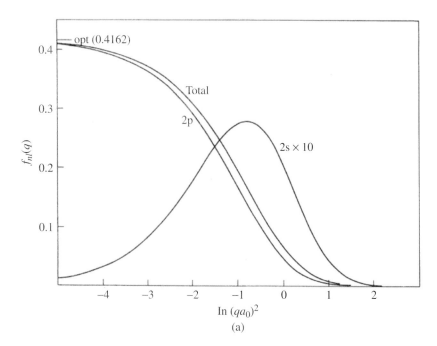

(a)

Figure 5.3 (a) Generalised oscillator strength (f_{nl}) for a hydrogen atom plotted as a function of momentum transfer for transitions to the second Bohr orbit. The momentum change is plotted as $\ln(qa_o)^2$, where q is the magnitude of the inelastic scattering vector and a_o is the Bohr radius. The total and individual final state oscillator strengths are indicated along with the 'optical' (i.e. $q \to 0$) value denoted by 'opt'. The oscillator strength df/dE for ionisation is shown in (b) for a range of energy losses E expressed in Rydberg energy 'R' units; (c) is the Bethe surface for hydrogen. From Inokuti (1971). Reproduced with permission; copyright American Physical Society.

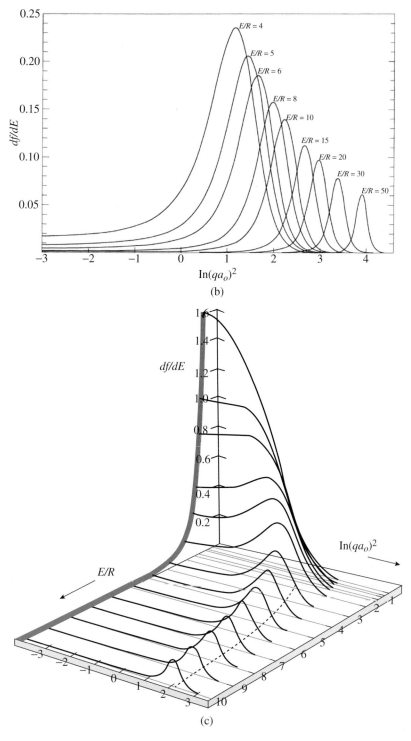

Figure 5.3 (*Continued*)

fixed momentum transfer while also becoming narrower. The physical origin of this behaviour is that for large energy transfers the atomic electron can effectively be treated as a free particle unbound to the nucleus, so that there is a unique correspondence between the energy and momentum ($\hbar q$) transfers, given by $E = (\hbar q)^2/(2m)$ (Inokuti, 1971). For small energy transfers, however, the atomic electron cannot be treated as a free particle, so that the momentum transfer $\hbar q$ is to the atom as a whole. The peak forms the so-called *Bethe ridge* of a three-dimensional Bethe surface, which plots the oscillator strength as a function of energy and momentum transfer (Figure 5.3a and b are two-dimensional slices through the Bethe surface at fixed energy transfer). The Bethe surface for hydrogen is shown in Figure 5.3c. Two distinct regions can be readily identified: (i) 'soft' collisions at small energy and momentum transfer that satisfy the dipole selection rule and have similar characteristics to optical absorption and (ii) 'hard' collisions that give rise to the Bethe ridge for large energy/momentum transfer. Soft collisions are widely used in EELS for chemical analysis, as well as ELNES and EXELFS. On the other hand, hard collisions have been used in Compton scattering experiments, where the momentum distribution of atomic electrons can be extracted by analysing the shape of the Bethe ridge (Williams *et al.*, 1984). There has been comparatively less work in this area, presumably due to complications from multiple scattering in the specimen, although the technique can potentially provide useful information on anisotropy of the electron wavefunction as well as charge transfer in atomic bonding (Schattschneider *et al.*, 1990).

Before concluding this section, anisotropy effects due to the transition matrix element will be discussed briefly. In graphite (an example of an anisotropic material) the C K-edge is due to a 1s electron being promoted to the unoccupied π or σ anti-bonding orbitals. The π-orbital is oriented perpendicular to the graphite basal planes (i.e. out of plane), while the σ-orbitals are in plane. At small scattering angles $\mathbf{q}$ is effectively anti-parallel to the incident electron wave vector (Figure 5.1), so that for graphite planes oriented perpendicular to the incident electron beam the σ-peak in the C K-edge is suppressed (i.e. the transition matrix element, Eq. (5.18), is approximately zero since $\mathbf{q}$ is largely perpendicular to the σ-orbital). Conversely, the intensity of the π-peak is maximised. Note that the anisotropy is due entirely to the final state, since the initial 1s state is isotropic. As shown in Figure 5.4 as the scattering angle is increased the relative intensity of the π-peak decreases while that of the σ-peak increases, owing to the larger transverse component of $\mathbf{q}$ (Leapman *et al.*, 1983).

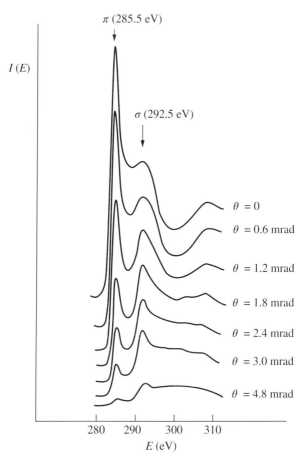

Figure 5.4 EELS spectra for graphite measured for different values of the scattering angle θ. The graphite basal planes are oriented perpendicular to the electron beam. From Leapman *et al.* (1983). Reproduced with permission; copyright American Physical Society.

5.1.2 Mixed Dynamic Form Factor

Thus far the discussion has focused on inelastic scattering of a single plane wave by an individual atom. However, if many coherent plane waves are involved their interference with one another is also important. Consider, for example, a focused STEM probe that consists of many partial plane waves originating from within the objective aperture. Each partial plane wave can give rise to inelastic scattering of the atom as described in the previous section. Intuitively, however, inelastic scattering should only take place when the atom is in the vicinity of the probe, that is, destructive interference of the partial plane waves outside the probe diameter must be taken into account. A similar scenario occurs for the electron wavefunction within a crystal, where a single incident plane wave can generate Bragg diffracted beams that interfere with one another as they propagate through the crystal. Dynamic diffraction in a crystal can be described using Bloch waves; for the straightforward case of a negative deviation parameter under two-beam conditions the Bloch state has intensity maxima along the atom columns, so that inelastic scattering is enhanced compared to positive deviation parameters, where the Bloch wave channels between the atom columns. Indeed, the ALCHEMI technique exploits this to determine site occupancy of atoms within a crystal (Chapter 4, Section 4.3.2; Spence and Taftø, 1983; Taftø and Krivanek, 1982).

To illustrate the effect of interference assume that a single atom is illuminated by not one, but two plane waves of the form $\exp(2\pi i k_1 \mathbf{n}_1 \cdot \mathbf{r})$ and $\exp[(2\pi i k_2 \mathbf{n}_2 \cdot \mathbf{r}) + i\phi]$, where $\mathbf{n}_1$, $\mathbf{n}_2$ are unit vectors along the direction of propagation, and the term $\exp(i\phi)$ represents the phase shift of the second plane wave with respect to the first. The two plane waves could represent (say) partial plane waves within a STEM probe or Bragg diffracted beams. Equation (5.9) then becomes

$$(\nabla_{\mathbf{r}}^2 + 4\pi^2 k_m^2)\psi_m(\mathbf{r})$$
$$= -\frac{2me}{\hbar^2} V_{m0}(\mathbf{r})[\exp(2\pi i k_1 \mathbf{n}_1 \cdot \mathbf{r}) + \exp(i\phi)\exp(2\pi i k_2 \mathbf{n}_2 \cdot \mathbf{r})]$$

$$(5.33)$$

Since the equation is linear with respect to ψ_m the solution follows directly from Eq. (5.10):

$$\psi_m(\mathbf{r}) = \frac{me}{2\pi\hbar^2} \int \frac{\exp(2\pi i k_m |\mathbf{r} - \mathbf{r}'|)}{|\mathbf{r} - \mathbf{r}'|}$$
$$\times V_{m0}(\mathbf{r}')[\exp(2\pi i k_1 \mathbf{n}_1 \cdot \mathbf{r}') + \exp(i\phi)\exp(2\pi i k_2 \mathbf{n}_2 \cdot \mathbf{r}')]d\mathbf{r}'$$

$$(5.34)$$

Proceeding along the lines of Eqs. (5.11) to (5.13), but with the modified solution for ψ_m (i.e. Eq. (5.34)), it follows that

$$d\sigma_m = \frac{m^2 e^2}{4\pi^2 \hbar^4} \left(\frac{k_m}{k_0}\right)$$
$$\times \left| \int [\exp(-2\pi i \mathbf{q}_1 \cdot \mathbf{r}') + \exp(i\phi)\exp(-2\pi i \mathbf{q}_2 \cdot \mathbf{r}')]V_{m0}(\mathbf{r}')d\mathbf{r}' \right|^2 d\Omega$$

$$(5.35)$$

The inelastic scattering vectors are $\mathbf{q}_1 = k_m \mathbf{n}_m - k_1 \mathbf{n}_1$ and $\mathbf{q}_2 = k_m \mathbf{n}_m - k_2 \mathbf{n}_2$. Applying Eq. (5.14) separately to the two integrands in Eq. (5.35) and expanding, we finally obtain

$$
\frac{\partial^2 \sigma_m}{\partial(\Delta E)\partial\Omega} = \left(\frac{1}{2\pi^2 a_o}\right)^2 \left(\frac{k_m}{k_0}\right)
$$

$$
\times \left[\frac{S(\mathbf{q}_1, \Delta E)}{q_1^4} + \frac{S(\mathbf{q}_2, \Delta E)}{q_2^4} + 2\frac{Re(e^{-i\phi}S(\mathbf{q}_1, \mathbf{q}_2, \Delta E))}{q_1^2 q_2^2} \right]
$$

(5.36)

$$
S(\mathbf{q}_1, \mathbf{q}_2, \Delta E) = \sum_{i,f,\alpha} p_i(\langle f | \exp(-2\pi i \mathbf{q}_1 \cdot \mathbf{r}_\alpha)|i\rangle)
$$

$$
\times (\langle i | \exp(2\pi i \mathbf{q}_2 \cdot \mathbf{r}_\alpha)|f\rangle)\delta(E_f - E_i - \Delta E)
$$

(5.37)

where $S(\mathbf{q}_1, \mathbf{q}_2, \Delta E)$ is known as the mixed dynamic form factor (Kohl and Rose, 1985) and the relationship $S(\mathbf{q}_1, \mathbf{q}_2, \Delta E) = S(\mathbf{q}_2, \mathbf{q}_1, \Delta E)^*$ has been used to simplify Eq. (5.36). $S(\mathbf{q}_1, \mathbf{q}_1, \Delta E)$ is equal to $S(\mathbf{q}_1, \Delta E)$, the dynamic form factor (Eq. (5.18)), and similarly for $S(\mathbf{q}_2, \Delta E)$. Comparing with Eqs. (5.17) and (5.18) it is clear that the first two terms in Eq. (5.36) represent direct inelastic scattering by the two plane waves, while the third term represents their interference. This is illustrated schematically in Figure 5.5a, where the degree of inelastic scattering is determined by the position of the target atom with respect to the interference pattern created by the two incoming plane waves. The inelastic scattered intensity consists of three contributions: two Lorentzians centred about the incident wave vectors representing direct scattering (i.e. the dynamic form factor terms in Eq. (5.36)) and an interference contribution due to the mixed dynamic form factor. Importantly, the relative contribution of each of the three terms depends on the position of the EELS spectrometer detector. For example, if the two plane waves are due to elastic Bragg scattering then interference effects can be expected to be important between the diffracted beams, since the Bragg angle is typically much larger than the characteristic scattering angle (θ_E), which determines the width of the Lorentzian.

In general, the mixed dynamic form factor $(\mathbf{q}_1 \neq \mathbf{q}_2)$ for a single atom is a complex quantity (compare with the dynamic form factor, which is always real). The imaginary part of $S(\mathbf{q}_1, \mathbf{q}_2, \Delta E)$ is dependent on

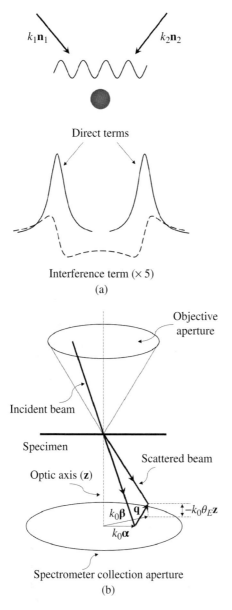

Figure 5.5 (a) Inelastic scattering from an atom (grey circle) due to two incident plane waves with wave vectors $k_1\mathbf{n}_1$ and $k_2\mathbf{n}_2$. The two waves give rise to an interference intensity pattern at the atom, so that the inelastic scattering consists of two direct Lorentzian contributions and an interference contribution. (b) Inelastic scattering in a STEM geometry; see text for details.

the magnetic properties of the atom, so that for non-magnetic atoms the mixed dynamic form factor is a real quantity (Schattschneider *et al.*, 2008; Hébert *et al.*, 2008). A further interesting property emerges for inelastic scattering from a crystal, where the form factors in Eq. (5.36) must be summed over all atoms. Considering the general case of the mixed dynamic form factor (Eq. (5.37))

$$\sum_{atoms} S(\mathbf{q}_1, \mathbf{q}_2, \Delta E)$$

$$= \sum_{n,i,f,j,\alpha} p_i(\langle f| \exp[-2\pi i\mathbf{q}_1 \cdot (\mathbf{r}_\alpha + n\mathbf{a}_j)]|i\rangle)$$

$$\times (\langle i| \exp[2\pi i\mathbf{q}_2 \cdot (\mathbf{r}_\alpha + n\mathbf{a}_j)]|f\rangle)\delta(E_f - E_i - \Delta E)$$

$$= \left\{ \sum_{n,j} \exp[2\pi i(k_1\mathbf{n}_1 - k_2\mathbf{n}_2) \cdot n\mathbf{a}_j] \right\} S(\mathbf{q}_1, \mathbf{q}_2, \Delta E) \qquad (5.38)$$

where n is an integer and $\mathbf{a}_j$ ($j = 1, 2, 3$) are the basis vectors of the crystal unit cell. The $S(\mathbf{q}_1, \mathbf{q}_2, \Delta E)$ term appearing in the right-hand side of the above equation is the mixed dynamic form factor for a single atom. For an infinite crystal (i.e. n unbounded) the term within the curly brackets reduces to a Dirac comb,[3] with non-zero values for $(k_1\mathbf{n}_1 - k_2\mathbf{n}_2)$, or equivalently $\mathbf{q}_2 - \mathbf{q}_1$, equal to a reciprocal lattice vector $\mathbf{g}$ (Schattschneider *et al.*, 2000). The physical significance of this result is as follows. The two incident plane waves, $\exp(2\pi ik_1\mathbf{n}_1 \cdot \mathbf{r})$ and $\exp[(2\pi ik_2\mathbf{n}_2 \cdot \mathbf{r}) + i\phi]$, give rise to an intensity $|\exp(2\pi ik_1\mathbf{n}_1 \cdot \mathbf{r}) + \exp[(2\pi ik_2\mathbf{n}_2 \cdot \mathbf{r}) + i\phi]|^2$ or $2\{1 + \cos[2\pi(k_1\mathbf{n}_1 - k_2\mathbf{n}_2) \cdot \mathbf{r} - \phi]\}$ at the object plane. If $(k_1\mathbf{n}_1 - k_2\mathbf{n}_2)$ is a reciprocal lattice vector, then the intensity distribution is identical for all unit cells (e.g. the intensity peaks at the atom columns or between them, depending on the value of ϕ). However, if this condition is not satisfied then inelastic scattering will be different for each unit cell, such that the individual contributions destructively interfere at the detector position (Figure 5.5a), resulting in a zero mixed dynamic form factor (Kohl and Rose, 1985). An alternative but useful way of stating this is that in inelastic scattering spatial information about the object is contained in the mixed dynamic form factor term, rather than the dynamic form factor (Eq. (5.36)).

---

[3]A Dirac comb is a series of periodically repeating delta functions, and is defined by $(1/T) \sum_{n=-\infty}^{\infty} \exp\left(\frac{2\pi inx}{T}\right)$, where T is the period along the x-axis. For Eq. (5.38) T can be taken as unity, so that the Dirac 'spikes' occur for all integer values of $(k_1\mathbf{n}_1 - k_2\mathbf{n}_2) \cdot \mathbf{a}_j$.

The mixed dynamic form factor will now be generalised to the more relevant case of inelastic scattering by a focused STEM probe. As shown in Chapter 3 (Section 3.2.2) the STEM probe wavefunction (ψ_0) is a coherent superposition of all partial plane waves within the objective aperture, such that at the specimen entrance surface $z = 0$:

$$\psi_0(\mathbf{R}, \mathbf{b}) = \gamma \int A(\boldsymbol{\alpha}) \exp[-i\chi(\boldsymbol{\alpha})] \exp[2\pi i k_0 \boldsymbol{\alpha} \cdot (\mathbf{R} - \mathbf{b})] d^2\boldsymbol{\alpha} \quad (5.39)$$

In order to aid the discussion the above equation is expressed using notation slightly different to Eq. (3.44). In particular, the target atom is placed at the origin, so that the probe incident position (i.e. $\mathbf{R}_p$ in Eq. (3.44)) is now equal to the impact parameter $\mathbf{b}$. Furthermore, the transverse wave vector component ($\mathbf{k}_t$ in Eq. (3.44)) is written as $k_0\boldsymbol{\alpha}$, where k_0 is the wave number and $\boldsymbol{\alpha}$ is a vector that depends on the incident angle of the partial plane wave (Figure 5.5b; here the small angle approximation is assumed). $A(\boldsymbol{\alpha})$, $\chi(\boldsymbol{\alpha})$ are aperture and aberration functions respectively, while $\mathbf{R}$ is a two-dimensional position vector in the plane perpendicular to the optic axis. γ is a normalisation constant for the probe intensity.

The energy loss spectrum is given by

$$\frac{dI}{d(\Delta E)} = \int_{\text{spectrometer}} \left(\frac{\partial^2 \sigma}{\partial(\Delta E)\partial\Omega} \right) d\Omega \quad (5.40)$$

where I is the intensity for energy losses $\leq \Delta E$ and the integration is carried out over the spectrometer entrance aperture. From the expression for the double differential cross-section (Eq. (5.36)) it follows that (Kohl and Rose, 1985)

$$\frac{dI}{d(\Delta E)} \propto \int A(\boldsymbol{\alpha}) A(\boldsymbol{\alpha}') D(\boldsymbol{\Theta}) \exp[i(\chi(\boldsymbol{\alpha}') - \chi(\boldsymbol{\alpha}))] \exp[2\pi i k_0 \mathbf{b} \cdot (\boldsymbol{\alpha}' - \boldsymbol{\alpha})]$$

$$\times \frac{S(\mathbf{q}, \mathbf{q}', \Delta E)}{q^2 q'^2} d^2\boldsymbol{\alpha} \, d^2\boldsymbol{\alpha}' \, d^2\boldsymbol{\Theta} \quad (5.41)$$

where $D(\boldsymbol{\Theta})$ is an aperture function for the spectrometer, such that $D(\boldsymbol{\Theta}) = 1$ for inelastic scattering angles within the collection aperture and $D(\boldsymbol{\Theta}) = 0$ otherwise (Figure 5.5b). The vector $\mathbf{q}$ (and similarly $\mathbf{q}'$) can be defined with the aid of Figure 5.5b in the small angle, small energy loss limit as

$$\mathbf{q} = -k_0[\theta_E \mathbf{z} + (\boldsymbol{\alpha} - \boldsymbol{\beta})] \quad (5.42)$$

where $\mathbf{z}$ is a unit vector parallel to the optic axis ($z > 0$ is into the sample) and $\boldsymbol{\beta}$ is the vector for the scattered wave, defined in a similar manner to $\boldsymbol{\alpha}$. In the dipole limit it can be shown that $S(\mathbf{q}_1, \mathbf{q}_2, \Delta E)$ for a non-magnetic atom is proportional to $\mathbf{q}_1 \cdot \mathbf{q}_2$ (Kohl and Rose, 1985). Assuming further a small spectrometer entrance aperture (i.e. negligible $\boldsymbol{\beta}$) from Eq. (5.42), it follows that

$$S(\mathbf{q}_1, \mathbf{q}_2, \Delta E) \propto [\theta_E^2 + \boldsymbol{\alpha} \cdot \boldsymbol{\alpha}'] \qquad (5.43)$$

Under these assumptions Eq. (5.41) can be simplified to (Kohl and Rose, 1985; Muller and Silcox, 1995)

$$\frac{dI}{d(\Delta E)} \propto |A_\parallel(b)|^2 + |A_\perp(b)|^2$$

$$A_\parallel(b) = \theta_E \int_0^{\alpha_0} \frac{J_0(2\pi k_0 b\alpha)}{\theta_E^2 + \alpha^2} \exp[-i\chi(\alpha)]\alpha \; d\alpha$$

$$A_\perp(b) = \int_0^{\alpha_0} \frac{J_1(2\pi k_0 b\alpha)}{\theta_E^2 + \alpha^2} \exp[-i\chi(\alpha)]\alpha^2 \; d\alpha \qquad (5.44)$$

where α_0 is the STEM probe semi-convergence angle and J_0, J_1 are Bessel functions of the first kind. $A_\parallel(b)$ and $A_\perp(b)$ are due to the first and second terms in Eq. (5.43), and represent dipole excitations parallel and perpendicular to the optic axis ($\mathbf{z}$) respectively. Figure 5.6a and b shows the intensity profiles for a single silicon atom illuminated by a 20 mrad, 200 kV STEM probe with zero electron optic aberrations. The θ_E values are 0.2 and 4.6 mrad respectively, and correspond to the energy loss of Si $L_{2,3}$ (99 eV) and K (1839 eV) edges. The intensity has a 'volcano' profile at the atom position, which is more pronounced for the smaller energy loss. The decrease in intensity at the atom is due to the $A_\perp(b)$ term in Eq. (5.44), since the $J_1(2\pi k_0 b\theta)$ Bessel term is zero at the origin. Large θ_E values at higher energy loss suppress the $A_\perp$ term more than $A_\parallel$ (Eq. (5.44)) and as a result the volcano effect is less evident (Figure 5.6b).

The underlying physical mechanism for the volcano profile can be qualitatively understood by considering the inelastic intensity distribution in the far field due to a single incident plane wave, which from Section 5.1.1 is given by a Lorentzian of the form q^{-2} or $1/(\theta_E^2 + \theta^2)$. The plane wave could represent one partial plane wave within the objective aperture of the STEM probe. Within the classical picture of the incident electron being treated as a particle, the intensity at large scattering angle θ is due to electrons with small impact parameter. Large angle deflected

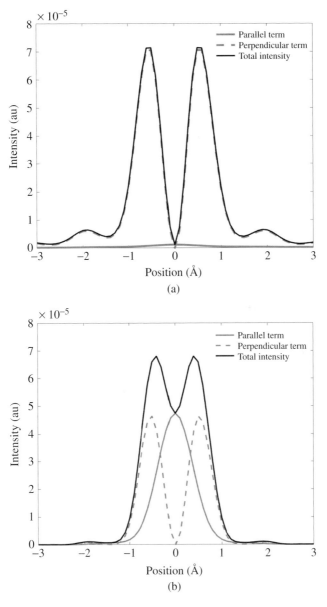

Figure 5.6 200 kV, aberration-free STEM EELS intensity profiles calculated for an individual Si atom. (a) and (b) correspond to the Si $L_{2,3}$ and K-edges for a 20 mrad STEM probe semi-convergence angle and point detector. The Si atom is at the origin and the parallel and perpendicular contributions to the total intensity (i.e. $A_\parallel$ and $A_\perp$ terms in Eq. (5.44)) are indicated. In (a) the curve for the total intensity largely overlaps with the perpendicular contribution. (c) shows the intensity profile for the Si $L_{2,3}$-edge with the detector collection semi-angle increased to 40 mrad, but keeping all other parameters unchanged. (d) is the intensity profile for the Si K-edge with 10 mrad probe convergence semi-angle and point detector.

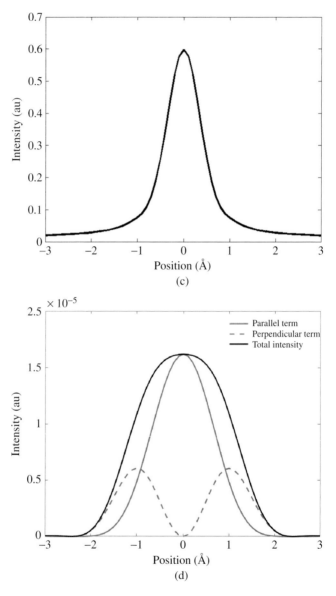

Figure 5.6 (*Continued*)

electrons are outside the (small) spectrometer entrance aperture and give rise to a decrease in intensity close to the atom, that is, the volcano profile. Figure 5.6c shows the low energy loss (i.e. $\theta_E = 0.2$ mrad) intensity profile for the same 20 mrad STEM probe, but now with a larger

40 mrad semi-collection angle, so that even the high angle scattered electrons may be detected by the spectrometer. Equation (5.41) was used for the calculation, since the assumption of a small collection angle no longer applies. As expected, the atom is now more localised with maximum intensity at the atom position. Conversely, rather than increase the size of the spectrometer aperture, the probe size can be decreased with respect to a fixed collection angle, resulting in a similar effect. This is illustrated in Figure 5.6d, which shows the high energy loss (i.e. $\theta_E = 4.6$ mrad) intensity profile for a 10 mrad, 200 kV aberration-free STEM probe with point detector, calculated using Eq. (5.44). The volcano structure is absent, although the atom is broader than Figure 5.6b owing to the larger width of the 10 mrad probe. These results are based on the dipole approximation for a single atom, but more detailed simulations by D'Alfonso *et al.* (2008) have highlighted the importance of non-dipole transitions, as well as the role of elastic scattering of the incident probe within a crystal.

The volcano structure is undesirable for chemical analysis of crystalline materials at the single atom level, since the delocalised intensity can overlap with neighbouring atom columns, thereby degrading the spatial resolution. Although the overall trends in Figure 5.6 are also valid for a thin crystal, they do not provide a quantitative measure of the delocalisation since dynamic elastic scattering of the STEM probe has not been taken into account. This is important since for atomic-scale chemical analysis the crystal must be tilted to a zone-axis orientation where diffraction is strong. Inelastic scattering in a crystalline environment is therefore the topic of the next section.

5.1.3 Yoshioka Equations and Inelastic Scattering within a Crystal

Many simulation methods for dynamic inelastic scattering within a crystal are derived from Yoshioka's theory (1957). In this section, implementation of the theory within the framework of multislice and Bloch wave methods will be discussed, and in Section 5.1.4 some general trends in atomic resolution chemical analysis are presented. Yoshioka's theory follows a similar approach to that outlined for a single atom (Section 5.1.1) so that Eq. (5.6), repeated below for convenience, is valid for a crystal as well:

$$\left(\nabla_r^2 + 4\pi^2 k_m^2 + \frac{2me}{\hbar^2} V_{mm}(\mathbf{r}) \right) \psi_m(\mathbf{r}) = -\frac{2me}{\hbar^2} \sum_{n \neq m} V_{mn}(\mathbf{r}) \psi_n(\mathbf{r}) \quad (5.45)$$

The potential $V_{mn}(\mathbf{r})$ is now an integration over all N atomic electrons within the solid and not just a single atom. It is defined in a similar manner to Eq. (5.7):

$$V_{mn}(\mathbf{r}) = \int a_m^*(\mathbf{r}_1, \ldots, \mathbf{r}_N) \left(\sum_i \frac{Z_i e}{4\pi\varepsilon_0 |\mathbf{r} - \mathbf{r}_i|} - \sum_\alpha \frac{e}{4\pi\varepsilon_0 |\mathbf{r} - \mathbf{r}_\alpha|} \right)$$
$$a_n(\mathbf{r}_1, \ldots, \mathbf{r}_N) \, d\mathbf{r}_1 \ldots d\mathbf{r}_N$$

$$= \left(\sum_i \frac{Z_i e}{4\pi\varepsilon_0 |\mathbf{r} - \mathbf{r}_i|} \right) \delta_{mn} - \int a_m^*(\mathbf{r}_1, \ldots, \mathbf{r}_N) \left(\sum_\alpha \frac{e}{4\pi\varepsilon_0 |\mathbf{r} - \mathbf{r}_\alpha|} \right)$$
$$a_n(\mathbf{r}_1, \ldots, \mathbf{r}_N) \, d\mathbf{r}_1 \ldots d\mathbf{r}_N$$
$$(5.46)$$

where Z_i is the atomic number of the ith nucleus located at position vector $\mathbf{r}_i$. $a_n(\mathbf{r}_1, \ldots, \mathbf{r}_N)$ is the electron wavefunction for the crystal. From Bloch's theorem for any given translation vector $\mathbf{R}_t$,

$$a_n(\mathbf{r}_1 + \mathbf{R}_t, \ldots, \mathbf{r}_N + \mathbf{R}_t) = \exp(2\pi i \mathbf{Q}_n \cdot \mathbf{R}_t) a_n(\mathbf{r}_1, \ldots, \mathbf{r}_N) \qquad (5.47)$$

where $\mathbf{Q}_n$ is the wave vector of the crystal in the nth state. $V_{mn}(\mathbf{r})$ for a perfect crystal can therefore be written in terms of its Fourier components (Howie, 1963):

$$V_{mn}(\mathbf{r}) = \exp(-2\pi i \mathbf{Q}_{mn} \cdot \mathbf{r}) \sum_\mathbf{g} V_\mathbf{g}^{mn} \exp(2\pi i \mathbf{g} \cdot \mathbf{r}) \qquad (5.48)$$

where $\mathbf{Q}_{mn} = \mathbf{Q}_m - \mathbf{Q}_n$. A more convenient method for evaluating the potential is to treat inelastic scattering from each atom in the crystal as being incoherent (Saldin and Rez, 1987), so that $V_{mn}(\mathbf{r})$ is the sum of all single atom transition potentials given by Eq. (5.7). The justification for this is that the initial core electron in the excited atom does not overlap significantly with neighbouring atoms, while the final level electron, although in the conduction band, typically interacts with only a few nearest neighbours, especially if the energy loss is just above the threshold.

Coene and Van Dyck (1990) have proposed an inelastic multislice simulation method based on Eq. (5.45). It is reasonably assumed that for a thin foil only the elastic wave ψ_0 has significant intensity, so that for inelastic scattering Eq. (5.45) simplifies to

$$\left(\nabla_\mathbf{r}^2 + 4\pi^2 k_m^2 + \frac{2me}{\hbar^2} V_{mm}(\mathbf{r}) \right) \psi_m(\mathbf{r}) = -\frac{2me}{\hbar^2} V_{m0}(\mathbf{r}) \psi_0(\mathbf{r}) \qquad (5.49)$$

The incident wave is expressed as $\psi_0(\mathbf{r}) = \Phi_0(\mathbf{r}) \exp(2\pi i \mathbf{k}_0 \cdot \mathbf{r})$ while the inelastic wave has the form $\psi_m(\mathbf{r}) = \Phi_m(\mathbf{r}) \exp(2\pi i \mathbf{k}_m \cdot \mathbf{r})$. For convenience it is assumed that $\mathbf{k}_m$ is parallel to $\mathbf{k}_0$ (i.e. small angle inelastic scattering, Figure 5.1), which is directed along the z-optic axis. The x, y-axes are parallel to the plane of the specimen. Substituting the above expressions for $\psi_0(\mathbf{r})$ and $\psi_m(\mathbf{r})$ in Eq. (5.49) and neglecting the $\partial^2 \Phi_m / \partial z^2$ term (since backscattering is weak for high energy electrons) results in

$$\frac{\partial \Phi_m}{\partial z} = D_m \Phi_m + i\sigma_m V_{m0} \Phi_0 \exp[-2\pi i(\mathbf{k}_m - \mathbf{k}_0) \cdot \mathbf{r}] \qquad (5.50)$$

where σ_m is the interaction constant for inelastic scattering and is equal to $(2\pi me/h^2 k_m)$. Furthermore,

$$D_m = \frac{i}{4\pi k_m} \nabla_{xy}^2 + i\sigma_m V_{mm}; \quad \nabla_{xy}^2 = \frac{\partial^2}{\partial x^2} + \frac{\partial^2}{\partial y^2} \qquad (5.51)$$

$V_{mm}(\mathbf{r})$ is the potential for elastic scattering of the inelastic wave while the crystal is in the mth excited state and, assuming the perturbation is small, is approximately equal to $V_{00}(\mathbf{r})$ (Howie, 1963). Comparing with the multislice equations for elastic scattering (i.e. Eq. (3.49)) it is clear that D_m represents propagation of the inelastic wave through the crystal. The second term in Eq. (5.50) represents inelastic scattering of the elastic wave. The solution for Φ_m is readily obtained as

$$\Phi_m(\mathbf{R}, z) = \int_0^z \exp\left[\int_{z'}^z D_m(\mathbf{R}, z'')dz''\right]$$
$$\times \{i\sigma_m V_{m0}(\mathbf{R}, z')\Phi_0(\mathbf{R}, z') \exp[-2\pi i(\mathbf{k}_m - \mathbf{k}_0)z']\}dz' \quad (5.52)$$

where $\mathbf{R}$ is a two-dimensional position vector in the xy-plane. The $\Phi_m(\mathbf{R}, 0) = 0$ boundary condition (i.e. zero inelastic wave amplitude at the specimen entrance surface) has been used to derive the above equation. The term within the curly brackets represents generation of the inelastic wave at depth z', while the exponential term represents free space propagation and elastic scattering through the remaining $(z - z')$ distance within the crystal (Coene and Van Dyck, 1990).

The multislice simulation method for inelastic scattering can be described as follows. The incident electron wavefunction is elastically propagated onto the slice at depth z_o, which contains the atom to be excited, using the standard multislice procedure (Chapter 3). Frozen phonons may be included for a more accurate estimate of the 'elastic' wavefunction (Dwyer, 2005). The inelastic wavefunction generated from within the slice of interest is from Eq. (5.50), given by

$$\Phi_m(\mathbf{R}, z_o) = i\sigma_m V_{m0}^{\mathrm{P}}(\mathbf{R})\Phi_0(\mathbf{R}, z_o)$$

$$V_{m0}^{p}(\mathbf{R}) = \int_{-\infty}^{\infty} V_{m0}(\mathbf{R}, z)\exp[-2\pi i(k_m - k_0)z]dz \qquad (5.53)$$

where $V_{m0}^{\mathrm{P}}(\mathbf{R})$ is the 'projected' potential (Allen *et al.*, 2015). The projection can effectively be carried out over the full range of z, since it is assumed that the slice thickness is sufficiently large to include a significant fraction of $V_{m0}(\mathbf{r})$. Furthermore, Φ_0 varies slowly over the slice thickness, so that it is treated as a constant. Following generation, the inelastic wave is propagated to the specimen exit surface using standard multislice methods, but with the modified D_m operator valid for the inelastic wave (Eq. (5.51)). Fourier transforming the exit wavefunction and integrating the square modulus over the detector collection angle gives the EELS signal. The EFTEM image intensity, on the other hand, is the square modulus of the exit wavefunction convolved with the point spread function of the imaging lens. The convolution simplifies to a multiplication in Fourier space as described in Section 3.1.3; note, however, that the lens aberration function (Eq. (3.28)) now contains an additional contribution to the defocus Δf term due to chromatic aberration. This is equal to $C_c(\Delta E/E_o)$, where C_c is the lens chromatic aberration coefficient and ΔE is the energy loss of the incident electrons of primary energy E_o (Lugg *et al.*, 2010).

In Section 5.1.1, the implications of an unobserved final state of the specimen on an inelastic scattering measurement was briefly discussed. The result is that transitions to all relevant final states must be incoherently summed. This includes degenerate energy levels of appropriate energy for all relevant atoms in the solid. The inelastic wave for each final state must be propagated to the specimen exit surface and the intensity in either real or reciprocal space incoherently summed to give the final image or diffraction pattern (Lugg *et al.*, 2010), which can result in long computation times. For EELS a significant reduction in computation time is achieved if the detector collection angle is large (Dwyer, 2005). In this case, elastic scattering following excitation can be ignored, so that the measured intensity is simply the square modulus of the inelastic wave (Eq. (5.53)), integrated over the position vector $\mathbf{R}$.

For completeness, it will also be shown how Eq. (5.49) can be solved using Bloch wave methods. The inelastic wavefunction ψ_m can be written as a sum of Bloch waves:

$$\psi_m(\mathbf{R}, z) = \sum_j \varepsilon_m^j(z) b_m^j(\mathbf{R}, z)$$

$$b_m^j(\mathbf{R}, z) = \sum_g C_g^j(\mathbf{k}_m) \exp(2\pi i \gamma_m^j z) \exp[2\pi i(\mathbf{k}_m + \mathbf{g}) \cdot \mathbf{r}] \qquad (5.54)$$

where $\varepsilon_m^j(z)$ represents the Bloch wave excitation at a given depth. The Schrödinger equation for $b_m^j(\mathbf{R}, z)$ is given by

$$\left(\nabla_r^2 + 4\pi^2 k_m^2 + \frac{2me}{\hbar^2} V_{mm}(\mathbf{r}) \right) b_m^j(\mathbf{R}, z) = 0 \qquad (5.55)$$

The procedure for calculating C_g^j and γ_m^j is similar to that described in Chapter 4 for elastic Bloch waves. Substituting Eq. (5.54) in (5.49) and simplifying with the aid of Eq. (5.55) gives

$$\sum_j 4\pi i[(\mathbf{k}_m)_z + \gamma_m^j] b_m^j \frac{\partial \varepsilon_m^j}{\partial z} = -\frac{2me}{\hbar^2} V_{m0} \psi_0 \qquad (5.56)$$

where $(\mathbf{k}_m)_z$ is the z-component of the wave vector $\mathbf{k}_m$. It has been assumed that the important reciprocal lattice vectors $\mathbf{g}$ in Eq. (5.54) have no z-component (i.e. projection approximation; Section 4.1.1) and that the $\partial^2 \varepsilon_m^j / \partial z^2$ term is negligible. Multiplying by b_m^{j*}, integrating over the xy-plane and using the fact that the Bloch waves are orthonormal results in

$$\frac{\partial \varepsilon_m^j}{\partial z} = \frac{ime}{2\pi \hbar^2 [(\mathbf{k}_m)_z + \gamma_m^j]} \int b_m^{j*} V_{m0} \psi_0 d\mathbf{R} \qquad (5.57)$$

The elastic wavefunction can also be written using Bloch waves:

$$\psi_0(\mathbf{R}, z) = \sum_i C_0^{i*}(\mathbf{k}_0) \left\{ \sum_h C_h^i(\mathbf{k}_0) \exp(2\pi i \gamma_0^i z) \exp[2\pi i(\mathbf{k}_0 + \mathbf{h}) \cdot \mathbf{r}] \right\} \qquad (5.58)$$

where C_0^{i*} represents excitation of the ith Bloch wave assuming the crystal is non-absorbing (Section 4.1.1). Substituting in Eq. (5.57):

$$\frac{\partial \varepsilon_m^j}{\partial z} = \frac{ime}{2\pi \hbar^2 [(\mathbf{k}_m)_z + \gamma_m^j]} \sum_i C_0^{i*}(\mathbf{k}_0) \exp(2\pi i \Delta^{ij} z) S^{ij}$$

$$S^{ij} = \sum_{g,h} C_g^{j\,*}(\mathbf{k}_m) C_h^i(\mathbf{k}_0) V_{g-h}^{m0}$$

$$\Delta^{ij} = \{ [(\mathbf{k}_0)_z + \gamma_0^i] - [(\mathbf{k}_m)_z + \gamma_m^j] \} - (\mathbf{Q}_{m0})_z \qquad (5.59)$$

where Eq. (5.48) has been used to expand V_{m0} and the general relationship

$$\int \exp(2\pi i\mathbf{u} \cdot \mathbf{R})d\mathbf{R} = \delta(\mathbf{u}) \qquad (5.60)$$

is used to simplify Eq. (5.59), where $\delta(\mathbf{u})$ is the Dirac delta function and $\mathbf{u}$ is a vector in reciprocal space. ε_m^j varies with depth z in the crystal due to intraband ($i = j$) and interband ($i \neq j$) scattering between the elastic and inelastic Bloch waves. If the potential is long range (e.g. plasmon excitation), such that V_{m0} is constant, then intraband inelastic scattering is dominant[4] (Howie, 1963), whereas for the more localised single electron inelastic scattering both mechanisms are allowed (Ishida, 1970). The Δ^{ij} term in Eq. (5.59) represents the fact that the wave vector may not be conserved in the z-direction. This is due to the finite foil thickness (t), so that from the Born–von Karman boundary conditions (Ashcroft and Mermin, 1976), the crystal wave vector along z (i.e. $(\mathbf{Q}_{m0})_z$ in Eq. (5.59)) has only discrete values corresponding to integer multiples of $(1/t)$. On the other hand, the crystal is effectively infinite in the x, y directions, meaning that $(\mathbf{Q}_{m0})_x$ and $(\mathbf{Q}_{m0})_y$ are continuous and can therefore be equal to $(\mathbf{k}_{0m})_x$ and $(\mathbf{k}_{0m})_y$ respectively, where $\mathbf{k}_{0m} = \mathbf{k}_0 - \mathbf{k}_m$, that is, the wave vector is conserved parallel to the surface foil during inelastic scattering.

For a perfect crystal S^{ij} is independent of depth, so that $\varepsilon_m^j(t)$ at the specimen exit surface can be found by integrating Eq. (5.59) with respect to z (Maslen and Rossouw, 1984);

$$\varepsilon_m^j(t) = \frac{me}{2\pi\hbar^2[(\mathbf{k}_m)_z + \gamma_m^j]} \sum_i C_0^{i\,*}(\mathbf{k}_0) S^{ij} \left[\frac{\exp(2\pi i\Delta^{ij}t) - 1}{2\pi\Delta^{ij}} \right] \qquad (5.61)$$

The inelastic exit wavefunction $\psi_m(\mathbf{R}, t)$, and hence EELS and EFTEM signals, can be calculated from Eqs. (5.54) and (5.61). In the multislice method it was necessary to incoherently sum over all final states of the excited atoms, because the specimen quantum state is not observed in

---

[4]For a constant potential V_{g-h}^{m0} is zero unless $\mathbf{g} = \mathbf{h}$. Therefore, S^{ij} reduces to $\sum C_g^{j\,*}(\mathbf{k}_m) C_g^i(\mathbf{k}_0)$. From the property of Bloch wave coefficients S^{ij} is therefore zero unless $i = j$ (Chapter 4), assuming $\mathbf{k}_m \approx \mathbf{k}_0$ (i.e. small energy and momentum change).

practice. Similarly, in the Bloch wave method the summation is carried out with respect to Δ^{ij} to account for all $(Q_{m0})_z$ values for the crystal. This has the effect of damping any 'resonance' effects that may occur for a particular value of Δ^{ij} (Young and Rez, 1975).

5.1.4 Coherence in Inelastic Scattering

Coherence of the inelastic signal in both EFTEM and STEM EELS will now be discussed. The intensity $I(\mathbf{R})$ of an EFTEM image is given by

$$I(\mathbf{R}) = |\psi_m(\mathbf{R}) \otimes H(\mathbf{R})|^2$$

$$= \iint \psi_m(\mathbf{R} - \mathbf{R}_1)\psi_m^*(\mathbf{R} - \mathbf{R}_2)H(\mathbf{R}_1)H^*(\mathbf{R}_2)d\mathbf{R}_1 \ d\mathbf{R}_2 \quad (5.62)$$

where $H(\mathbf{R})$ is the point spread function of the objective lens for the energy loss of interest and $\otimes$ denotes convolution. The intensity is governed by interference of the wavefunction from different specimen points so that the final image is (partially) coherent. This means that EFTEM is in principle similar to HREM lattice images formed from elastically scattered electrons and is prone to contrast reversals due to changes in defocus, specimen thickness, etc. However, ψ_m is proportional to the projected potential (Eq. (5.53)), which becomes more localised as the energy loss increases. The spatial coherence length therefore decreases with increasing energy loss; indeed, the coherence length has been shown to decrease from a few nanometres for low loss features (e.g. plasmons) to a few Angstroms for core loss edges above 100 eV (Kimoto and Matsui, 2003). EFTEM maps of core loss edges therefore show very little phase contrast.

On the other hand, the intensity (I) of the STEM EELS signal for a given probe position is

$$I = \int |\tilde{\psi}_m(\mathbf{K})|^2 \tilde{D}(\mathbf{K})d\mathbf{K} \quad (5.63)$$

where $\tilde{\psi}_m(\mathbf{K})$ is the Fourier transform of the inelastic exit wavefunction ψ_m and $\tilde{D}(\mathbf{K})$ is the detector function, which has a value of unity for all reciprocal vectors $\mathbf{K}$ within the collection aperture and zero outside it. It is instructive to express the above equation in real space. To do so note that the integrand can be multiplied by $\tilde{D}(\mathbf{K})^*$ without affecting its value, so that by Parseval's theorem (Lupini and Pennycook, 2003)

$$I = \int |\tilde{\psi}_m(\mathbf{K})\tilde{D}(\mathbf{K})|^2 d\mathbf{K} = \int |\psi_m(\mathbf{R}) \otimes D(\mathbf{R})|^2 d\mathbf{R} \quad (5.64)$$

where $D(\mathbf{R})$ is the inverse Fourier transform of $\tilde{D}(\mathbf{K})$. Consider first the case of a point detector located at the reciprocal point $\mathbf{K}_o$ (i.e. $\tilde{D}(\mathbf{K}) = \delta(\mathbf{K} - \mathbf{K}_o)$, where δ is a Dirac delta function), so that from Eq. (5.63) the measured intensity is

$$I = |\tilde{\psi}_m(\mathbf{K}_o)|^2 = \left| \int \psi_m(\mathbf{R}) \exp(-2\pi i \mathbf{K}_o \cdot \mathbf{R}) d\mathbf{R} \right|^2 \qquad (5.65)$$

It is clear that the intensity of a point detector depends on interference of the wavefunction from different specimen points $\mathbf{R}$ and is therefore coherent. Conversely, for a large detector where $\tilde{D}(\mathbf{K})$ is equal to unity everywhere in reciprocal space, or equivalently a delta function in real space, Eq. (5.64) gives

$$I = \int |\psi_m(\mathbf{R})|^2 d\mathbf{R} \qquad (5.66)$$

In other words, the intensity for a large EELS detector is incoherent. Note that this also true for EDX, since characteristic X-ray emission is isotropic and hence there is no restriction on the detected $\mathbf{K}$-values. In typical EELS experiments, however, the detector size is somewhere in between the two extremes discussed above and the measured intensity is therefore partially coherent.

Consider now incoherent imaging in more detail. Rewriting Eq. (5.53) in terms of the complete wavefunctions ψ_0 and ψ_m,

$$\psi_m(\mathbf{R}, z_o) = i\sigma_m V_{m0}^{\mathrm{P}}(\mathbf{R})\psi_0(\mathbf{R}, z_o) \exp[2\pi i(\mathbf{k}_m - \mathbf{k}_0) \cdot \mathbf{r}] \qquad (5.67)$$

Since the detector angle is large, elastic scattering of ψ_m within the crystal is not important. Substituting ψ_m in Eq. (5.66), the measured intensity is

$$I = \sigma_m^2 \int |V_{m0}^{\mathrm{P}}(\mathbf{R})|^2 |\psi_0(\mathbf{R}, z_o)|^2 d\mathbf{R} \qquad (5.68)$$

The measured signal therefore depends on the probe intensity as well as the magnitude of the projected potential. In high spatial resolution chemical analysis delocalisation of the inelastic signal is potentially an important artefact (Muller and Silcox, 1995). If delocalisation is strong then the signal recorded from a probe incident along a given atom column will have contributions from neighbouring atom columns, making direct interpretation either difficult or impossible. Examples will now be presented illustrating the effect of the projected potential and probe intensity on delocalisation of the Si $L_{2,3}$-edge.

Figure 5.7 shows the projected potential 1 eV above the Si $L_{2,3}$-edge threshold for 100 keV incident electrons (Allen et al., 2015). The initial

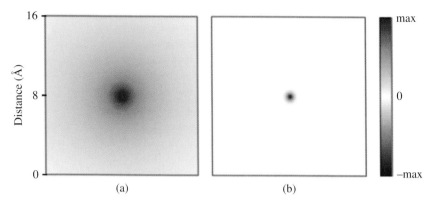

Figure 5.7 Projected potential 1 eV above the Si $L_{2,3}$-edge onset for transition to the (a) $l = 0$, $m = 0$ and (b) $l = 3$, $m = 0$ final states, where l, m are the orbital and magnetic quantum numbers respectively. The former is a dipole transition, while the latter is a quadrupole transition. The incident electron energy is 100 keV. The potential is plotted over a 16 Å × 16 Å area with the silicon atom in the centre. From Allen *et al.* (2015). Reproduced with permission; copyright Elsevier.

core electron has (l, m) quantum numbers of (1, 0) while the quantum numbers for the final state are (0, 0) in Figure 5.7a (dipole transition) and (3, 0) for Figure 5.7b (quadrupole transition). The projected potential is real for the former and imaginary for the latter. The dipole transition potential extends over distances larger than the typical inter-atomic spacings of a few Angstroms, so that the Si $L_{2,3}$ signal can be delocalised even when the incident probe is as small as the atom column of interest. This is demonstrated in the experimental STEM EELS results for a [0001]-Si_3N_4 sample shown in Figure 5.8 (Kimoto *et al.* 2008). The high angle annular dark field (HAADF) image and summed EELS spectrum from the entire spectrum image area are shown in Figure 5.8a and b respectively. Only the heavier Si atoms are visible in the HAADF image. Background subtracted intensity maps extracted from the energy windows labelled in Figure 5.8b are shown in Figure 5.8c–g respectively. Figure 5.8c–f are for the Si $L_{2,3}$-edge while Figure 5.8g is for the N K-edge. Intensity maps have been plotted on a common greyscale so that they can be directly compared with one another. The Si $L_{2,3}$ intensity becomes more localised at the atom columns as the energy integration window is shifted further above the edge onset. Since the intensity maps were extracted from the same spectrum image elastic wave propagation is identical, so that the change in delocalisation must be due to the nature of the projected potential. Just above the edge onset the potential is broad (e.g. Figure 5.7) and can overlap with neighbouring Si atom columns at

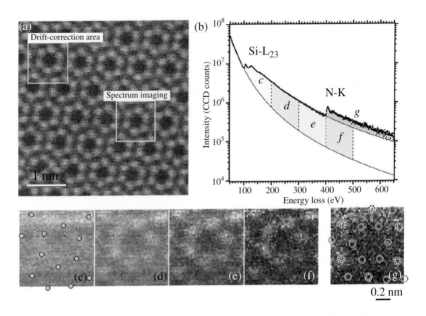

Figure 5.8 (a) HAADF image of [0001]-Si$_3$N$_4$. The summed EELS spectrum extracted from the spectrum imaging area in (a) is shown in (b). The background subtracted Si L$_{2,3}$ intensity maps in figures (c) to (f) are extracted from the corresponding energy windows labelled in (b). Figure (g) is the background subtracted N K-edge intensity. From Kimoto *et al.* (2008). Reproduced with permission; copyright Elsevier.

a distance of 2.7 Å in [0001]-Si$_3$N$_4$. However, at higher energy losses the potential becomes more localised (Muller and Silcox, 1995). From a classical picture as the energy loss increases a larger fraction of inelastic events undergo high angle scattering (Eqs. (5.17) and (5.19)), so that the impact parameter of the incident electron effectively decreases, resulting in stronger localisation.

100 kV STEM EELS results for a relatively thick (91 nm) [110]-Si sample is shown in Figure 5.9 (Wang *et al.*, 2008). Figure 5.9a is the HAADF image, while background subtracted Si L$_{2,3}$ intensity maps extracted from 20 eV wide energy loss windows centred at 153 and 290 eV are shown in Figure 5.9b and c respectively. Simulation results are shown overlaid on the experimental data. Comparing with atomic positions in the HAADF image the 153 eV intensity map shows a contrast reversal, such that the open channels between the atom columns appear 'brighter'. This is because at this specimen thickness there is significant dechannelling of the STEM probe. A probe incident along

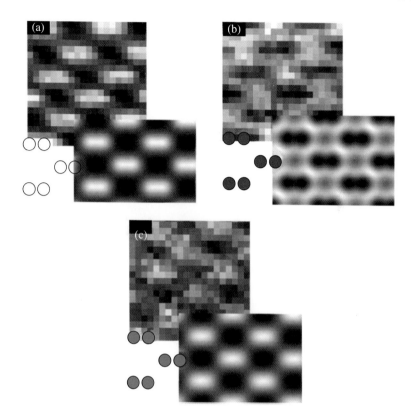

Figure 5.9   (a) HAADF image of a 91 nm thick, [110]-Si specimen acquired at 100 kV beam energy. The background subtracted Si $L_{2,3}$ intensity maps extracted from a 20 eV wide energy window centred about 153 and 290 eV are shown in (b) and (c) respectively. The HAADF and EELS data were acquired simultaneously from the same spectrum image. For each 'image' the atomic structure is overlaid along with the simulation result. From Wang *et al.* (2008). Reproduced with permission; copyright American Physical Society.

the open channels can therefore spread to many neighbouring atom columns and, combined with a delocalised projected potential, give rise to a higher inelastic intensity. The contrast for the 290 eV intensity map, however, is the same as the HAADF image and is due to the projected potential being more localised at the higher energy loss, so that the probe intensity along the excited atom columns becomes the dominant factor. This phenomenon, where the chemical map is influence by dynamic scattering of the electron probe, is called 'preservation of elastic contrast'. This is observed not only in STEM but in EFTEM as well. As an example Figure 5.10a shows the Ti $L_{2,3}$ EFTEM map acquired from

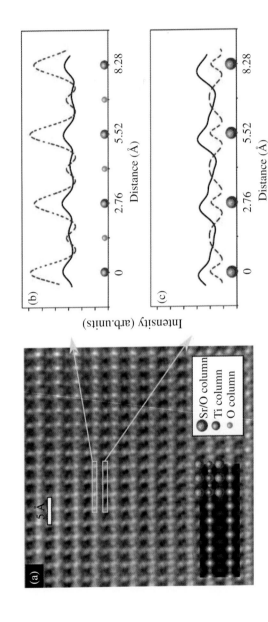

Figure 5.10 (a) Ti L$_{2,3}$ EFTEM map acquired from a 30 nm thick, [110]-SrTiO$_3$ specimen using a C$_s$- and C$_c$-corrected microscope operating at 200 kV. The underlying atomic structure and simulation result are overlaid on the experimental data. Intensity profiles for the Ti–O and Sr/O row of atoms are shown in (b) and (c) respectively. The solid black line and red dashed line are extracted from the experimental and simulated images respectively. From Forbes *et al.* (2014). Reproduced with permission; copyright Elsevier.

a 30 nm thick, [110]-SrTiO$_3$ specimen using a C_{s-} and C_c-corrected microscope at 200 kV (Forbes *et al.*, 2014). Intensity profiles extracted across the Ti–O and Sr/O rows of atoms are shown in Figure 5.10b and c respectively, along with the simulation results. Clear peaks in Ti L$_{2,3}$-intensity are observed at the O and Sr—O atom columns and is due to a combination of the delocalised projected potential as well as channelling of the incident plane wave along the atom columns. Furthermore, atomic resolution STEM EDX maps of [100]-SrTiO$_3$ are found to give inaccurate values for the chemical composition due to channelling (Kothleitner *et al.*, 2014). Thin specimens and high energy losses are therefore required for the chemical map to be directly interpretable.

5.2 FINE STRUCTURE OF THE ELECTRON ENERGY LOSS SIGNAL

5.2.1 Origin of Fine Structure

The fine structure in an EELS core loss edge is a powerful technique for probing the local bonding environment of the excited atom (see, e.g. Garvie *et al.*, 1994; Krivanek and Paterson, 1990; Paxton *et al.*, 2000; surveying minerals, transition metal oxides, carbides and nitrides). The cross-section for inelastic scattering is given by Eqs. (5.17) and (5.18); the specimen information is contained in the square modulus of the transition matrix element:

$$\left| \sum_{\alpha} \langle f| \exp(-2\pi i \mathbf{q} \cdot \mathbf{r}_{\alpha})|i\rangle \right|^2$$

$$= \left| \sum_{\alpha} \int \varphi_f^*(\mathbf{r}_1, \ldots, \mathbf{r}_N) \exp(-2\pi i \mathbf{q} \cdot \mathbf{r}_{\alpha})\varphi_i(\mathbf{r}_1, \ldots, \mathbf{r}_N)\, d\mathbf{r}_1 \ldots d\mathbf{r}_N \right|^2$$

$$(5.69)$$

where φ_i, φ_f denote initial and final state wavefunctions respectively.[5] The transition matrix element is dependent on the energy loss due to the scattering vector $\mathbf{q}$ (Eq. (5.19)), although in most cases this dependence is weak (Rez *et al.*, 1999). From Eq. (5.18) the inelastic scattering cross-section also depends on the unoccupied density of states, through the $\delta(E_f - E_i - \Delta E)$ term. As has been shown previously in the dipole

---

[5]In Section 5.1.1, '*u*' was used to denote the atomic wavefunction. φ in Eq. (5.69), however, is more general and covers the case of solids as well as a free atom.

approximation only $\Delta l = \pm 1$ transitions are allowed. The inelastic scattering cross-section therefore contains information on the local, angular momentum resolved density of unoccupied states.

Evaluating Eq. (5.69) is a formidable task, since it involves many electrons in the solid or atom. The standard approach is to find solutions to the wavefunction under a one electron approximation, that is, by solving an effective one particle Schrödinger equation for each electron. The Hamiltonian, which consists of the Coulomb and exchange interactions with the surrounding electron density, must be self-consistent, since it is also a function of the electron wavefunction that is being solved. Within the one electron approximation the transition matrix element simplifies to

$$\left| \sum_{\alpha} \langle f | \exp(-2\pi i q \cdot \mathbf{r}_{\alpha}) | i \rangle \right|^2 = \left| \int \phi_f^*(\mathbf{r}) \exp(-2\pi i q \cdot \mathbf{r}) \phi_i(\mathbf{r}) d\mathbf{r} \right|^2 \quad (5.70)$$

where ϕ_i, ϕ_f are the initial and final one electron wavefunctions for the electron undergoing the transition; the one electron wavefunctions for all other electrons are assumed to be unchanged during the transition. A derivation of this result can be found in Appendix C.

Equation (5.70) can be solved at varying levels of complexity. The simplest case is a free atom, which, although considerably different from a solid, can still nevertheless reproduce the gross shape of the experimental EELS edge. For spherically symmetric atoms the potential $V(r)$ is a function only of the radial distance r, so that the one electron wavefunction can be expressed as the product of a radial and angular term, that is, $\phi(\mathbf{r}) = L_{nl}(r) Y_{lm}(\theta, \phi)$, where Y_{lm} is the spherical harmonic given by Eq. (5.22) (Rae, 2008). As discussed previously, the angular term governs the selection rules for the transition at small momentum transfer; the magnitude of an allowed transition is, however, also determined by the radial function $L_{nl}(r)$, which satisfies the following differential equation (see Eq. (2.55)):

$$\left[\frac{\hbar^2}{2m} \frac{d^2}{dr^2} + eV(r) - \frac{l(l+1)\hbar^2}{2mr^2} + \varepsilon_{nl} \right] rL_{nl} = 0 \quad (5.71)$$

where ε_{nl} is the energy of the electron eigenstate and $l(l+1)\hbar^2/2mr^2$ is the so-called *centrifugal potential*. In practice, due to the finite collection angle of the EELS spectrometer the l quantum number for the final state need not be restricted by the dipole selection rule (i.e. $\Delta l = \pm 1$), although this will be the dominant contribution at small scattering angles. However, for strong overlap between the initial and final states the energy ε_{nl} for the latter must be larger than its centrifugal potential (Leapman *et al.*, 1980); final states with large angular momentum therefore do not contribute significantly to the transition matrix element. Clearly, the number of final state l-values that need to be considered increases with energy loss above the edge onset.

The centrifugal potential has important implications for the overall EELS edge shape; it determines whether the onset is sharp (e.g. a hydrogenic edge) or displays a delayed maximum. An example of the latter is the Si $L_{2,3}$-edge, which is shown in Figure 5.11a. The solid line is the experimental edge for pure silicon, while the dashed line is the simulation result for a single atom (Leapman *et al.*, 1980; the former is measured using photoabsorption, so that dipole selection rules apply). The similarity of the two curves is striking despite ignoring solid state bonding in the calculation. Note that the maximum intensity for the Si $L_{2,3}$-edge is at a higher energy loss compared to the onset. The reason for the delayed maximum is apparent from the generalised oscillator strength (GOS; Eq. (5.32)) diagram shown in Figure 5.11b. The GOS is plotted for different final state energy values (zero energy being arbitrarily assigned to the edge onset) as a function of the scattering vector magnitude q. At small q the GOS first increases with energy loss, reaching a maximum at $\sim$15 eV, before starting to decrease; the peak at larger q-values observed for energies greater than $\sim$100 eV is the Bethe ridge. The dipole transition channels are from the silicon 2p-core state to continuum states of s and d-angular momentum. The latter with $l = 2$ has a relatively large centrifugal potential barrier, which can only be overcome at energy losses sufficiently above the edge onset. Although there is no centrifugal barrier for the s-channel its GOS is relatively small and does not contribute any appreciable intensity at the edge onset. This is the origin of the delayed maximum for Si $L_{2,3}$. The Si L_1-edge, however, is sharp and has a saw tooth profile, since here the optical channel is from 2s to p-states, which has a smaller centrifugal barrier.

The EELS edge shape for a solid frequently consists of intensity oscillations (i.e. fine structure), which are otherwise not observed in a free atom calculation. The physical origin of this fine structure can be described as follows: the photoelectron[6] emitted from the excited atom propagates outwards as a spherical wave with wave number determined by its kinetic energy (i.e. the energy above the edge onset). The spherical wave can

---

[6]The term 'photoelectron' is strictly valid when the excitation source is a photon, such as an X-ray, but is used here for convenience.

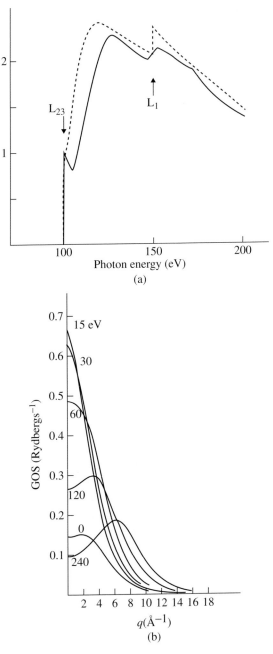

Figure 5.11 (a) Si $L_{2,3}$-edge photoabsorption spectrum (solid line) and simulated result for a silicon atom (dashed line). (b) Generalised oscillator strength for different values of the Si $L_{2,3}$ energy loss plotted as a function of the magnitude of the inelastic scattering vector q. The energy loss at the Si $L_{2,3}$-edge onset is arbitrarily assigned a zero value. From Leapman *et al.* (1980). Reproduced with permission; copyright AIP Publishing LLC.

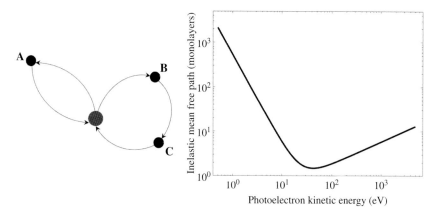

Figure 5.12 (a) Potential scattering pathways for an outgoing photoelectron that is redirected towards the emitting atom (grey circle). The neighbouring atom labelled 'A' is involved in single scattering, while atoms 'B' and 'C' give rise to multiple scattering. (b) The inelastic mean free path for a photoelectron plotted as a function of its kinetic energy; note the log scale for both axes. The graph is a 'universal curve' based on measurements for several different elements. From Seah and Dench 1979.

be *elastically* scattered by neighbouring atoms such that it is redirected towards the emitter (Figure 5.12a). This could be via a single scattering event (i.e. backscattering) or multiple scattering involving several neighbouring atoms. The final state of the excited atom therefore consists of an outgoing wave and incoming wave(s), which interfere with one another. The phase shift between the two waves depends on the scattering path length as well as the phase change due to the potential of the emitting and scattering atoms (Rehr and Albers, 2000; Sayers *et al.*, 1971). For a given scattering pathway, the interference will vary with the photoelectron kinetic energy, such that maxima in the EELS edge intensity are produced for constructive interference and minima for destructive interference. These oscillations in intensity constitute the EELS edge fine structure.

The fine structure is typically divided into two regions. The first is the ELNES, which extends up to ~20 eV above the edge onset. Beyond that is the EXELFS. The physical origin of ELNES and EXELFS is as described above and their demarcation is somewhat arbitrary. Nevertheless, ELNES is a long range effect where the photoelectron undergoes multiple scattering events, while EXELFS is short range and typically involves only a single scattering event. This can be explained by considering the inelastic mean free path of the photoelectron within the solid. A typical curve for the inelastic mean free path as a function of photoelectron energy is shown in Figure 5.12b[7] (Seah and Dench, 1979). A photoelectron can lose energy in a number of ways, including phonon scattering, electron–hole pair generation, plasmon excitation, etc. The mean free path displays a minimum with respect to photoelectron energy. The gradual increase at high energies is due to decreasing inelastic scattering cross-sections. On the other hand, the very rapid increase in mean free path at low energy is due to the fact that only phonon scattering becomes possible at these energies, the other scattering mechanisms being relatively high energy processes (phonon energies are only a few million electron volts). In low energy ELNES the photoelectron can therefore traverse large distances, undergoing scattering multiple times by the atoms in the solid before returning to the emitting atom. ELNES is therefore a long range phenomenon, although if the excited atom is surrounded by strongly backscattering atoms then much of the fine structure is due to the nearest neighbours. These tend to be the light elements for the low photoelectron energies encountered in ELNES (Ravel, 2005). This is an especially useful feature for ELNES of oxides, since tetrahedral, octahedral and other coordination environments, as well as crystal field splitting, can be probed by examining the EELS edge of the respective cation (Brydson, 1996; Garvie and Buseck, 1999; Stoyanov et al., 2007).

In higher energy EXELFS, however, the photoelectron can travel only relatively short distances (Figure 5.12b), typically undergoing only one scattering event from the first few coordination shells surrounding the excited atom. In the method of Sayers et al. (1971), originally developed for X-ray absorption edges, single scattering is assumed, so that each neighbouring atom will contribute a unique EXELFS oscillation, from which inter-atomic distances, number of atoms within a coordination shell, etc. can be extracted. This information is complementary to the radial distribution function that can be obtained from diffraction experiments, since in EXELFS the coordination environment is probed around a specific atom of interest. The extension of Sayers et al. (1971) method to EELS fine structure can be found in Egerton (1996).

5.2.2 Core Hole Effects

In some cases, there might be deviations from the expected one-to-one correspondence between the ELNES fine structure and the partial density of unoccupied states. This is because the final state φ_f in Eq. (5.69)

---

[7]Strictly speaking, Figure 5.12b plots the photoelectron attenuation length, since this is what is measured, rather than the inelastic mean free path.

does not correspond to the ground state of the solid, but rather an excited state containing a vacancy (i.e. core hole) at the initial core energy level of the 'photoelectron'. The core hole will eventually be filled by an electron from a higher energy level, either as a radiative or an Auger recombination event, but its lifetime can be sufficiently long such that it is still present upon arrival of the photoelectron at the excited atom after undergoing elastic scattering from neighbouring atom shells. For example, the core hole lifetime for Al is estimated to be $\sim 10^{-15}$ s, based on the tabulated value for the core hole energy broadening and by applying the uncertainty principle (Seabourne et al., 2009). On the other hand, a photoelectron 10 eV above the EELS edge onset will take $\sim 10^{-16}$ s to be backscattered by the neighbouring atoms, assumed to be at a distance of 1 Å away from the excited atom. During this time, the positively charged core hole may be fully or partially screened by other electrons in the solid. In the Thomas–Fermi model of screening, valid for a free electron gas, the screened potential decays as $\exp(-k_{TF}r)/r$, where k_{TF} is the Thomas–Fermi screening wave number and r is the distance (Ashcroft and Mermin, 1976; recall that an unscreened Coulomb potential has the form $1/r$). For efficient core hole screening $(1/k_{TF})$ must therefore be smaller than the inter-atomic distance. In EELS the core hole is found to be important for insulators and semiconductors and less so for solids with delocalised, metallic-type bonding (Rez et al., 1999; Duscher et al., 2001). If the core hole is not fully screened then the net positive charge localises the final state electron wavefunction around the excited atom and lowers its energy. This causes the EELS edge onset to shift to a lower energy loss and increase in intensity, due to greater spatial overlap with the initial state φ_i (Eq. (5.69)). This is evident in Figure 5.13, which shows the C K-edge in TiC simulated with and without a core hole; the core hole increases the relative intensity of peak '1' at the edge onset, and slightly shifts the peak to lower energy loss (Scott et al., 2001).

The core hole can be introduced in a number of ways in EELS simulations (Duscher et al., 2001; Shirley, 2006). Each one of these is an approximation and so has varying degrees of complexity and ease of implementation. The simplest method is the so-called $(Z + 1)$ approximation, where the excited atom is replaced with the next element in the periodic table, while keeping all other atoms in the solid unchanged. A physically more realistic model would be to keep all atoms the same, but introduce a core hole in the excited atom by promoting a core electron to the lowest unoccupied level (e.g. for the Si $L_{2,3}$-edge the electronic configuration of silicon would therefore be $1s^2\ 2s^2\ 2p^5\ 3s^2\ 3p^3$; note the extra electron in the 3p level and the missing 2p electron). This is also known

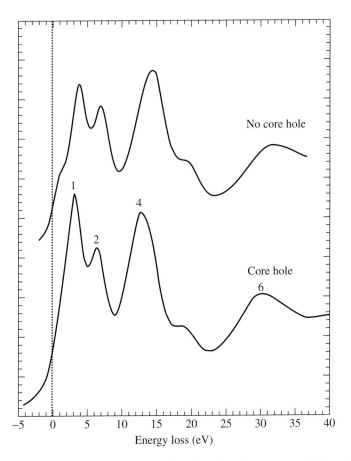

Figure 5.13 C K-edge in TiC simulated with and without a core hole, using the FEFF multiple scattering code. The results are plotted on an arbitrary energy scale. The core hole was modelled using the Z^* approximation, such that the electronic configuration of the excited C atom is $1s^1\,2s^2\,2p^3$. The core hole increases the relative intensity of the peak labelled '1' at the edge onset, as well as shifts it to lower energy loss. From Scott *et al.* (2001). Reproduced with permission; copyright American Physical Society.

as the Z^* approximation. A still higher level of accuracy would be to take into account the interaction between the final state electron and the core hole, that is, the core 'exciton' effect (Mizoguchi *et al.*, 2010). This requires solving the computationally expensive Bethe–Salpeter equation, which is based on a two-body Hamiltonian. Core exciton effects are important for solids with large exciton binding energy (i.e. small dielectric constant) and are typically observed in the K-edges of light elements, such as lithium. Peak positions as well as peak intensities can be modified by the core exciton (Mizoguchi *et al.*, 2010). Partial screening can also be approximated by considering fractional core holes (typically half a core hole is assumed), although this is not possible to implement in all electron codes, and requires the use of excited state pseudopotentials (Seabourne *et al.*, 2009).

For band structure calculations, such as density functional theory, an important consideration is the size of the supercell containing the core hole (Seabourne *et al.*, 2009; Mizoguchi *et al.*, 2010). The supercell must be sufficiently large to minimise the interaction between neighbouring core holes imposed by periodic boundary conditions in the calculation. As noted by Seabourne *et al.* (2009), the incident electron beam current for a typical EELS measurement is sufficiently small so as to allow complete relaxation of the core hole between successive inelastic scattering events, so that any interaction between core holes is physically unrealistic. Finally, the uncertainty principle predicts that the finite lifetime of the core hole leads to an energy broadening that could potentially 'smear' the fine structure in an EELS spectrum. This is modelled by convolving the simulated EELS spectrum with a Lorentzian; a Lorentzian width of ~0.2 eV is appropriate for most EELS edges (Hébert, 2007). Other sources of broadening include the energy spread of the electron source and energy resolution of the EELS spectrometer (modelled as a single Gaussian) as well as an energy-dependent broadening term due to the lifetime of the excited photoelectron, which can be estimated from the photoelectron inelastic mean free path (Figure 5.12b; Seabourne *et al.*, 2009). The latter becomes dominant (i.e. several electronvolts broadening) for photoelectron energies close to the mean free path minimum, and limits the useful width of the ELNES region.

5.2.3 Magnetic Circular Dichroism

A more recent development, inspired by X-ray absorption spectroscopy, is the use of ELNES as a probe for the magnetic properties of the material (Schattschneider *et al.*, 2006, 2008; Hébert *et al.*, 2008). The technique is known as *electron magnetic circular dichroism* (EMCD) and in this section the basic principles will first be described using X-rays, followed by the modifications required for conducting a similar experiment using electrons.

In Section 5.1.1, it was shown that for single electron excitation in the dipole limit the electric field ε of photons (e.g. X-rays) is equivalent to the inelastic scattering vector $\mathbf{q}$ for electrons. Consider circular polarised

X-rays where the electric field vector is of the form $(\boldsymbol{\varepsilon}_1 \pm i\boldsymbol{\varepsilon}_2)$, where $\boldsymbol{\varepsilon}_1$, $\boldsymbol{\varepsilon}_2$ are the electric field vector components in two orthogonal directions and the magnitudes of $\boldsymbol{\varepsilon}_1$ and $\boldsymbol{\varepsilon}_2$ are both equal. The $\pm i$ term indicates that $\boldsymbol{\varepsilon}_2$ is phase shifted by $\pm\pi/2$ radians with respect to $\boldsymbol{\varepsilon}_1$. They represent the two scenarios where the electric field vector rotates in a right- or left-hand helix (Fowles, 1989; note that for convenience the $\exp(-i\omega t)$ time dependence of the electric field has been omitted). Absorption of a circular polarised X-ray therefore gives rise to a $\Delta m = \pm 1$ transition with rotating electric dipole, as illustrated in Figure 5.2b. By comparing the absorption spectra acquired from circular polarised X-rays with different helicity any asymmetry between the $\Delta m = +1$ and $\Delta m = -1$ transitions can be probed. For non-magnetic materials there will be no difference, but for magnetic materials there will be an asymmetry due to the difference in spin up and spin down electron populations (Stöhr, 1995).

To perform a similar experiment with EELS the $(\boldsymbol{\varepsilon}_1 \pm i\boldsymbol{\varepsilon}_2)$ X-ray electric field must be replaced by an inelastic scattering vector of the form $(\mathbf{q}_1 \pm i\mathbf{q}_2)$, where $\mathbf{q}_1$ is perpendicular to $\mathbf{q}_2$ and $|\mathbf{q}_1| = |\mathbf{q}_2|$. Furthermore, $\mathbf{q}_1$ and $\mathbf{q}_2$ must also be phase shifted by $\pm\pi/2$ radians. The successful implementation of EMCD uses kinematic scattering within a crystal to generate two beams: the unscattered incident electron beam (O) and a Bragg scattered beam (G) phase shifted by $\pi/2$ radians. As shown in Figure 5.14a, the EELS spectrometer entrance aperture is positioned at points marked '+' and '−' in the diffraction plane, that is, along the perpendicular bisector to the imaginary line joining O to G and on the circumference of the circle with diameter OG. This ensures that the conditions $|\mathbf{q}_1| = |\mathbf{q}_2|$ and $\mathbf{q}_1$ perpendicular to $\mathbf{q}_2$ are satisfied. Furthermore, for position '+' the inelastic scattering vector is $(\mathbf{q}_1 + i\mathbf{q}_2)$ with $\pi/2$ radian phase shift, while for position '−' the scattering vector is $(-i\mathbf{q}_1 - \mathbf{q}_2)$ or $[(-i\mathbf{q}_1) - i(-i\mathbf{q}_2)]$ with $-\pi/2$ radian phase shift. The two scattering vectors therefore have opposite helicity.

For EMCD the double differential inelastic scattering cross-section (Eq. (5.36)) with inelastic scattering vector $(\mathbf{q}_1 \pm i\mathbf{q}_2)$, $|\mathbf{q}_1| = |\mathbf{q}_2|$ reduces to

$$\frac{\partial^2 \sigma_m}{\partial(\Delta E)\partial\Omega} = \left(\frac{1}{2\pi^2 a_o}\right)^2 \left(\frac{k_m}{k_0}\right)$$

$$\times \frac{[S(\mathbf{q}_1, \Delta E) + S(\mathbf{q}_2, \Delta E) \pm 2\text{Im}\{S(\mathbf{q}_1, \mathbf{q}_2, \Delta E)\}]}{q^4} \quad (5.72)$$

where q is the magnitude of $\mathbf{q}_1$ (or equivalently $\mathbf{q}_2$) and 'Im' denotes the imaginary part of a complex number; the other terms in Eq. (5.72) are as

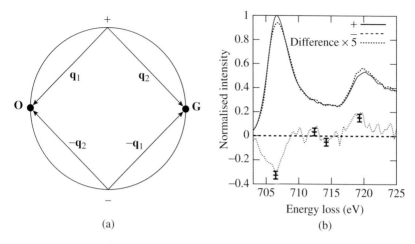

(a)   (b)

Figure 5.14 (a) Experimental set-up for electron magnetic circular dichroism (EMCD) measurements. The unscattered and Bragg diffracted beams in the diffraction plane are denoted by O and G respectively. An EELS detector at the '+' or '−' positions has an inelastic scattering vector that is equivalent to circular polarised X-rays of opposite helicity. (b) EMCD data for the Fe $L_{2,3}$-edge measured from a pure iron specimen. The EELS spectra recorded at the '+' and '−' detector positions are shown along with the difference spectrum. The 002 reflection is used for the Bragg diffracted beam. From Schattschneider *et al.* (2006). Reproduced with permission; copyright Nature Publishing Group.

defined previously. Consider the difference between the two EELS spectra that are acquired with inelastic scattering vectors of opposite helicity (i.e. positions '+' and '−' in Figure 5.14a). In Eq. (5.72) the dynamic form factor terms vanish, so that the only remaining contribution to the difference spectrum is due to the imaginary part of the mixed dynamic form factor, which in the dipole limit can be shown to depend on the magnetic properties of the atom (Schattschneider *et al.*, 2008). An example of an EMCD measurement using the Fe $L_{2,3}$-edge in a pure iron specimen is shown in Figure 5.14b.

5.3 SUMMARY

Single electron excitation from a core energy level within an atom is an important inelastic scattering process in both EELS and EDX analysis. The scattering cross-section for a single atom illuminated by an electron plane wave was shown to depend on the dynamic form factor. In the limit of small scattering angles, the so-called *dipole approximation*, selection rules for atomic transitions apply, due to the symmetry of the initial and final electronic states. For dipole transitions, the change in orbital quantum number must be ± 1 and the change in magnetic quantum number must be either zero or ± 1. When the atom is illuminated by more than one plane wave the scattering cross-section also contains a mixed dynamic form factor, which is a generalisation of the dynamic form factor and takes into account interference between the plane waves. The mixed dynamic form factor has important implications for STEM.

Chemical mapping of atoms within a crystal is more complicated. Here, it is important to also take into account the dynamic elastic scattering of the probe electrons, via Yoshioka's equations. Artefacts such as 'preservation of elastic contrast' and delocalisation of inelastic scattering complicate atomic resolution chemical mapping in the electron microscope. The effects can be minimised by using thinner samples, larger signal collection angles (for EELS) as well as by using higher energy loss edges for analysis.

Finally, the fine structure of the EELS edge can also contain valuable information about the partial unoccupied density of states and coordination environment. The fine structure is due to the photoelectron being elastically scattered by surrounding atoms before returning to the excited atom, causing constructive and destructive interference between incoming and outgoing waves. Close to the edge onset the inelastic mean free path of the photoelectron is large, so that it undergoes multiple elastic scattering by the atoms in the solid over a large distance. The ELNES region therefore contains information about the angular momentum resolved, unoccupied density of states. In contrast, higher energy photoelectrons have shorter inelastic mean free path, and are typically elastically scattered only once. Consequently, the EXELFS region contains information about the coordination environment surrounding the excited atom. An important artefact, especially for ELNES, is the core hole created by the photoelectron. If the core hole is not fully screened by conduction electrons in the solid then the final state of the excited atom cannot be described by ground state electronic properties. A straightforward interpretation in terms of unoccupied density of states may therefore not always be possible.

REFERENCES

Allen, L.J., D'Alfonso, A.J. and Findlay, S.D. (2015) *Ultramicroscopy* **151**, 11.
Ashcroft, N.W. and Mermin, N.D. (1976) *Solid State Physics*, Holt-Saunders Japan Ltd, Tokyo.

Auerhammer, J.M. and Rez, P. (1989) *Phys. Rev. B* **40**, 2024.

Bosman, M., Keast, V.J., García-Muñoz, J.L., D'Alfonso, A.J., Findlay, S.D. and Allen, L.J. (2007) *Phys. Rev. Lett.* **99**, 086102.

Brydson, R. (1996) *J. Phys. D: Appl. Phys.* **29**, 1699.

Chu, M.W., Liou, S.C., Chang, C.P., Choa, F.S. and Chen, C.H. (2010) *Phys. Rev. Lett.* **104**, 196101.

Coene, W. and Van Dyck, D. (1990) *Ultramicroscopy* **33**, 261.

D'Alfonso, A.J., Findlay, S.D., Oxley, M.P. and Allen, A.J. (2008) *Ultramicroscopy* **108**, 677.

D'Alfonso, A.J., Freitag, B., Klenov, D. and Allen, L.J. (2010) *Phys. Rev. B* **81**, 100101R.

Duscher, G., Buczko, R., Pennycook, S.J. and Pantelides, S.T. (2001) *Ultramicroscopy* **86**, 355.

Dwyer, C. (2005) *Ultramicroscopy* **104**, 141.

Egerton, R.F. (1996) *Electron Energy-Loss Spectroscopy in the Electron Microscope*, 2nd edition, Plenum Press, New York.

Forbes, B.D., Houben, L., Mayer, J., Dunin-Borkowski, R.E. and Allen, L.J. (2014) *Ultramicroscopy* **147**, 98.

Fowles, G.R. (1989) *Introduction to Modern Optics*, Dover edition, USA.

Garvie, L.A.J., Craven, A.J. and Brydson, R. (1994) *Am. Mineral.* **79**, 411.

Garvie, L.A.J. and Buseck, P.R. (1999) *Am. Mineral.* **84**, 946.

Gonis, A. and Butler, W.H. (2000) *Multiple Scattering in Solids*, Springer-Verlag, New York.

Gradshteyn, I.S. and Ryzhik, I.M. (1980) *Tables of Integrals, Series and Products*, Academic Press, New York.

Haider, M., Uhlemann, S., Schwan, E., Rose, H., Kabius, B. and Urban, K.W. (1998) *Nature* **392**, 768.

Hébert, C. (2007) *Micron* **38**, 12.

Hébert, C., Schattschneider, P., Rubino, S., Novak, P., Rusz, J. and Stöger-Pollach, M. (2008) *Ultramicroscopy* **108**, 277.

Howie, A. (1963) *Proc. Roy. Soc. A* **271**, 268.

Inokuti, M. (1971) *Rev. Mod. Phys.* **43**, 297.

Ishida, K. (1970) *J. Phys. Soc. Japan* **28**, 450.

Kabius, B., Hartel, P., Haider, M., Müller, H., Uhlemann, S., Loebau, U., Zach, J. and Rose, H. (2009) *J. Electron Microsc.* **58**, 147.

Kimoto, K. and Matsui, Y. (2003) *Ultramicroscopy* **96**, 335.

Kimoto, K., Ishizuka, K. and Matsui, Y. (2008) *Micron* **39**, 257.

Kohl, H. and Rose, H. (1985) *Adv. Electron. Electron Phys.* **65**, 173.

Kothleitner, G., Neish, M.J., Lugg, N.R., Findlay, S.D., Grogger, W., Hofer, F. and Allen, L.J. (2014) *Phys. Rev. Lett.* **112**, 085501.

Krivanek, O.L. and Paterson, J.H. (1990) *Ultramicroscopy* **32**, 313.

Krivanek, O.L., Corbin, G.J., Dellby, N., Elston, B.F., Keyse, R.J., Murfitt, M.F., Own, C.S., Szilagyi, Z.S. and Woodruff, J.W. (2008) *Ultramicroscopy* **108**, 179.

Leapman, R.D., Rez, P. and Mayers, D.F. (1980) *J. Chem. Phys.* **72**, 1232.

Leapman, R.D., Fejes, P.L. and Silcox, J. (1983) *Phys Rev B* **28**, 2361.

Levine, I.R. (2006) *Quantum Chemistry*, 5th Edition, Prentice-Hall of India, New Delhi.

Lugg, N.R., Freitag, B., Findlay, S.D. and Allen, L.J. (2010) *Ultramicroscopy* **110**, 981.

Lupini, A.R. and Pennycook, S.J. (2003) *Ultramicroscopy* **96**, 313.

Martin, R.M. (2004) *Electronic Structure- Basic Theory and Practical Methods*, Cambridge University Press, UK.

Maslen, V.W. and Rossouw, C.J. (1984) *Phil. Mag. A* **49**, 735.

McDaniel, E.W. (1989) *Atomic Collisions: Electron and Photon Projectiles*, John Wiley & Sons, USA.

Mizoguchi, T., Olovsson, W., Ikeno, H. and Tanaka, I. (2010) *Micron* **41**, 695.

Moreno, M.S., Jorissen, K. and Rehr, J.J. (2007) *Micron* **38**, 1.

Muller, D.A. and Silcox, J. (1995) *Ultramicroscopy* **59**, 195.

Muller, D.A., Fitting Kourkoutis, L., Murfitt, M., Song, J.H., Hwang, H.Y., Silcox, J., Dellby, N. and Krivanek, O.L. (2008) *Science* **319**, 1073.

Paxton, A.T., van Schilfgaarde, M., MacKenzie, M. and Craven, A.J. (2000) *J. Phys.: Condens. Matter* **12**, 729.

Rae, A.I.M. (2008) *Quantum Mechanics*, 5th edition, Taylor & Francis, USA.

Ravel, B. (2005) *J. Alloys Cmpds.* **401**, 118.

Rehr, J.J. and Albers, R.C. (2000) *Rev. Mod. Phys.* **72**, 621.

Rez, P., Alvarez, J.R. and Pickard, C. (1999) *Ultramicroscopy* **78**, 175.

Saldin, D.K. and Rez, P. (1987) *Phil. Mag. B* **55**, 481.

Sayers, D.E., Stern, E.A. and Lytle, F.W. (1971) *Phys. Rev. Lett.* **27**, 1204.

Schattschneider, P., Pongratz, P. and Hohenegger, H. (1990) *Scanning Microsc.* **4**, 35.

Schattschneider, P., Nelhiebel, M., Souchay, H. and Jouffrey, B. (2000) *Micron* **31**, 333.

Schattschneider, P., Rubino, S., Hébert, C., Rusz, J., Kunes, J., Novák, P., Carlino, E., Fabrizioli, M., Panaccione, G. and Rossi, G. (2006) *Nature* **441**, 486.

Schattschneider, P., Hébert, C., Rubino, S., Stöger-Pollach, M., Rusz, J. and Novák, P. (2008) *Ultramicroscopy* **108**, 433.

Scott, A.J., Brydson, R., MacKenzie, M. and Craven, A.J. (2001) *Phys. Rev. B* **63**, 245105.

Seabourne, C.R., Scott, A.J., Brydson, R. and Nicholls, R.J. (2009) *Ultramicroscopy* **109**, 1374.

Seah, M.P. and Dench, W.A. (1979) *Surf. Interface Anal.* **1**, 2.

Shirley, E. (2006) *Ultramicroscopy* **106**, 986.

Spence, J.C.H. and Taftø, J. (1983) *J. Microsc.* **130**, 147.

Stöhr, J. (1995) *J. Electron Spectrosc. Related Phenom.* **75**, 253.

Stoyanov, E., Langenhorst, F. and Steinle-Neumann, G. (2007) *Am. Mineral.* **92**, 577.

Taftø, J. and Krivanek, O.L. (1982) *Phys. Rev. Lett.* **48**, 560.

Urban, K.W., Mayer, J., Jinschek, J.R., Neish, M.J., Lugg, N.R. and Allen, L.J. (2013) *Phys. Rev. Lett.* **110**, 185507.

Wang, P., D'Alfonso, A.J., Findlay, S.D., Allen, L.J. and Bleloch, A.L. (2008) *Phys. Rev. Lett.* **101**, 236102.

Williams, B.G., Sparrow, T.G. and Egerton, R.F. (1984) *Proc. R. Soc. Lond. A* **393**, 409.

Yoshioka, H. (1957) *J. Phys. Soc. Japan* **12**, 618.

Young, A.P. and Rez, P. (1975) *J. Phys. C* **8**, L1.

6

Electrodynamic Theory of Inelastic Scattering

In the previous chapter, on single electron excitation, the incident high energy electron transfers part of its energy to a core electron lying deep within an atom. There are, however, examples where energy is transferred to not one but many electrons in the solid, such as, collective oscillations of valence electrons, which are also known as plasmons (Kittel, 2005; Pitarke *et al.*, 2007). Furthermore, for low energy loss inelastic scattering the impact parameter of the incident electron can be several nanometres in length (Section 2.1.2), such that the electric field of the incident electron, as seen by the target atom, is modified by screening and polarisation of the surrounding atoms. In such cases, the one electron approximation of the transition matrix element (Section 5.2.1) is no longer valid, and an alternative theory is required. The method adopted is due to Fermi (1940). It was originally proposed to calculate the stopping power in solids and is based on the electrodynamics of the incident electron, with the material response characterised via its dielectric function. While it can be shown that this is equivalent to the quantum mechanical description of the previous chapter (Kohl and Rose, 1985), the electrodynamic approach is nevertheless conceptually simpler, since the many body interactions of the valence electrons are all contained in a single physical property, the dielectric function.

The low energy loss region contains many interesting features, such as the optical properties of a material, bulk and surface plasmon modes, interband transitions, band gaps, and excitons. These can be characterised using electron energy loss spectroscopy (e.g. Erni and Browning, 2005) and cathodoluminescence (CL) for the radiative transitions

Electron Beam-Specimen Interactions and Simulation Methods in Microscopy,
First Edition. Budhika G. Mendis.
© 2018 John Wiley & Sons Ltd. Published 2018 by John Wiley & Sons Ltd.

(e.g. Yamamoto *et al.*, 1996a). This chapter first considers energy loss in an 'infinite' solid; key concepts in electrodynamic theory are introduced at this stage. Some background knowledge of Maxwell's equations is required and can be found at the start of Appendix D or one of the many textbooks on the subject (e.g. Jackson, 1998; Griffiths, 1999; note that SI units are used throughout this chapter). In transmission electron microscopy, the material examined has finite dimensions, and hence the effect of free surfaces and internal interfaces is also discussed. This is followed by a section on radiative energy losses, namely Cerenkov and transition radiation. For certain measurements, such as band gaps, radiative energy losses can lead to unwanted artefacts and it is therefore important to understand their origin as well as devise strategies to minimise its effects. In the final part of the chapter, computational methods based on the electrodynamic theory are introduced.

6.1 BULK AND SURFACE ENERGY LOSS

6.1.1 Energy Loss in an 'Infinite' Solid

The stopping power of a high energy electron in an 'infinite' medium is derived in this section. It is assumed that the energy loss is sufficiently small, so that the trajectory (i.e. both speed and direction) of the incident electron is unchanged to first approximation. Furthermore, the so-called *non-retarded* limit is assumed, where the solid can respond instantaneously to a perturbation in the electric charge density or current density, which from Maxwell's equations are the source of electric and magnetic fields respectively. This effectively means that there is no time delay for the fields to propagate between any two given points (i.e. speed of light is infinite). Neglecting retardation effects is not entirely appropriate in most electron microscopy measurements, where the incident electron may be travelling at a significant fraction of the speed of light. Nevertheless, it considerably simplifies the calculations and is therefore used here. The derivation for bulk energy loss in the retarded regime can be found in Appendix D.

Consider a high energy incident electron travelling at velocity **v** along the z-axis. The x,y- axes form a right-handed coordinate system with z. The divergence of the electric displacement **D** is given by Maxwell's equations (see Appendix D):

$$\vec{\nabla} \cdot \mathbf{D}(\mathbf{r}, t) = \rho_f(\mathbf{r}, t) \tag{6.1}$$

where $\mathbf{r}$ is the position vector, t is time and ρ_f is the free charge density. The incident electron is treated classically as a point particle, so that (Ritchie, 1957)

$$\rho_f(\mathbf{r}, t) = -e\delta(x)\delta(y)\delta(z - vt) \tag{6.2}$$

Here $-e$ is the charge of the electron, v its speed and δ represents Dirac's delta function. Consider the relationship between the electric displacement $\mathbf{D}$ and the electric field $\mathbf{E}$. For a homogeneous, isotropic solid the two quantities are related by (Schattschneider, 1986)

$$\mathbf{D}(\mathbf{r}, t) = \varepsilon_o \iint \varepsilon(\mathbf{r} - \mathbf{r}', t - t')\mathbf{E}(\mathbf{r}', t')d\mathbf{r}'\, dt' \tag{6.3}$$

where ε is the dielectric function of the material and ε_o is the permittivity of free space. The dielectric function characterises the response of the electrons and nuclei in the system to an applied electric field. $\mathbf{D}$ is a convolution of ε and $\mathbf{E}$ in spatial and temporal domains. Fourier transforming and making use of the convolution theorem gives

$$\tilde{\mathbf{D}}(\mathbf{q}, \omega) = \varepsilon_o \varepsilon(\mathbf{q}, \omega)\tilde{\mathbf{E}}(\mathbf{q}, \omega) \tag{6.4}$$

where $\mathbf{q}$ is the reciprocal vector and ω the frequency. A tilde is used to denote a Fourier transform with respect to both $\mathbf{r}$ and t; it is omitted from $\varepsilon(\mathbf{q}, \omega)$ as a matter of convenience. In order to simplify the calculations it is common to make use of the 'local approximation' (Wang, 1996), which states that the local permittivity does not depend on the surrounding environment, that is, $\varepsilon(\mathbf{r} - \mathbf{r}', t - t') = \varepsilon(t - t')\delta(\mathbf{r} - \mathbf{r}')$. Making this approximation in Eq. (6.3) leads to Eq. (6.5a) and taking its Fourier transform, which is now required only in the time domain, leads to Eq. (6.5b)

$$\mathbf{D}(\mathbf{r}, t) = \varepsilon_o \int \varepsilon(t - t')\mathbf{E}(\mathbf{r}, t')dt' \tag{6.5a}$$

$$\mathbf{D}(\mathbf{r}, \omega) = \varepsilon_o \varepsilon(\omega)\mathbf{E}(\mathbf{r}, \omega) \tag{6.5b}$$

In the local approximation ε is therefore a function of ω only. The assumption is valid for small electron energy loss (EELS) collection angles, that is, $|\mathbf{q}| \approx 0$.

Let us now return to Eq. (6.1). It is advantageous to express this in Fourier space. The Fourier transform of a function $f(\mathbf{r}, t)$ and its inverse transform are defined as:

$$\tilde{f}(\mathbf{q}, \omega) = \iint f(\mathbf{r}, t) \exp[-2\pi i \mathbf{q} \cdot \mathbf{r} + i\omega t] d\mathbf{r} \, dt$$

$$f(\mathbf{r}, t) = \iint \tilde{f}(\mathbf{q}, \omega) \exp[2\pi i \mathbf{q} \cdot \mathbf{r} - i\omega t] d\mathbf{q} \, d\omega \qquad (6.6)$$

Therefore:

$$\vec{\nabla} f(\mathbf{r}, t) = \iint 2\pi i \mathbf{q} \tilde{f}(\mathbf{q}, \omega) \exp[2\pi i \mathbf{q} \cdot \mathbf{r} - i\omega t] d\mathbf{q} \, d\omega \qquad (6.7)$$

Comparing Eqs. (6.6) and (6.7) it is clear that the Fourier transform of the $\vec{\nabla}$ operator is $2\pi i \mathbf{q}$. Using this fact, the Fourier transform of Eq. (6.1) is

$$2\pi i \mathbf{q} \cdot \tilde{\mathbf{D}}(\mathbf{q}, \omega) = 2\pi i \varepsilon_o \varepsilon(\mathbf{q}, \omega) \mathbf{q} \cdot \tilde{\mathbf{E}}(\mathbf{q}, \omega) = \tilde{\rho}_f(\mathbf{q}, \omega) \qquad (6.8)$$

while that of Eq. (6.2) is

$$\tilde{\rho}_f(\mathbf{q}, \omega) = -e\delta(2\pi q_z v - \omega) \qquad (6.9)$$

where q_z is the z-component of the reciprocal vector $\mathbf{q}$. In Eq. (6.9) use has been made of the standard result that the Fourier transform of $\exp(2\pi i u)$ is the Dirac delta function $\delta(u)$, where u is an arbitrary variable. Note that Eq. (6.8) indicates the broadband nature of the electric field of the incident electron (i.e. a continuous spectrum of frequencies ω are generated). In the non-retarded limit, $\mathbf{E}(\mathbf{r}, t) = -\vec{\nabla}\phi(\mathbf{r}, t)$ or in Fourier space $\tilde{\mathbf{E}}(\mathbf{q}, \omega) = -2\pi i \mathbf{q} \tilde{\phi}(\mathbf{q}, \omega)$, where ϕ is the electric potential.[1] Substituting in Eqs. (6.8) and (6.9) it then follows that

$$\tilde{\phi}(\mathbf{q}, \omega) = -\frac{e\delta(2\pi q_z v - \omega)}{4\pi^2 \varepsilon_o \varepsilon(\mathbf{q}, \omega) q^2} \qquad (6.10)$$

The energy loss dW over an infinitesimal distance dz along the electron trajectory is given by $-F_z \, dz$, where F_z is the force component along the z-axis; a negative sign is included so that dW is a positive quantity. The force due to the electric field is $-e\mathbf{E}(\mathbf{r}, t)$. The instantaneous magnetic induction $\mathbf{B}$ of the electron does not affect the energy loss, since

---

[1]In fact $\mathbf{E} = -\vec{\nabla}\phi - \partial\mathbf{A}/\partial t$, where $\mathbf{A}$ is the magnetic vector potential. For a Lorentz gauge $\vec{\nabla} \cdot \mathbf{A} = -(1/c^2)\partial\phi/\partial t$, so that $\mathbf{A} \to 0$ in the non-retarded limit where the speed of light $c \to \infty$.

the Lorentz force ($= -e\mathbf{v} \times \mathbf{B}$) has zero component along the electron trajectory.[2] The stopping power dW/dz is therefore

$$\frac{dW}{dz} = eE_z(\mathbf{r}, t)|_{x,y=0,z=vt} = -e\frac{\partial\phi}{\partial z}\bigg|_{x,y=0,z=vt} \tag{6.11}$$

where E_z is the z-component of the electric field. From Eqs. (6.6) and (6.10) it follows that

$$\frac{dW}{dz} = \frac{e^2 i}{2\pi\varepsilon_o} \iint \frac{\delta(2\pi q_z v - \omega)}{\varepsilon(\mathbf{q}, \omega)q^2} q_z \exp[2\pi i\mathbf{q} \cdot \mathbf{r} - i\omega t]d\mathbf{q}\, d\omega\bigg|_{x,y=0,z=vt} \tag{6.12}$$

The integration over q_z can be carried out using the relationship:

$$\int g(q_z)\delta(2\pi q_z v - \omega)dq_z = \frac{1}{2\pi v}\int g(q_z)\delta(2\pi q_z v - \omega)d(2\pi q_z v)$$

$$= \frac{g(\omega/2\pi v)}{2\pi v} \tag{6.13}$$

where $g(q_z)$ is an arbitrary function of q_z. Setting $g(q_z) = q_z \exp(2\pi i q_z z)/[\varepsilon(\mathbf{q}_\perp, q_z, \omega) \cdot (q_\perp^2 + q_z^2)]$ and noting that $z = vt$, the stopping power in Eq. (6.12) simplifies to

$$\frac{dW}{dz} = \frac{e^2 i}{8\pi^3 v^2 \varepsilon_o} \int d\mathbf{q}_\perp \int_{-\infty}^{\infty} \frac{\omega}{\varepsilon(\mathbf{q}, \omega)[q_\perp^2 + (\omega/2\pi v)^2]}$$

$$\times \exp[2\pi i\mathbf{q}_\perp \cdot \mathbf{R}]d\omega\bigg|_{x,y=0}$$

$$= \frac{e^2 i}{8\pi^3 v^2 \varepsilon_o} \int d\mathbf{q}_\perp \int_{-\infty}^{\infty} \frac{\omega}{\varepsilon(\mathbf{q}, \omega)[q_\perp^2 + (\omega/2\pi v)^2]}d\omega \tag{6.14}$$

where $\mathbf{q}_\perp$ and $\mathbf{R}$ are the components of $\mathbf{q}$ and $\mathbf{r}$ in the xy-plane (i.e. perpendicular to the z-axis) respectively. $\varepsilon(\mathbf{q}, \omega)$ is evaluated for the reciprocal vector components $\mathbf{q} = (\mathbf{q}_\perp, \omega/2\pi v)$. In practice, the measured energy loss $\hbar\omega$ ($= h\omega/2\pi$, where h is the Planck's constant) can have only positive values, so that the integral over all frequencies in Eq. (6.14) must be simplified. Since $\varepsilon(t)$ is a real quantity from Eq. (6.6) its Fourier transform satisfies the relationship $\varepsilon(-\omega) = \varepsilon^*(\omega)$. The final expression for the

---

[2] In the retarded limit, however, the time derivative of the magnetic field can affect the energy loss by inducing an electric field component (see Footnote 1).

stopping power is therefore

$$\frac{dW}{dz} = \frac{e^2}{4\pi^3 v^2 \varepsilon_o} \int dq_\perp \int_0^\infty \frac{\omega}{[q_\perp^2 + (\omega/2\pi v)^2]} \mathrm{Im} \left[-\frac{1}{\varepsilon(\mathbf{q}, \omega)} \right] d\omega \quad (6.15)$$

where 'Im' denotes the imaginary part of a complex quantity. Note the modified integration limits for the frequency integral. The integrand can also be expressed as $(\hbar\omega) \cdot [\partial^2 P/\partial\mathbf{q}_\perp \partial\omega]$, where P is the energy loss probability per unit path length. Hence,

$$\frac{\partial^2 P}{\partial\mathbf{q}_\perp \partial\omega} = \frac{e^2}{4\pi^3 v^2 \hbar \varepsilon_o} \frac{1}{[q_\perp^2 + (\omega/2\pi v)^2]} \mathrm{Im} \left[-\frac{1}{\varepsilon(\mathbf{q}, \omega)} \right] \quad (6.16)$$

In practice, the experimentally measured parameter is the double differential scattering cross-section $\partial^2\sigma/\partial\Omega\partial(\Delta E)$, where σ is the scattering cross-section, Ω the solid angle and $\Delta E = \hbar\omega$ is the energy loss. The stopping power can be expressed as (see Section 2.1.2):

$$\frac{dW}{dz} = \iint n\Delta E \frac{\partial^2\sigma}{\partial\Omega\partial(\Delta E)} d\Omega d(\Delta E) \quad (6.17)$$

where n is the number of atoms per unit volume. The integrands in Eqs. (6.15) and (6.17) must be equal. For small energy loss and scattering angles $d\Omega = 2\pi(\sin\theta)d\theta \approx (2\pi\theta)d\theta$ and $q_\perp \approx k_0\theta$ (see Figure 5.1), so that $d\mathbf{q}_\perp = 2\pi q_\perp(dq_\perp) \approx (2\pi k_0^2\theta)d\theta$, where θ is the scattering angle and k_0 is the incident electron wave number. Substituting in Eqs. (6.15) and (6.17) and comparing the integrands gives

$$\frac{\partial^2\sigma}{\partial\Omega\partial(\Delta E)} = \frac{e^2}{n\pi v^2 \varepsilon_o \hbar^2} \frac{\mathrm{Im}[-1/\varepsilon(\omega)]}{\theta^2 + \theta_E^2} = \frac{\mathrm{Im}[-1/\varepsilon(\omega)]}{n\pi^2 m v^2 a_o[\theta^2 + \theta_E^2]} \quad (6.18)$$

where $a_o = 4\pi\varepsilon_o \hbar^2/me^2$ is the Bohr radius and m the mass of the electron. Furthermore, the characteristic scattering angle θ_E follows from $k_0\theta_E = (\omega/2\pi v)$.[3] Equation (6.18) is derived in the local approximation, that is, $\varepsilon(\mathbf{q}_\perp, \omega) \approx \varepsilon(\omega)$, and is therefore only valid at small scattering angles. Furthermore, the result assumes single inelastic scattering, while

---

[3] By definition $k_0\theta_E = k_0(\Delta E/2E_o)$, where $\Delta E = \hbar\omega$ is the energy loss and E_o is the incident electron energy. Substituting $E_o = \frac{1}{2}mv^2$ and using de Broglie's relationship $k_0 = mv/h$ gives the desired result, $k_0\theta_E = \omega/2\pi v$.

in practice the finite thickness of the specimen means that multiple inelastic scattering, as well as elastic scattering, will give rise to a redistribution of intensity.

From Eq. (6.16), it is clear that there is no energy loss when the imaginary part of the dielectric function is zero; this is also true for absorption of an electromagnetic wave in a solid (Hecht, 2002). Recall that Fermi's golden rule for inelastic scattering (Section 5.1.1) also showed a close relationship between electrons and photons. By comparing the double differential cross-section, Eq. (6.18), with Eq. (5.17), it is clear that the loss function $Im[-1/\varepsilon(\mathbf{q}, \omega)]$ in classical electrodynamics is equivalent to the dynamic form factor $S(\mathbf{q}, \Delta E = \hbar\omega)$ in quantum mechanics. A formal derivation of this result can be found in Kohl and Rose (1985). Similarly, the Lorentzian distribution $1/(\theta^2 + \theta_E^2)$ for single scattering is also common to both approaches (see Eq. (6.18) and Section 5.1.1).

Equation (6.16) can be simplified further by assuming the local approximation, so that $\varepsilon(\mathbf{q}, \omega) = \varepsilon(\omega)$. Integrating over $\mathbf{q}_\perp$ and recalling that $d\mathbf{q}_\perp = 2\pi q_\perp (dq_\perp)$,

$$
\begin{aligned}
\frac{dP}{d\omega} &= \frac{e^2}{4\pi^3 v^2 \hbar \varepsilon_o} Im\left[-\frac{1}{\varepsilon(\omega)}\right] \int_0^{q_c} \frac{2\pi q_\perp\ dq_\perp}{q_\perp^2 + (\omega/2\pi v)^2} \\
&= \frac{e^2}{4\pi^2 v^2 \hbar \varepsilon_o} Im\left[-\frac{1}{\varepsilon(\omega)}\right] \ln[1 + (2\pi q_c v/\omega)^2] \\
&\approx \frac{e^2}{2\pi^2 v^2 \hbar \varepsilon_o} Im\left[-\frac{1}{\varepsilon(\omega)}\right] \ln(2\pi q_c v/\omega)
\end{aligned}
\tag{6.19}
$$

where q_c is the largest reciprocal vector component $q_\perp$ detected by the EELS spectrometer. The final approximation in Eq. (6.19) is valid for large q_c.

From Eq. (6.3) it is apparent that the electric displacement $\mathbf{D}$ at any given time t is determined by the temporal evolution of the electric field $\mathbf{E}$ and the response of the solid, as expressed by the dielectric function. The dielectric function must obey the causality condition, such that $\varepsilon(\mathbf{r} - \mathbf{r}', t - t')$ is zero for $t' > t$ (Eq. (6.3)). This means that only the electric field prior to time t can influence $\mathbf{D}(\mathbf{r}, t)$. It can be shown that causality implies that the real and imaginary parts of the dielectric function are related to one another through the Kramers–Kronig equations (see Jackson, 1998 for a derivation). In Eq. (6.19) the EELS spectrum is proportional to $Im[-1/\varepsilon(\omega)]$. The real part $Re[1/\varepsilon(\omega)]$ can then be extracted from the measured values of $Im[-1/\varepsilon(\omega)]$ via the relevant Kramers–Kronig relationship (Egerton, 1996):

$$\mathrm{Re}\left[\frac{1}{\varepsilon(\omega)}\right] = 1 - \frac{2}{\pi}P\int_0^\infty \frac{\omega'\mathrm{Im}[-1/\varepsilon(\omega')]}{(\omega')^2 - (\omega)^2}d\omega' \qquad (6.20)$$

where 'P' indicates the Cauchy principal value of the integral.[4] The procedure for performing a Kramers–Kronig analysis on experimental EELS spectra can be found in Egerton (1996) and Stöger-Pollach (2008) (see also Section 6.2.1 for a discussion of Cerenkov radiation artefacts). The dielectric function is fully determined from $\mathrm{Re}[1/\varepsilon(\omega)]$ and $\mathrm{Im}[-1/\varepsilon(\omega)]$; this enables the optoelectronic properties of a material to be analysed, since the complex refractive index $n(\omega)$ is given by $\sqrt{\varepsilon(\omega)}$.

Conversely, if the dielectric function, or equivalently complex refractive index, of a solid is known then it is possible to calculate the EELS spectrum using Eq. (6.19) or the more accurate Eq. (6.16). The dielectric function can also be expressed analytically using models with varying degrees of complexity or alternatively directly calculated from density functional theory (Keast, 2013). Here the Drude–Lorentz model (Jackson, 1998) for the dielectric function is used to explain two important features in the EELS spectrum, namely interband transitions and plasmons. Although not the most accurate, the Drude–Lorentz model is nevertheless sufficiently detailed to capture much of the underlying physics. The method relies on calculating the polarisation $\mathbf{P}$, which is related to the electric displacement through the equation $\mathbf{D}(\mathbf{r},\ t) = \varepsilon_o\mathbf{E}(\mathbf{r},\ t) + \mathbf{P}(\mathbf{r},\ t)$. For a spatially uniform electric field in the local approximation, $\mathbf{D}(\omega) = \varepsilon_o\varepsilon(\omega)\mathbf{E}(\omega)$ (Eq. (6.5b)), so that $\mathbf{P}(\omega) = \varepsilon_o[\varepsilon(\omega)\ -\ 1]\mathbf{E}(\omega)$. Furthermore, for a dielectric material, consisting of bound electrons within the constituent atoms or molecules, the polarisation is also equal to the electric dipole density, that is, $\mathbf{P}(t) = -ne\mathbf{x}(t)$, where n is the bound electron density and $\mathbf{x}$ is the dipole separation due to displacement of the bound electrons from their equilibrium positions by the electric field $\mathbf{E}$. The equation of motion for a bound electron is given by

$$m\frac{d^2\mathbf{x}}{dt^2} + m\gamma\frac{d\mathbf{x}}{dt} + K\mathbf{x} = -e\mathbf{E} \qquad (6.21)$$

---

[4]For a function $f(x)$, which has a singularity at c the Cauchy principal value of the integral $\int_a^b f(x)dx$ is defined as $\lim\limits_{\gamma\to 0}\left[\int_a^{c-\gamma} f(x)dx + \int_{c+\gamma}^b f(x)dx\right]$ where $a < c < b$ and γ is positive; Dass and Sharma, (1998).

where m is the electron mass, γ is a damping constant and K is a restoring force constant. The damping of the electron motion is due to energy dissipation from electron–electron collisions, while the restoring force is due to the Coulomb attraction between the bound electron and the atomic nucleus. Consider an electric field of the form $\mathbf{E}(t) = \mathbf{E}(\omega)\exp(-i\omega t)$, which consists of only a single frequency ω. A potential solution for the dipole separation is then $\mathbf{x}(t) = \mathbf{x}(\omega)\exp(-i\omega t)$ and substituting this in Eq. (6.21) gives

$$\mathbf{x}(\omega) = \frac{-e\mathbf{E}(\omega)}{-m\omega^2 - im\gamma\omega + K} \tag{6.22}$$

Substituting Eq. (6.22) in $\mathbf{P}(\omega) = -ne\mathbf{x}(\omega)$ and using the fact that $\mathbf{P}(\omega) = \varepsilon_0[\varepsilon(\omega) - 1]\mathbf{E}(\omega)$ gives

$$\varepsilon(\omega) = 1 - \frac{\omega_p^2}{(\omega^2 - \omega_0^2) + i\gamma\omega} \tag{6.23}$$

where $\omega_0 = \sqrt{(K/m)}$ is the resonant frequency of the bound electron and $\omega_p = \sqrt{(ne^2/m\varepsilon_0)}$ is the so-called *plasmon frequency*. The latter is equal to the oscillation frequency of a plasma where the polarisation of *free* electrons with volume density n completely cancels the applied electric field (Kittel, 2005). This follows directly from Eq. (6.23) since $\omega_0 = 0$ for free electrons, so that $\varepsilon(\omega_p) = 0$ assuming negligible damping (i.e. $\gamma \to 0$). Since $\mathbf{D} = \varepsilon_0\varepsilon(\omega)\mathbf{E} = \varepsilon_0\mathbf{E} + \mathbf{P} = 0$ the induced polarisation must oppose the applied electric field. The term 'plasmon' is used to describe the free electron oscillation.

In practice, the electrons in a solid consist of free (or nearly free valence) electrons as well as bound electrons of different resonant frequencies. It is customary to express the dielectric function of this ensemble as (Fox, 2008)

$$\varepsilon(\omega) = 1 - \omega_p^2 \sum_j \frac{f_j}{(\omega^2 - \omega_{0j}^2) + i\gamma_j\omega} \tag{6.24}$$

where ω_{0j} and γ_j are the resonant frequency and damping constant for the jth oscillator electron, which has an oscillator strength f_j. The oscillator strength phenomenologically takes into account the relative contribution of the different electrons and satisfies the sum rule $\Sigma_j f_j = 1$. For simplicity, assume that a solid consists of free electrons (oscillator strength f) and only one type of bound electron. From Eq. (6.24) the dielectric function is therefore

$$\varepsilon(\omega) = 1 - \frac{f\omega_p^2}{\omega^2 + i\gamma\omega} - \frac{(1-f)\omega_p^2}{(\omega^2 - \omega_0^2) + i\gamma\omega} \qquad (6.25)$$

where a constant damping term has been assumed for both free and bound electrons. Figure 6.1a shows the calculated energy loss spectrum, $Im[-1/\varepsilon(\omega)]$, for the above dielectric function assuming values of $\hbar\omega_p = 15$ eV, $\hbar\omega_0 = 10$ eV, $\hbar\gamma = 1$ eV (Keast, 2013) and $f = 0.9$. Two peaks are observed at 9.2 and 15.5 eV. By comparing with the values of $\hbar\omega_0$ and $\hbar\omega_p$ these can be attributed to an interband transition of the bound electrons and plasmon oscillation of the free electrons respectively. Note, however, that the peaks have shifted from their expected values owing to the presence of more than one energy loss mechanism in the dielectric function (Eq. (6.25); Egerton, 1996). The peak shift is small when $\hbar\omega_0$ and $\hbar\omega_p$ are far apart in energy, but becomes more apparent for similar energy values.

Figure 6.1b shows the real (ε_1) and imaginary (ε_2) parts of the dielectric function corresponding to Figure 6.1a. The peak in ε_2 and characteristic 'anomalous dispersion' in ε_1 (i.e. decrease in refractive index with increasing energy) are indicative of the bound electron resonance at $\hbar\omega_0$ (Fox, 2008). The plasmon is typically characterised by a zero crossing of ε_1 with positive slope, with the value of ε_2 being small close to the crossing (Egerton, 1996). This follows from Eq. (6.23) and the fact that for an undamped plasma the dielectric function is given by $\varepsilon(\omega) = 1 - (\omega_p/\omega)^2$. The dielectric function is therefore negative for $\omega < \omega_p$ and positive for $\omega > \omega_p$. The zero crossing of ε_1 is indicated by the vertical arrow in Figure 6.1b; note that it has shifted to a slightly higher energy than $\hbar\omega_p$ owing to the interband transition. By performing a Kramers–Kronig analysis and comparing the real and imaginary parts of the dielectric function it is possible to determine if a given feature in an energy loss spectrum is due to an interband transition or plasmon, although in some cases the presence of more than one transition may complicate the analysis (Egerton, 1996).

There are many applications in microanalysis that involve interband transitions and plasmons. Since these features appear in the high intensity, low loss region of the EELS spectrum, the acquisition time is comparatively short, making it ideal for beam damaging materials. Interband transitions can be used to identify specific chemical elements and quantify their composition. For interband transitions that overlap with the plasmon peak, Gass et al. (2004) have demonstrated that quantification can

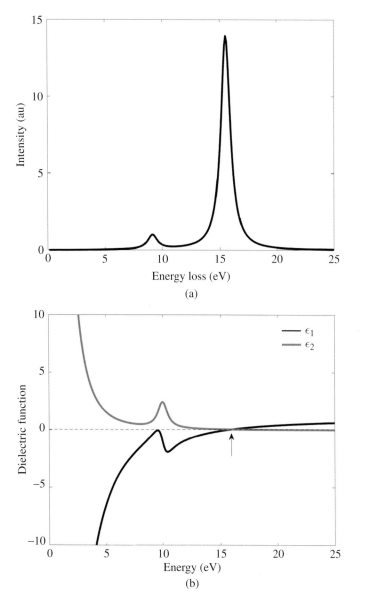

Figure 6.1 (a) Calculated energy loss function $\text{Im}[-1/\varepsilon(\omega)]$ for a solid containing bound electrons with 10 eV resonant energy. The plasmon energy of the free electrons (0.9 oscillator strength) is 15 eV, while the damping energy for all electrons is 1 eV. The dielectric function $\varepsilon(\omega)$ is modelled using Eq. (6.25). (b) Plots of the real (ε_1) and imaginary (ε_2) parts of the dielectric function as a function of energy for the solid in (a).

be improved by using $\mathrm{Im}[\varepsilon(\omega)]$ derived from a Kramers–Kronig analysis, since then the contribution from the plasmon is minimised. Plasmons are also useful in microanalysis since the plasmon energy, in its most general form, is given by $\omega_p^2 = (ne^2/m^*\varepsilon_0\varepsilon_\infty)$, where m^* is the effective electron mass and ε_∞ is a high frequency 'background' permittivity due to the ion cores in the material (Fox, 2008; compare with the expression for ω_p^2 in Eq. (6.23)). In favourable cases, such as sp-valence metals (e.g. Al–Mg alloys; Cundy $et\ al.$, 1968), alloying will cause a change in the valence electron density n, so that the shift in the plasmon peak position can be used to detect solute segregation at grain boundaries etc. Similarly, the plasmon energy has been used to estimate strain (Sanchez $et\ al.$, 2006) and effective mass (Gass $et\ al.$, 2006) of semiconductor heterostructures. It is also useful in the analysis of layered carbon-based materials, such as nanotubes (Kociak $et\ al.$ 2000) and graphene (Eberlain $et\ al.$, 2008).

6.1.2 Phonon Spectroscopy

Infrared (IR) spectroscopy is an optical technique widely used to characterise molecules, and is based on absorption of low energy, IR radiation via vibronic excitations in molecules (Harris and Bertolucci, 1989). Similar measurements can also be performed using electrons, in the form of reflection EELS from surfaces, where the energy resolution is $\sim$1 meV, although the spatial resolution is a less impressive $\sim$1 mm (Egerton, 2014). The recent development of stable, high energy resolution (i.e. $\sim$10 meV) monochromators for transmission electron microscopes (e.g. Krivanek $et\ al.$, 2009) has, however, made phonon spectroscopy at high spatial resolution a distinct possibility. Phonon losses in polar solids can be described using dielectric theory and is the topic of this section; for a quantum mechanical theory of phonon excitation see Forbes and Allen (2016).

Consider a crystal with two distinct atoms per primitive cell. An analysis of the phonon modes for such a crystal can be found in Kittel (2005) and is illustrated schematically in Figure 6.2a. With acoustic mode phonons the atom vibrations are in phase, while for optical modes the vibrations of the two distinct atoms are out of phase. Furthermore, for each mode both longitudinal and transverse phonon waves are possible, depending on whether the vibrations are parallel or perpendicular to the phonon wave vector. For each wave number (k) there corresponds a unique angular frequency (ω) or energy $\hbar\omega$ of the phonon. An ω–k diagram plots the phonon dispersion, and a schematic for acoustic and optical phonons is shown in Figure 6.2b (due to crystal

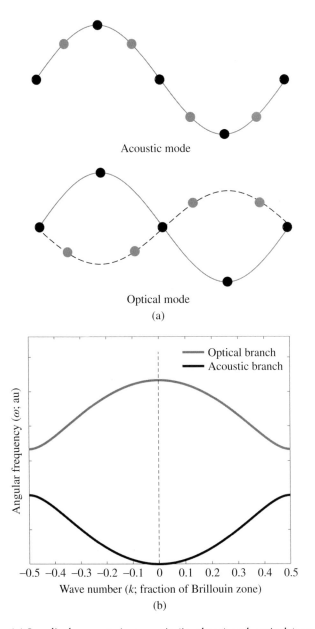

Acoustic mode

Optical mode

(a)

Optical branch
Acoustic branch

Angular frequency (ω; au)

−0.5 −0.4 −0.3 −0.2 −0.1   0   0.1   0.2   0.3   0.4   0.5
Wave number (k; fraction of Brillouin zone)

(b)

Figure 6.2 (a) Ion displacements in acoustic (in-phase) and optical (out-of-phase) transverse phonon modes. The positive and negative ions are depicted by black and grey circles. (b) ω–k dispersion plots for the acoustic and optical phonon modes. The wave number k is expressed as a fraction of the first Brillouin zone. (c) Real (ε_1) and imaginary (ε_2) parts of the dielectric function $\varepsilon(\omega)$ given by Eq. (6.31), with $\hbar\omega_T = 41.25$ meV, $\hbar\omega_L = 45.40$ meV and $\gamma = 10^{11}$ s^{-1}. The energy loss function Im$[-1/\varepsilon(\omega)]$ is shown in (d).

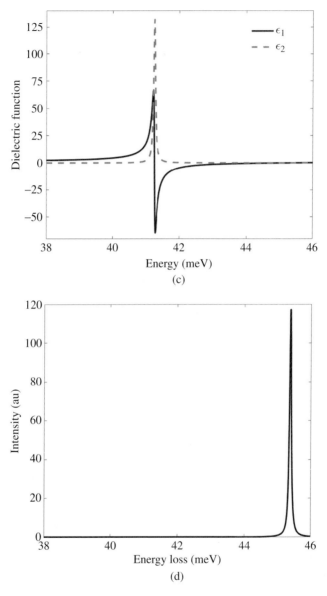

Figure 6.2 (*Continued*)

periodicity the dispersion needs only be plotted within the first Brillouin zone). Longitudinal and transverse modes are typically non-degenerate, but for simplicity this is not indicated in the diagram. At $k = 0$ the group velocity, expressed as $d\omega/dk$, of an acoustic phonon is equal to the speed of sound through the solid (Kittel, 2005), while the optical phonon shows no dispersion.

It is of interest to derive the dielectric function $\varepsilon(\omega)$ due to phonons, since this determines the energy loss function $\mathrm{Im}[-1/\varepsilon(\omega)]$ of a charged particle, such as an incident electron. Recall that $\varepsilon(\omega)$ characterises the response of a solid to a time varying electric field with angular frequency ω. It is clear that optical phonons in a *polar* solid can couple to such an electric field, since the oppositely charged ions are displaced in opposite directions by the electric field. However, there can be no coupling to acoustic phonons, since the displacements are all in phase (Figure 6.2a).

The equations of motion for the positive and negative ions in the presence of an electric field E can be written in a form similar to Eq. (6.21):

$$m^+ \frac{d^2 \mathbf{x}^+}{dt^2} + m^+ \gamma \frac{d\mathbf{x}^+}{dt} + K(\mathbf{x}^+ - \mathbf{x}^-) = q_{\text{charge}} \mathbf{E} \qquad (6.26)$$

$$m^- \frac{d^2 \mathbf{x}^-}{dt^2} + m^- \gamma \frac{d\mathbf{x}^-}{dt} + K(\mathbf{x}^- - \mathbf{x}^+) = -q_{\text{charge}} \mathbf{E} \qquad (6.27)$$

where t is time and $m^\pm$, $\mathbf{x}^\pm$ are the mass and displacement from equilibrium position for the positive and negative ions, which have charge $\pm q_{\text{charge}}$. The second term in each equation represents damping, while the third term is the restoring force; γ and K are the damping and restoring force constants. Since the phonon is a wave (Figure 6.2a) the displacements $\mathbf{x}^\pm$ are, strictly speaking, functions of position as well as time. The spatial dependence is, however, omitted from Eqs. (6.26) and (6.27) owing to the fact that the phonons that are excited by the incident electron typically have long wavelengths, so that $\mathbf{x}^\pm$ is slowly varying. From Section 5.1.1, the inelastic intensity follows a $1/[\theta^2 + \theta_E^2]$ Lorentzian distribution with respect to scattering angle θ, where θ_E is the characteristic scattering angle that is proportional to the energy loss. In semiconductors a typical phonon energy loss is ~40 meV, so that for 200 kV incident electrons the inelastic intensity distribution decreases to 50% of its maximum value within only 0.1 μrad, compared to 0.25 mrad for 100 eV energy loss from core level excitation. The magnitude of the inelastic scattering vector $\mathbf{q}$ is therefore small (see Eq. (5.19)), and since momentum is conserved this means that the wave number of the excited phonon is also small, that is, the phonon has long wavelength.

Dividing Eq. (6.26) by m^+, Eq. (6.27) by m^- and subtracting gives

$$\frac{d^2(\mathbf{x}^+ - \mathbf{x}^-)}{dt^2} + \gamma \frac{d(\mathbf{x}^+ - \mathbf{x}^-)}{dt} + \frac{K}{m_{\text{red}}}(\mathbf{x}^+ - \mathbf{x}^-) = \frac{q_{\text{charge}} \mathbf{E}}{m_{\text{red}}} \qquad (6.28)$$

where $1/m_{red} = (1/m^+ + 1/m^-)$ is the reciprocal of the reduced mass m_{red}. The quantity $q_{charge}(\mathbf{x}^+ - \mathbf{x}^-)$ is the polarisation of a single pair of oppositely charged ions. Note that the polarisation is zero when there is no displacement from the equilibrium lattice positions, that is, the crystal has no permanent electric dipole. Oscillation of permanent dipoles alter the dielectric function $\varepsilon(\omega)$ at much lower frequencies (e.g. microwaves; Kittel, 2005) compared to the IR frequencies for phonons and will therefore not be considered here. Following the methods outlined in Section 6.1.1 (see Eqs. (6.21)–(6.23)) the polarisation can be determined by solving Eq. (6.28) for a time varying electric field of the form $E(t) = E(\omega)\exp(-i\omega t)$. Relating polarisation to the dielectric function gives

$$\varepsilon(\omega) = 1 + \frac{(nq_{charge}^2/\varepsilon_o m_{red})}{\omega_T^2 - \omega^2 - i\gamma\omega} \tag{6.29}$$

where $\omega_T = K/m_{red}$. For a system with negligible damping (i.e. $\gamma \to 0$) there exists an angular frequency ω_L such that $\varepsilon(\omega_L) = 0$. From Eq. (6.29) the following relationship is obtained:

$$\omega_L^2 = \omega_T^2 + \left(\frac{nq_{charge}^2}{\varepsilon_o m_{red}}\right) \tag{6.30}$$

Equation (6.29) can therefore be expressed in the alternative form:

$$\varepsilon(\omega) = 1 + \frac{\omega_L^2 - \omega_T^2}{\omega_T^2 - \omega^2 - i\gamma\omega} \tag{6.31}$$

It can be shown that $\hbar\omega_T$ and $\hbar\omega_L$ correspond to the energies of transverse and longitudinal optical phonon modes respectively.[5] Figure 6.2c shows the real and imaginary parts of $\varepsilon(\omega)$ for parameters that are typical of phonons in semiconductor materials, that is, $\hbar\omega_T = 41.25$ meV, $\hbar\omega_L = 45.40$ meV and $\gamma = 10^{11}$ s^{-1} (Fox, 2008). The real part of $\varepsilon(\omega)$ shows a pole at ω_T, as well as a zero crossing close to ω_L. Between ω_T

---

[5]When no external charges are present Maxwell's equation for the electric displacement gives $\vec{\nabla} \cdot \mathbf{D} = \varepsilon_o \varepsilon(\omega)\vec{\nabla} \cdot \mathbf{E} = \varepsilon_o \vec{\nabla} \cdot \mathbf{E} + \vec{\nabla} \cdot \mathbf{P} = 0$. Assuming no damping in Eq. (6.31) it follows that $\varepsilon(\omega_T) = \infty$, so that $\vec{\nabla} \cdot \mathbf{E}$ and hence $\vec{\nabla} \cdot \mathbf{P}$ are both zero. If the polarisation has the form $\mathbf{P}(\mathbf{r}, t) = \mathbf{P}_0\exp(2\pi i\mathbf{k}\cdot\mathbf{r} - i\omega t)$, where $\mathbf{k}$ is the wave vector and $\mathbf{r}$ a position vector, this means that the polarisation is perpendicular to $\mathbf{k}$, that is, the ion displacements are transverse. Similarly, $\varepsilon(\omega_L) = 0$ implies $\varepsilon_o\vec{\nabla} \cdot \mathbf{E} = -\vec{\nabla} \cdot \mathbf{P}$. This is valid when the polarisation is parallel to $\mathbf{k}$, that is, the ion displacements are longitudinal.

and ω_L the real part of the dielectric function is negative; in this region, the semiconductor is highly reflective to electromagnetic radiation (Fox, 2008), since the complex refractive index $\sqrt{\varepsilon(\omega)}$ is pure imaginary in the limit of no damping, and hence radiation cannot penetrate deep into the material. The imaginary part of $\varepsilon(\omega)$ shows a single peak close to ω_T. Figure 6.2d shows the energy loss function $\text{Im}[-1/\varepsilon(\omega)]$ plotted on the same energy scale as Figure 6.2c. The phonon loss peak is at 45.37 meV, close to the value of $\hbar\omega_L$. This behaviour is readily explained by examining the functional form of $\varepsilon(\omega)$, as expressed by Eq. (6.31). Phonon spectroscopy of MgO by Lagos et al. (2017) has confirmed the longitudinal optical mode to be the dominant contribution at small inelastic scatting vector, although the relatively large spectrometer collection angle in that experiment meant that the longitudinal acoustic signal was also observed. Furthermore, surface phonon modes were detected at the surfaces of MgO smoke cubes, which are excited at the expense of the bulk phonon modes, that is, the 'Begrenzung' effect (see Section 6.1.3).

An important practical consideration is the cross-section for phonon excitation compared to the intensity of the zero loss peak *tail* at the phonon energy. Despite a narrow full width at half maximum the tails of the zero loss peak can extend to relatively large energy losses due to aberrations in the monochromator and/or spectrometer, as well as scattering of the high energy electrons within the spectrometer (Egerton, 2014). Scattering cross-sections have been calculated by Rez (2014) for the stretching mode in diatomic molecules. The energy of the stretching mode increases as the reduced mass of the constituent atoms in the molecule decreases, such that for molecules containing hydrogen the phonon energy is typically several 100 meV. Significantly, the cross-section for this type of molecular scattering can be greater than core loss scattering from a carbon K-edge (Rez, 2014). Together with a comparatively weak zero loss peak intensity at energy losses >100 meV, this means that detecting hydrogen, at least indirectly, is possible (Krivanek et al., 2014), although beam damage of the specimen can be an issue.

Dwyer (2014) has carried out image simulations on diatomic molecules of H_2 and CO to determine the spatial resolution achievable with phonon inelastic scattering. Despite increased delocalisation at low energy losses (see Section 6.1.3), atomic resolution is still possible provided the spectrometer entrance aperture is large enough, such as, for example, an 80 mrad collection semi-angle for a 30 mrad, 100 kV electron probe. Using classical physics this can be explained by the fact that electrons inelastically scattered out to large angles have small impact parameter and are therefore more sensitive to the intense, rapidly varying part of the molecular potential, which is more localised (see also Section 5.1.4 for the effect of collection angle on coherence of inelastic images). Similar conclusions are also valid for the CO molecule, although it has a permanent electric dipole with a long range electrostatic potential field. The dipole could therefore potentially degrade the spatial resolution, although its contribution to inelastic scattering turns out to be sufficiently weak, such that atomic resolution can still be obtained for the largest collection apertures (Dwyer, 2014).

6.1.3 *Interface and Surface Contributions*

In reality transmission electron microscopy requires thin specimens, so that the bulk energy loss formula (Eq. (6.16)) is only an approximation. The change in the dielectric properties outside the free surfaces of the specimen modifies the electric field of the incident electron and consequently its energy loss and stopping power. A similar situation arises with internal interfaces, such as, for example, a multi-layer thin-film structure typical of most electronic devices. In this section, the energy loss in the presence of free surfaces and internal interfaces is discussed.

Assume a bulk solid consisting of two materials 'A' and 'B', which have dielectric functions ε_A and ε_B respectively (Figure 6.3a). The incident electron travels close to, and parallel to, the internal interface between the two materials. The solid is assumed to be sufficiently large such that the effect of free surfaces can be ignored. The boundary conditions for the electric field of the incident electron are therefore determined only by the internal interface. This example serves to highlight the general methodology used to analyse energy loss in the presence of interfaces and/or free surfaces. For simplicity, the non-retarded limit is once again assumed. The results will be used to discuss several physical phenomena, such as the 'Begrenzung' effect (Howie, 1999) and delocalisation. Energy loss in a uniform, thin-film material containing top and bottom free surfaces can be calculated in a similar manner to that outlined here, although only the final result will be presented, since no new concepts are involved.

The energy loss is governed by the electric field of the incident electron, which is travelling along the z-axis in material 'B', with impact parameter x_o from the interface (Figure 6.3a). In Section 6.1.1, the electric field was determined by Fourier transforming Eq. (6.1) in both the spatial and time domains. Here, however, only a partial Fourier transform is carried out, with the **r** position vector component normal to the interface (i.e. along

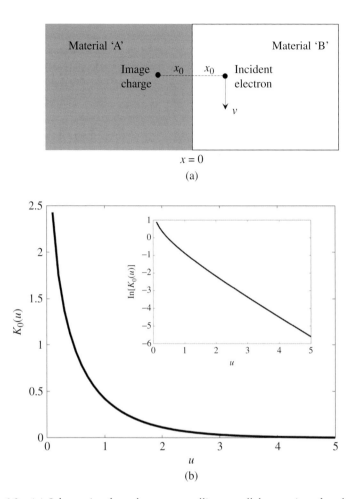

Material 'A'   Material 'B'

Image x_0 x_0 Incident
charge   electron

                                v

$x = 0$

(a)

(b)

Figure 6.3 (a) Schematic of an electron travelling parallel to an interface between two materials 'A' and 'B'. The interface is arbitrarily positioned at $x = 0$ and the impact parameter of the electron, which is travelling in material 'B' with speed v, is x_0. An image charge is 'induced' on the opposite side of the interface. (b) Plot of the modified Bessel function $K_0(u)$ with respect to the dimensionless variable u. The inset shows $\ln[K_0(u)]$ versus u. (c) Shows the measured EELS signal for different energy losses of a 60 kV electron probe as a function of distance from a graphene sheet edge. The EELS signal has been normalised for direct comparison. The normalised annular dark field (ADF) signal is also shown superimposed. From Zhou *et al.* (2012). Reproduced with permission; copyright Elsevier Science. (d) Shows the calculated energy loss probability per unit path length for a 200 kV electron incident 2 nm either side of a vacuum–specimen interface. The specimen dielectric function is the same as that in Figure 6.1 and the EELS collection semi-angle is 5 mrad. For comparison the energy loss probability per unit path length for the bulk specimen is also shown superimposed.

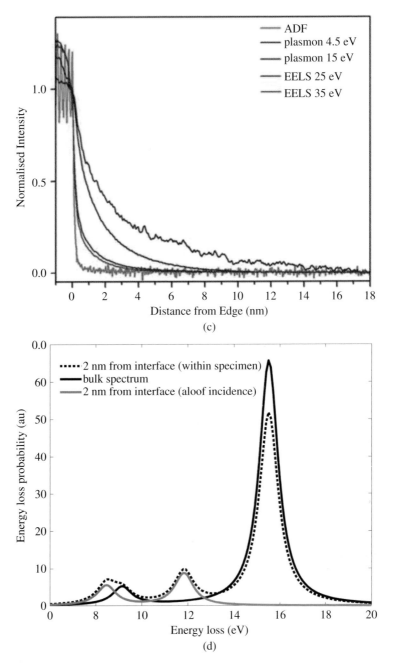

Figure 6.3 (*Continued*)

the x-axis) kept untransformed. As will be shown later, this enables an analytical solution for the potential that satisfies the boundary conditions at the interface. Fourier transforming Eq. (6.1) in the time domain and invoking the local approximation (i.e. Eq. (6.5)) gives

$$\varepsilon_o \varepsilon(\omega) \vec{\nabla} \cdot \mathbf{E}(\mathbf{r}, \omega) = -\varepsilon_o \varepsilon(\omega) \nabla^2 \phi(\mathbf{r}, \omega) = \rho_f(\mathbf{r}, \omega) \qquad (6.32)$$

where use has been made of $\mathbf{E}(\mathbf{r}, t) = -\vec{\nabla}\phi(\mathbf{r}, t)$ or equivalently $\mathbf{E}(\mathbf{r}, \omega) = -\vec{\nabla}\phi(\mathbf{r}, \omega)$. Equation (6.32) is essentially Poisson's equation. Fourier transforming with respect to components of $\mathbf{r}$ along y and z-axes,

$$\varepsilon_o \varepsilon(\omega) \left[\frac{\partial^2}{\partial x^2} - 4\pi^2 q_{yz}^2 \right] \phi(x, \mathbf{q}_{yz}, \omega) = -\rho_f(x, \mathbf{q}_{yz}, \omega) \qquad (6.33)$$

where $\mathbf{q}_{yz}$ is the reciprocal vector component in the yz-plane (i.e. parallel to the interface, Figure 6.3a). The charge density of the incident electron is

$$\rho_f(\mathbf{r}, t) = -e\delta(x - x_o)\delta(y)\delta(z - vt) \qquad (6.34)$$

so that from Eq. (6.6) its partial Fourier transform is given by

$$\rho_f(x, \mathbf{q}_{yz}, \omega) = -e\delta(x - x_o)\delta(2\pi q_z v - \omega) \qquad (6.35)$$

Substituting in Eq. (6.33),

$$\left[\frac{\partial^2}{\partial x^2} - 4\pi^2 q_{yz}^2 \right] \phi_B(x, \mathbf{q}_{yz}, \omega) = \frac{e\delta(x - x_o)\delta(2\pi q_z v - \omega)}{\varepsilon_o \varepsilon_B(\omega)} \qquad (6.36)$$

where the subscript 'B' has been added to indicate that the relevant parameters are evaluated in material 'B'. Since there is no free charge density in material 'A' it follows that

$$\left[\frac{\partial^2}{\partial x^2} - 4\pi^2 q_{yz}^2 \right] \phi_A(x, \mathbf{q}_{yz}, \omega) = 0 \qquad (6.37)$$

Equations (6.36) and (6.37) determine the electric potential in the two materials and can be solved using standard methods for linear differential equations (e.g. Dass and Sharma, 1998). In particular, solutions for the latter are of the form $\exp(\pm 2\pi q_{yz} x)$. The potential must asymptotically decay to zero as $x \to \pm\infty$. This rules out solutions of the form $\exp(-2\pi q_{yz}x)$, since material 'A' extends over negative values of x (Figure 6.3a). Therefore,

$$\phi_A(x, \mathbf{q}_{yz}, \omega) = a \exp(2\pi q_{yz}x) \tag{6.38}$$

where 'a' is a constant to be determined. The solution for Eq. (6.36) is

$$\phi_B(x, \mathbf{q}_{yz}, \omega) = b \exp(-2\pi q_{yz}x) - \frac{e}{4\pi\varepsilon_o\varepsilon_B(\omega)q_{yz}}$$
$$\times \exp(-2\pi q_{yz}|x - x_o|)\delta(2\pi q_z v - \omega) \tag{6.39}$$

The first term is the homogeneous solution, obtained by setting the right-hand side of Eq. (6.36) to zero (solutions of the form $\exp(2\pi q_{yz}x)$ are omitted due to divergence of the potential as $x \to \infty$). 'b' is a constant to be determined. The second term is the inhomogeneous solution; the | | symbol represents the absolute value of the term within it. The constants 'a' and 'b' can be determined from the boundary conditions at the interface, namely that the potential must be continuous and that the displacement field components normal to the interface are equal, that is,

$$[\phi_A(x, \mathbf{q}_{yz}, \omega)]_{x=0} = [\phi_B(x, \mathbf{q}_{yz}, \omega)]_{x=0}$$
$$\varepsilon_A(\omega)\left[\frac{\partial \phi_A(x, \mathbf{q}_{yz}, \omega)}{\partial x}\right]_{x=0} = \varepsilon_B(\omega)\left[\frac{\partial \phi_B(x, \mathbf{q}_{yz}, \omega)}{\partial x}\right]_{x=0} \tag{6.40}$$

Note that the first condition guarantees identical electric field components parallel to the interface. The resulting expressions for 'a' and 'b' are

$$a = -\left(\frac{2\varepsilon_B}{\varepsilon_A + \varepsilon_B}\right)\frac{e}{4\pi\varepsilon_o\varepsilon_B q_{yz}}\exp(-2\pi q_{yz}x_o)\delta(2\pi q_z v - \omega)$$
$$b = -\left(\frac{\varepsilon_B - \varepsilon_A}{\varepsilon_A + \varepsilon_B}\right)\frac{e}{4\pi\varepsilon_o\varepsilon_B q_{yz}}\exp(-2\pi q_{yz}x_o)\delta(2\pi q_z v - \omega) \tag{6.41}$$

The stopping power dW/dz from Eq. (6.11) is given by

$$\frac{dW}{dz} = -e\frac{\partial \phi_B(\mathbf{r}, t)}{\partial z}\bigg|_{x=x_o, y=0, z=vt}$$
$$= -2\pi i e \iint \phi_B(x, \mathbf{q}_{yz}, \omega)q_z \exp[2\pi i \mathbf{q}_{yz} \cdot \mathbf{r}_{yz} - i\omega t]d\mathbf{q}_{yz}\, d\omega\bigg|_{x=x_o, y=0, z=vt} \tag{6.42}$$

where r_{yz} is the position vector component in the yz-plane (Figure 6.3a). First consider the stopping power dW_1/dz due to the first term in Eq. (6.39), that is, the homogeneous solution part of ϕ_B. Substituting in Eq. (6.42) and after some standard algebra as described in Section 6.1.1, it follows that

$$
\frac{dW_1}{dz} = \frac{e^2 i}{8\pi^2 v^2 \varepsilon_o} \int_{-\infty}^{\infty} \int_{-\infty}^{\infty} \omega \frac{\exp[-4\pi(q_y^2 + (\omega/2\pi v)^2)^{1/2} x_o]}{[q_y^2 + (\omega/2\pi v)^2]^{1/2}}
$$
$$
\times \left[\frac{\varepsilon_B - \varepsilon_A}{\varepsilon_B(\varepsilon_B + \varepsilon_A)} \right] d\omega dq_y
$$
$$
= \frac{e^2}{2\pi^2 v^2 \varepsilon_o} \int_0^{\infty} \omega \text{Im} \left[-\frac{\varepsilon_B - \varepsilon_A}{\varepsilon_B(\varepsilon_B + \varepsilon_A)} \right]
$$
$$
\times \left(\int_0^{\infty} \frac{\exp[-4\pi(q_y^2 + (\omega/2\pi v)^2)^{1/2} x_o]}{[q_y^2 + (\omega/2\pi v)^2]^{1/2}} dq_y \right) d\omega \qquad (6.43)
$$

where the property $\varepsilon(-\omega) = \varepsilon^*(\omega)$ has been made use of; note also the change in the integration limits. The integral over q_y can be simplified by noting that it is equivalent to

$$
\int_0^{\infty} \frac{\exp[-(2\omega x_o/v)(1 + (vu/\omega)^2)^{1/2}]}{[1 + (vu/\omega)^2]^{1/2}} d(vu/\omega)
$$
$$
= \int_0^{\infty} \exp[-(2\omega x_o/v) \cosh t] dt \qquad (6.44)
$$

where $u = 2\pi q_y$ and $\cosh t = [1 + (vu/\omega)^2]^{1/2}$. The integral over t is the definition of the modified Bessel function of the second kind $K_0(2\omega x_o/v)$ (Gradshteyn and Ryzhik, 1980). The energy loss probability per unit path length due to the first term in ϕ_B is therefore

$$
\frac{dP_1}{d\omega} = \frac{e^2}{2\pi^2 v^2 \hbar \varepsilon_o} \text{Im} \left[-\frac{\varepsilon_B - \varepsilon_A}{\varepsilon_B(\varepsilon_B + \varepsilon_A)} \right] K_0(2\omega x_o/v) \qquad (6.45)
$$

Similarly, it can be shown that the stopping power dW_2/dz due to the second term in ϕ_B (Eq. (6.39)) is

$$
\frac{dW_2}{dz} = \frac{e^2}{2\pi^2 v^2 \varepsilon_o} \int_0^{\infty} \omega \text{Im} \left(\frac{-1}{\varepsilon_B} \right) \left(\int_0^{\infty} \frac{dq_y}{[q_y^2 + (\omega/2\pi v)^2]^{1/2}} \right) d\omega \qquad (6.46)
$$

The energy loss probability per unit path length is therefore given by

$$\frac{dP_2}{d\omega} = \frac{e^2}{2\pi^2 v^2 \hbar \varepsilon_o} \mathrm{Im} \left(\frac{-1}{\varepsilon_B}\right) \int_0^\infty \frac{dq_y}{[q_y^2 + (\omega/2\pi v)^2]^{1/2}} \tag{6.47}$$

The integration over q_y is divergent. In practice, however, energy loss is measured only up to a finite scattering vector magnitude q_c; assuming a large value for q_c, Eq. (6.47) simplifies to

$$\frac{dP_2}{d\omega} \approx \frac{e^2}{2\pi^2 v^2 \hbar \varepsilon_o} \mathrm{Im} \left[-\frac{1}{\varepsilon_B}\right] \ln(2\pi q_c v/\omega) \tag{6.48}$$

Equation (6.48) is identical to the bulk energy loss formula, Eq. (6.19). The total energy loss probability per unit path length is the sum of Eqs. (6.45) and (6.48), that is,

$$\boxed{\begin{aligned} \frac{dP}{d\omega} \approx \frac{e^2}{2\pi^2 v^2 \hbar \varepsilon_o} & \left\{ \mathrm{Im} \left[\frac{-1}{\varepsilon_B}\right] [\ln(2\pi q_c v/\omega) - K_0(2\omega x_o/v)] \right. \\ & \left. + \mathrm{Im} \left[\frac{-2}{\varepsilon_A + \varepsilon_B}\right] K_0(2\omega x_o/v) \right\} \end{aligned}} \tag{6.49}$$

For an electron impact parameter x_o from the interface the bulk energy loss is reduced by an amount proportional to $K_0(2\omega x_o/v)$, while a second energy loss term of the form $\mathrm{Im}[-2/(\varepsilon_A + \varepsilon_B)]$ appears in Eq. (6.49). This is known as the 'Begrenzung' effect (Howie, 1999) and the physical origin for this 'interface loss' term is illustrated in Figure 6.3a. The electric field in material 'B' is equivalent to the sum of two terms: (i) the electric field of the incident electron as if it were in an infinitely large volume of material 'B' and (ii) the electric field of an image charge of magnitude $[(\varepsilon_A - \varepsilon_B)/(\varepsilon_A + \varepsilon_B)]e$ located at distance x_o from the interface in material 'A' (Jackson, 1998). The interface can be shown to be polarised and therefore has a net (bound) surface charge density (Jackson, 1998). The interface loss term is due to the fact that the incident electron is moving under the influence of a (fictitious) image charge or equivalently a (real) interface charge density.

The modified Bessel function $K_0(u)$ appearing in Eq. (6.49), where 'u' is a dimensionless parameter, is plotted in Figure 6.3b. $K_0(u)$ diverges as $u \to 0$, but very rapidly approaches an exponential decay for larger values of u (Jackson, 1998), as confirmed by the logarithm plot in the figure inset. The long tails of the exponential decay means that the incident electron can excite the interface loss term from impact parameters that are relatively large compared to atomic dimensions. For example, $u = 1$ corresponds to $\sim 10\%$ of the $\ln(2\pi q_c v/\omega)$ term (Eq. (6.49)), assuming a 5 mrad collection semi-angle for 200 kV incident electrons that have excited a 15 eV energy plasmon. The impact parameter under these conditions is 5 nm. Clearly, the smaller the energy loss the larger the delocalisation of the inelastic scattering. A standard test for delocalisation involves scanning a focused electron probe from the vacuum region towards the edge of the specimen, while simultaneously acquiring energy loss spectra and the annular dark field (ADF) intensity at each scan position. The latter is used to estimate the edge of the specimen, assuming that the detector inner angle is sufficiently large to produce an incoherent image. Delocalisation can be estimated by plotting the intensity of the energy loss signal as a function of position and comparing the resulting profile with the ADF signal. Results for a monolayer graphene specimen (Zhou *et al.*, 2012) are shown in Figure 6.3c and confirm the expected trend of more delocalisation for smaller energy loss. It is also interesting that the measured delocalisation is of the same order of magnitude as that predicted by Eq. (6.49), despite the fact that the theoretical result was derived for an 'infinitely' thick specimen.

Consider the interface loss function $\mathrm{Im}[-2/(\varepsilon_A + \varepsilon_B)]$. Comparing with the bulk loss formula $\mathrm{Im}[-1/\varepsilon]$ an equivalent dielectric function ε' can be defined such that $\varepsilon' = (\varepsilon_A + \varepsilon_B)/2$. For an ideal vacuum–specimen interface (i.e. no surface layers) and a specimen dielectric function given by Eq. (6.25) it follows that

$$\varepsilon'(\omega) = 1 - \frac{f\omega_s^2}{\omega^2 + i\gamma\omega} - \frac{(1-f)\omega_s^2}{(\omega^2 - \omega_0^2) + i\gamma\omega} \;\; ; \;\; \omega_s = \frac{\omega_p}{\sqrt{2}} \qquad (6.50)$$

ε' has the same form as the bulk specimen dielectric function, with an interband resonance at ω_0 and modified plasmon frequency ω_s. It will be shown that $\hbar\omega_s$ is the energy of a surface plasmon at the vacuum–specimen interface. The interface therefore affects the long-range collective plasmon oscillations, but not the short-range single electron excitations at $\hbar\omega_0$.

Figure 6.3d shows the energy loss probability per unit path length, calculated from Eq. (6.49), for a 200 kV incident electron moving parallel to a vacuum–specimen interface, with 2 nm impact parameter on either side of the interface. The specimen dielectric function is given by Eq. (6.25) where the interband transition energy $\hbar\omega_0 = 10.0$ eV, the bulk plasmon energy $\hbar\omega_p = 15.0$ eV and the oscillator strength $f = 0.9$. A surface plasmon is therefore expected at 10.6 eV ($=15/\sqrt{2}$) energy and corresponds to the additional peak at 11.8 eV in Figure 6.3d (the energy shift is due to the interband transition; Section 6.1.1). The energy loss probability per unit path length for the bulk material (Eq. (6.19)) is also shown superimposed in Figure 6.3d. The Begrenzung effect is particularly noticeable for the bulk plasmon peak. Furthermore, the bulk plasmon peak is absent for the electron travelling in vacuum, although surface plasmon excitation is evident. Equation (6.49) and its predictions have been verified for MgO smoke cubes (Walls and Howie, 1989). Krivanek *et al.* (2014) have used the 'aloof' incidence geometry, where the electron probe is placed outside the specimen, to measure vibrational EELS modes in beam-sensitive materials containing hydrogen. These measurements rely on strong delocalisation of the low energy loss phonon signal; with the electron beam away from the specimen, beam damage is avoided. Furthermore, Maclean *et al.* (2001) have tested the theory for the interface between a Mg_2Si precipitate and Al-rich matrix in a 6061 aluminium alloy. It is also possible to extend the theory to multi-layer systems with multiple internal interfaces (Bolton and Chen, 1995). Couillard *et al.* (2007, 2008) have applied this theory to gate dielectric stacks found in field effect transistors.

In a thin-foil specimen the free surfaces result in similar 'artefacts' to the internal interface discussed previously. Derivation of the energy loss for a thin foil proceeds along similar lines, that is, Poisson's equation is Fourier transformed with respect to the x,y-axes in the plane of the specimen only and the unique solution is determined by matching the boundary conditions at the free surface (Ritchie, 1957; Kroger, 1970). The (non-retarded) result for a specimen of thickness t is (Ritchie, 1957)

$$
\frac{\partial^2 P}{\partial q_\perp \partial\omega} = \frac{e^2}{4\pi^3 v^2 \hbar\varepsilon_o} \left\{ \frac{t}{[q_\perp^2 + (\omega/2\pi v)^2]} \operatorname{Im}\left(\frac{-1}{\varepsilon}\right) + \frac{2q_\perp}{[q_\perp^2 + (\omega/2\pi v)^2]^2} \right.
$$

$$
\left. \times \operatorname{Im}\left[-\frac{1-\varepsilon}{\varepsilon} \cdot \frac{2(\varepsilon-1)\cos(\omega t/v) + (\varepsilon-1)\exp(-q_\perp t) + (1-\varepsilon^2)\exp(q_\perp t)}{(\varepsilon-1)^2 \exp(-q_\perp t) - (\varepsilon+1)^2 \exp(q_\perp t)}\right]\right\}
$$
$$(6.51)$$

Note that P here is an energy loss probability, rather than an energy loss probability per unit path length as in Eqs. (6.16) and (6.49). The first term is the bulk energy loss for the electron traversing the specimen, while the second term is due to surface loss contributions. First, consider the form of Eq. (6.51) in the limit of large t:

$$\frac{\partial^2 P}{\partial q_\perp \partial \omega} = \frac{e^2}{4\pi^3 v^2 \hbar \varepsilon_o} \left\{ \frac{t}{[q_\perp^2 + (\omega/2\pi v)^2]} \text{Im}\left(\frac{-1}{\varepsilon}\right) + \frac{2q_\perp}{[q_\perp^2 + (\omega/2\pi v)^2]^2} \right.$$
$$\left. \times \text{Im}\left[\frac{(1-\varepsilon)^2}{\varepsilon(1+\varepsilon)}\right]\right\}$$

$$= \frac{e^2}{4\pi^3 v^2 \hbar \varepsilon_o (q_\perp^2 + (\omega/2\pi v)^2)}$$
$$\times \left[\left(t - \frac{2q_\perp}{q_\perp^2 + (\omega/2\pi v)^2}\right)\text{Im}\left(\frac{-1}{\varepsilon}\right) + \frac{8q_\perp}{q_\perp^2 + (\omega/2\pi v)^2}\text{Im}\left(\frac{-1}{1+\varepsilon}\right)\right]$$

$$(6.52)$$

where the expansion $[(1-\varepsilon)^2/\varepsilon(1+\varepsilon)] = 1 + (1/\varepsilon) - [4/(1+\varepsilon)]$, has been used to simplify Eq. (6.52). For a free electron metal with dielectric function $\varepsilon(\omega) = 1 - \omega_p^2/(\omega^2 + i\gamma\omega)$, the surface loss term $\text{Im}[-1/(1+\varepsilon)]$ has a maximum value or 'resonance' when $\omega = \omega_p/\sqrt{2}$ and negligible damping (i.e. $\gamma \to 0$). This corresponds to the frequency of a surface plasmon, which is due to oscillations in the charge density at the specimen free surfaces (Rocca, 1995). A schematic of the charge density and electric field lines for the surface plasmon at the interface between a semi-infinite solid and vacuum is illustrated in Figure 6.4a. The Begrenzung effect is also evident in Eq. (6.52), that is, the surface loss contribution is accompanied by a diminishing of the bulk loss from its expected value for a specimen of thickness t. Furthermore, from Eqs. (6.51) and (6.52) the surface loss scattering distribution has the form $q_\perp/[q_\perp^2 + (\omega/2\pi v)^2]^2$, which is narrower than the Lorentzian distribution for bulk loss scattering (Eq. (6.16)). The surface loss contribution can therefore be reduced by measuring the EELS spectrum using a slightly off-axis detector (Egerton, 1996). Mkhoyan et al. (2007) have proposed that surface loss contributions can be identified by examining EELS spectra acquired at different specimen thicknesses, since only the bulk loss contributions increase with thickness.

For thin specimens, Eq. (6.52) is no longer valid and the full expression, that is, Eq. (6.51), must be used instead. The resonance condition for the surface loss is then given by

$$|(\varepsilon - 1)^2 \exp(-q_\perp t) - (\varepsilon + 1)^2 \exp(q_\perp t)| = 0 \qquad (6.53)$$

where '| |' denotes the amplitude of a complex number. For a free electron metal with negligible damping, this leads to the condition

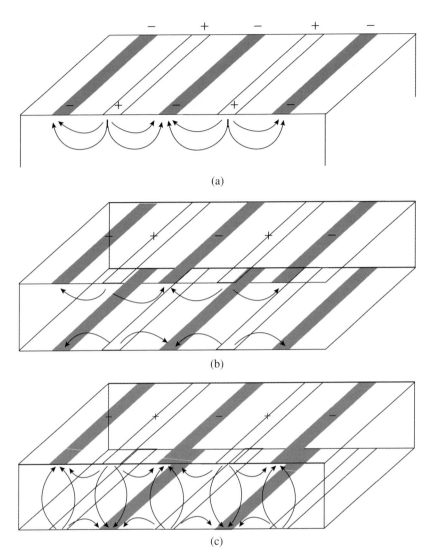

Figure 6.4 (a) Charge density and electric field lines for the surface plasmon at the interface between a semi-infinite solid and vacuum. Regions of negative charge are displayed as filled strips while positive charge regions are displayed as open strips. Only the electric field lines within the solid are shown. (b) and (c) are the equivalent diagrams for the symmetric and anti-symmetric surface plasmon modes in a thin-foil respectively. (d) Intensity of 1.75, 2.70 and 3.20 eV surface plasmon modes as excited by a 100 kV STEM electron probe scanned over a triangular shaped silver nano-prism. The specimen boundary is indicated in the figure. From Nelayah *et al.* (2007). Reproduced with permission; copyright Nature Publishing Group.

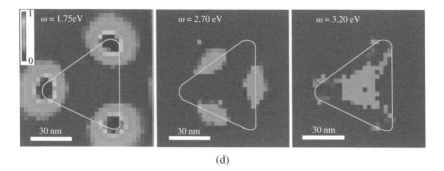

(d)

Figure 6.4   (*Continued*)

$(\varepsilon - 1/\varepsilon + 1) = \pm \exp(q_\perp t)$ or equivalently $\omega_s = (\omega_p/\sqrt{2})[1 \mp \exp(-q_\perp t)]^{1/2}$. The two surface plasmon frequencies ω_s are due to the symmetric and anti-symmetric distribution of charge at the two opposing free surfaces as illustrated in Figure 6.4b and c respectively (Ferrell, 1958). The symmetric mode has the higher frequency and hence larger energy. The individual frequencies are a function of the specimen thickness t, although for thick specimens the interaction between the two surfaces is negligible, so that they converge to the asymptotic value $(\omega_p/\sqrt{2})$. A similar dependence of ω_s with respect to momentum transfer $q_\perp$ is also predicted and has been confirmed experimentally (Vincent and Silcox, 1973; surface plasmons with large wave number $q_\perp$ have rapidly oscillating charge densities, so that coupling between the two free surfaces is suppressed).

The surface plasmon modes can be determined analytically for relatively simple specimen geometries, such as, spheres and cylinders (Wang, 1996). The high spatial resolution in an electron microscope enables excitation of these modes to be mapped as a function of position, even for nano-sized objects. As an example, Figure 6.4d shows the intensity of three surface plasmon modes at 1.75, 2.70 and 3.20 eV respectively in a triangular shaped Ag nano-prism (Nelayah *et al.*, 2007). EELS spectrum imaging was used to map the energy loss, and hence excitation of a given plasmon mode, as a function of position. Similar results can also be obtained using energy filtered TEM (EFTEM) with the advantage that the entire region of interest is mapped simultaneously, albeit at a single energy loss window (Schaffer *et al.*, 2009). It is also useful to combine EELS with CL, since the former will detect all plasmon modes while the latter is only sensitive to radiative modes.[6] There are also

---

[6] A key criterion for a surface plasmon to be radiative is that its phase velocity must be larger than the speed of light in vacuum.

strong parallels between EELS and CL and optically measured absorption and scattering cross-sections respectively (Losquin *et al.*, 2015). For the Ag nano-prism in Figure 6.4d, CL has been used to demonstrate that only the dipolar mode at 1.75 eV energy loss is radiative (Losquin *et al.*, 2015).

6.2 RADIATIVE PHENOMENA

6.2.1 *Cerenkov Radiation and Band Gap Measurement*

The energy lost by the incident electron is transferred to the electrons in the solid (collective plasmon oscillations, interband transitions, etc.) as well as any electromagnetic radiation emitted into the far field. For the latter, the instantaneous power (P_{rad}) escaping a closed surface S' surrounding the incident electron is given by (Griffiths, 1999)

$$P_{rad} = \frac{1}{\mu_o} \oint_S (\mathbf{E} \times \mathbf{B}) \cdot d\mathbf{S}' \tag{6.54}$$

where μ_o is the permeability of free space and $d\mathbf{S}'$ is an infinitesimal (vector) area element on the surface S', which is assumed to be embedded in a non-magnetic medium. For an incident electron within a bulk solid two regimes can be identified; the first is when the electron is moving slower than the phase velocity of light in the medium. Then, the $\mathbf{E}$ and $\mathbf{B}$-fields can be shown to be short range, so that no radiation is emitted (Jackson, 1998). Indeed, at these relatively slow speeds the electrons in the solid have sufficient time to respond to the electromagnetic field of the incident electron and all energy transfer occurs via this process. On the other hand, when the incident electron exceeds the phase velocity of light the limited response time means that some energy transfer occurs via emission of the so-called *Cerenkov* radiation. The criterion for its generation is expressed as

$$v > \frac{c}{n(\omega)} \tag{6.55}$$

where v is the speed of the incident electron, c is the speed of light in vacuum and $n(\omega)$ is the refractive index; the right-hand side of Eq. (6.55) is equal to the phase velocity of light in the solid.

Cerenkov radiation is emitted in a cone about the forward direction of the incident electron. This can be explained with the aid of Figure 6.5. Initially, the incident electron is at the origin O and emits Cerenkov radiation at an angle θ_c. After a time t the incident electron has travelled

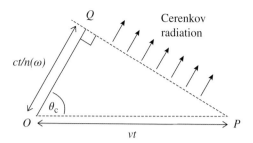

Figure 6.5 Schematic of Cerenkov radiation emitted by an electron travelling faster than the speed of light in a given medium. From the origin O the electron moves a distance vt to the new point P, where v is the electron speed and t is the time elapsed. Cerenkov radiation is emitted at an angle θ_c. Radiation emitted from the origin O travels a distance $ct/n(\omega)$, where c is the speed of light in vacuum and $n(\omega)$ is the refractive index.

a distance vt to the new position 'P', continuously emitting Cerenkov radiation along its path. It is assumed that the incident electron is of sufficiently high energy, so that its speed is largely unaffected by Cerenkov losses. From the triangle OPQ it follows that

$$\cos \theta_c = \frac{1}{\beta n(\omega)} \qquad (6.56)$$

where $\beta = v/c$. CL detection of Cerenkov radiation can be suppressed by the size of the emission angle θ_c. For example, in the case of CL in the TEM, assuming a parallel-sided foil and normal beam incidence, the angle θ_c is often larger than the critical angle $(=\sin^{-1}[1/n(\omega)])$ for total internal reflection, especially in semiconductor materials. Thus, the Cerenkov radiation cannot escape through the surface of the specimen (Yamamoto et al., 1996a). This, however, may not be the case if the specimen/beam is tilted or the specimen has a shape different to that of a parallel-sided foil (e.g. nanoparticles). On the other hand, Cerenkov losses are detected by EELS and this has important consequences for band gap measurements[7] as discussed below.

---

[7]The band gap in an EELS spectrum corresponds to the energy loss onset. This follows directly from the loss function (i.e. Im$[-1/\varepsilon(\omega)]$; Eq. 6.19) or by recognising that the band gap energy is the minimum energy that must be transferred to promote an electron into an unoccupied state. For indirect band gap materials, however, the momentum change during the transition must be provided by the $q_\perp$ component of the inelastically scattered incident electron and consequently a sufficiently large EELS spectrometer aperture is required (Rafferty and Brown, 1998).

Consider first the bulk energy loss in the retarded regime, which is derived in Appendix D and is given by

$$\frac{\partial^2 P}{\partial q_\perp \, \partial \omega} = \frac{e^2}{4\pi^3 v^2 \hbar \varepsilon_o} \text{Im} \left\{ -\frac{1 - \varepsilon(\omega)\beta^2}{\varepsilon(\omega)[q_\perp^2 + (\omega/2\pi v)^2(1 - \varepsilon(\omega)\beta^2)]} \right\}$$
(6.57)

Note that in the non-retarded limit (i.e. $\beta \to 0$) the above expression simplifies to Eq. (6.16) as required. The 'resonant' condition for energy loss is given by the following criterion:

$$|q_\perp^2 + (\omega/2\pi v)^2(1 - \varepsilon(\omega)\beta^2)| \to 0 \qquad (6.58)$$

Since $q_\perp^2$ is always real and positive this requires that $\text{Re}[\varepsilon(\omega)\beta^2] > 1$ and $\text{Im}[\varepsilon(\omega)] \to 0$. The latter is valid below the band gap (assuming no significant band gap states); the former condition is then equivalent to Eq. (6.55), the criterion for Cerenkov radiation, since $n(\omega) \approx \sqrt{\varepsilon(\omega)}$ when the imaginary part of $\varepsilon(\omega)$ is negligible. Cerenkov radiation therefore shifts the energy loss onset in an EELS spectrum below the band gap value (Erni and Browning, 2005; Gu et al., 2007; Stöger-Pollach and Schattschneider, 2007; Stöger-Pollach, 2008). Experiment as well as theory (i.e. Eq. (6.57)), however, reveals that Cerenkov radiation is present at only small values of $q_\perp$. This is illustrated in Figure 6.6a and b for GaN, which has a band gap of 3.4 eV (Gu et al., 2007). Figure 6.6a shows the experimental $\omega - q_\perp$ diagram, where the bulk plasmon peak at 19.6 eV and interband transitions at ~7–12 eV are evident (the latter is more clearly visible in Figure 6.6b). In Figure 6.6b, individual EELS spectra are extracted from three separate values of $q_\perp$, corresponding to 0, 20 and 60 μrad scattering angle respectively. At $q_\perp = 0$ there is significant intensity in the EELS signal below the 3.4 eV band gap due to Cerenkov losses, although it decreases rapidly as the scattering angle increases.

Confinement of Cerenkov losses to small values of $q_\perp$ provides two methods to minimise its effect. The first involves measuring the EELS signal in diffraction mode in the TEM whilst positioning the EELS spectrometer entrance aperture slightly away from the unscattered beam. The second method, also valid for TEM diffraction mode, is to record two EELS spectra for two different collection angles β_1 and β_2 ($\beta_2 > \beta_1$) and subtract the two to obtain a difference spectrum. The spectrometer entrance aperture is placed symmetrically about the unscattered beam. Provided β_1 is greater than the maximum scattering angle for Cerenkov radiation, the difference spectrum will be free of Cerenkov artefacts. The validity of these methods has been tested on a range of different materials (see, e.g. Gu *et al.*, 2007; Stöger-Pollach and Schattschneider, 2007). Although Cerenkov losses are largely removed, the fact that the spectrum is acquired at $q_\perp \neq 0$ means that the measurement can be affected by anisotropy of $\varepsilon(\mathbf{q}, \omega)$. It may not always be possible to compare with optical band gap measurements, which are valid for $\mathbf{q} = 0$. This is especially true if the band structure of the valence and conduction bands are not flat, such as, for example, in silicon (Stöger-Pollach and Schattschneider, 2007). Furthermore, from an experimental point of view, removing the small $\mathbf{q}$ contribution dramatically reduces the signal, resulting in long acquisition times. STEM probe geometries are also not feasible, so that if high spatial resolution is required the microscope must be set up in nano-beam, parallel illumination mode.

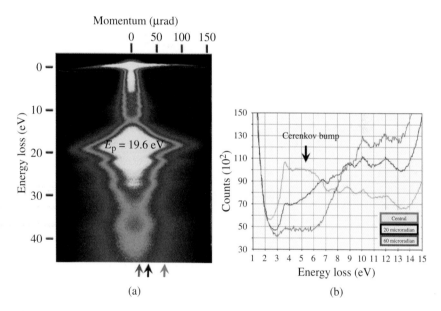

(a)   (b)

Figure 6.6   (a) Experimentally measured $\omega - q_\perp$ diagram for GaN, where $\hbar\omega$ is the EELS energy loss and $q_\perp$ is the transverse momentum change of the incident electron expressed as a scattering angle. The feature at 19.6 eV corresponds to the plasmon energy (labelled E_p). The three arrows at the bottom indicate the locations where individual EELS spectra were extracted; these correspond to 0, 20 and 60 μrad scattering angle respectively and are shown superimposed in (b). From Gu *et al.* (2007). Reproduced with permission; copyright American Physical Society. (c) Shows the low energy loss region of GaP EELS spectra measured at different specimen thicknesses with a 2.8 mrad collection semi-angle. From Stöger-Pollach (2008). Reproduced with permission; copyright Elsevier. In all cases, the incident electron energy is 200 keV.

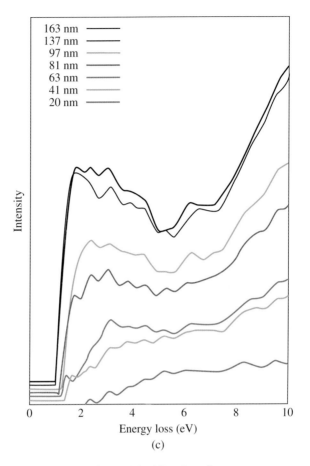

Figure 6.6   (*Continued*)

Reducing the incident electron energy is another way to minimise Cerenkov radiation (Eq. (6.55)), although this technique is more suitable for wide band gap materials (Gu *et al.*, 2007), since then the refractive index $n(\omega)$ is comparatively smaller. Alternatively, the specimen thickness could also be reduced; here destructive interference of the Cerenkov radiation takes place due to total internal reflection, apart from guided modes[8] that are present at only discrete values of $q_\perp$ for a given energy loss (García de Abajo *et al.*, 2004). The double differential energy loss probability for a thin, parallel-sided specimen in the retarded regime

---

[8]Guided modes are electromagnetic waves within a confined volume. Multiple internal reflection at the boundaries result in standing waves, which restrict the allowed wave vector in the confinement directions. For more details see, for example, Griffiths (1999).

was first derived by Kroger (1968, 1970) and this (rather complicated) expression can be used to determine the effect of any Cerenkov losses on the EELS spectrum for a given specimen thickness. On the other hand, Figure 6.6c shows how the EELS spectrum measured using a 200 keV electron beam varies with specimen thickness in GaP (Stöger-Pollach, 2008); as the specimen thickness is reduced to 20 nm Cerenkov losses are suppressed, so that the energy loss onset is close to the true band gap value of ~2 eV.

6.2.2 Transition Radiation

The electric field of an incident electron is determined by the dielectric properties of the medium in which it is travelling (Section 6.1.1). It follows that when the electron crosses an interface between two dielectric media the electric field must alter its form, especially if the dielectric functions of the two media are very different. Such a scenario arises at the free surfaces of electron microscopy specimens. On crossing a free surface energy is therefore dissipated as electromagnetic radiation, also known as *transition* radiation. A physical explanation of this phenomenon is provided in Figure 6.7a, which shows an incident electron

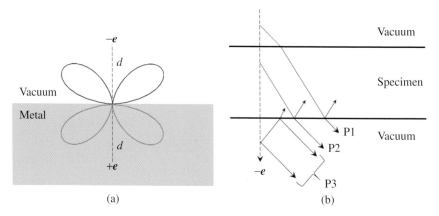

(a)   (b)

Figure 6.7 (a) Polar diagram representation of the emission pattern for transition radiation at a vacuum–metal boundary. The distance of the butterfly wing-shaped contour from the origin is proportional to the intensity emitted along that particular direction. (b) Radiation generated by an incident electron traversing a thin-foil specimen. 'P1' and 'P3' represent transition radiation generated at the electron entrance and exit surfaces respectively, while 'P2' represents any Cerenkov radiation generated from within the specimen. It is assumed that the radiation can undergo both reflection and transmission at the vacuum–specimen boundaries.

near a vacuum–specimen interface, with the specimen being an earthed metal (García de Abajo, 2010). A positive image charge is 'induced' within the metal; the image charge and incident electron form a dipole, whose separation decreases continuously as the electron approaches the vacuum–specimen boundary. The emission pattern of the transition radiation is therefore similar to an oscillating electric dipole (Figure 6.7a). If the specimen is a dielectric instead of a metal then the transition radiation is due to the change in specimen polarisation, as induced by the incident electron, with respect to time (Jackson, 1998). Each electric dipole in the specimen can emit radiation, but for appreciable intensity the radiation from neighbouring dipoles must be coherent. A 'formation zone' for transition radiation can be defined, which extends from the dielectric boundary into each medium, and is of the order of $(\beta\lambda/2\pi)$ in size, where λ is the wavelength of emitted radiation (Yamamoto et al., 1996a). The formation zone for 500 nm transition radiation is therefore ~50 nm in size for a 200 kV electron beam (strictly speaking the size also depends on the refractive index of the material as well as emission angle of the transition radiation; Yamamoto et al., 1996a). For normal electron incidence the transition radiation is p-polarised, that is, the electric field of the electromagnetic wave lies in the plane of incidence (Kroger, 1970).

For TEM thin-foil specimens there are two free surfaces (i.e. electron beam entrance and exit surfaces) from which transition radiation can be emitted. Furthermore, Cerenkov radiation can also be generated from within the specimen, provided that Eq. (6.55) is satisfied. The net intensity detected in the far field is determined by interference between all three sources of radiation. Figure 6.7b illustrates the interference mechanisms for forward emission (Yamamoto et al., 1996a). 'P1' represents transition radiation emitted in the forward direction at the electron entrance surface, 'P3' the transition radiation generated at the exit surface (emitted in both forward and backward directions) and 'P2' the Cerenkov radiation generated from within the specimen. In Figure 6.7b it has been assumed that a ray can undergo both transmission and reflection at the vacuum–specimen interface; there are, however, exceptions to this rule, such as, total internal reflection above the critical incidence angle. A similar diagram can be drawn for back-emitted radiation (Yamamoto et al., 1996b). Analytical expressions for the radiation cross-section can be found in Kroger (1970) and Yamamoto et al. (1996a, b). The transition radiation spectrum varies with the incident electron energy, since this determines the electric field (Section 6.1.1), as well as the specimen thickness due to interference effects. As an example, Figure 6.8 shows CL spectra for mica of different thicknesses, acquired in the backward emission direction, so that Cerenkov radiation is virtually absent. The spectra consist of multiple maxima and minima, which vary with specimen thickness due to the interference effect.

Stöger-Pollach *et al.* (2017) have used aluminium, a strong emitter of transition radiation, as an example to demonstrate that transition radiation losses are a relatively minor contribution in an EELS spectrum. Nevertheless, transition radiation can be an unwanted artefact in TEM–CL, particularly when analysing defects such as grain boundaries and dislocations (Mendis *et al.*, 2016). Here, defect contrast is due to the incoherent luminescence generated by electron–hole pair recombination, that is, non-radiative recombination is typically stronger at a defect due to electronic states within the band gap, so that there is a corresponding

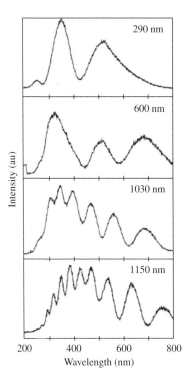

Figure 6.8   TEM-cathodoluminescence (CL) spectra measured in the backward emission direction (i.e. anti-parallel to the incident electron direction) for mica of thickness 290, 600, 1030 and 1150 nm respectively at 200 keV beam energy. From Yamamoto *et al.* (1996a). Reproduced with permission; copyright The Royal Society of London.

decrease in the luminescence intensity. Transition radiation, on the other hand, is generated at the vacuum–specimen interface and is therefore not sensitive to any underlying defect structure. Panchromatic CL images can therefore exhibit lower defect contrast due to the transition radiation 'background' (Mendis *et al.*, 2016). This is illustrated in Figure 6.9 for a 280 nm thick CdTe specimen at 200 keV beam energy. Figure 6.9a shows the CL spectrum, which consists of a peak at ~820 nm wavelength due to incoherent luminescence and a much broader peak centred at ~600 nm wavelength due to transition radiation. In Figure 6.9b, the intensity within a 760–850 nm wavelength range is used to map the

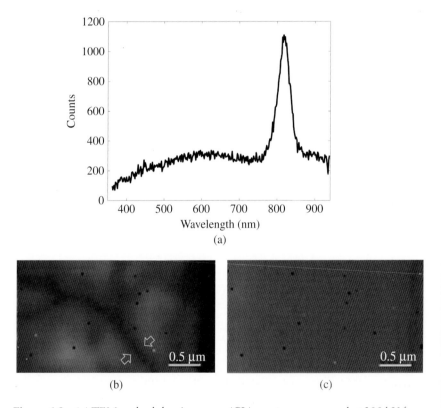

Figure 6.9 (a) TEM-cathodoluminescence (CL) spectrum measured at 200 kV from a ~280 nm thick CdTe specimen. The peak at ~820 nm is due to electron–hole pair recombination and the broad peak centred at ~600 nm is due to transition radiation. (b) and (c) show intensity maps extracted from a CL spectrum image over the wavelength windows 760–850 nm and 400–700 nm respectively. Grain boundaries are visible in the former, as indicated by the arrows. The anomalous intensity in some of the pixels in (b) and (c) are due to light collection artefacts. From Mendis *et al.* (2016). Reproduced with permission; copyright Elsevier.

CdTe specimen and reveal the presence of grain boundaries. However, the same area mapped with CL intensity in the 400–700 nm wavelength range shows no such contrast, owing to the nature of transition radiation (Figure 6.9c). In Figure 6.9a, the transition radiation contributes ~30% to the overall CL intensity, so that it can be an important factor when interpreting TEM–CL data, especially if the radiative recombination efficiency of the material is low. Similarly, Brenny *et al.* (2014) have shown that transition radiation is important in SEM–CL of Si but not GaAs, owing to the incoherent luminescence signal being much larger for the latter (GaAs has a direct band gap, while Si is an indirect band gap semiconductor).

6.3 SIMULATING LOW ENERGY LOSS EELS SPECTRA

Analytical methods for calculating the energy loss are only tractable for relatively simple specimen shapes, such as spheres, spherical cavities and cylinders (Wang, 1996). More complex geometries require numerical methods, the two most widely used being the discrete dipole approximation (DDA) and the boundary element method (BEM),[9] which are summarised below.

6.3.1 Discrete Dipole Approximation (DDA)

In DDA, the specimen is modelled as a collection of dipoles. The polarisation (p_i) of the ith dipole at position vector r_i is proportional to the local electric field (E_{local}), that is,

$$p_i(\omega) = \alpha(\omega)E_{local}(r_i, \omega) \qquad (6.59)$$

The proportionality constant α is the polarisability of the dipole. For anisotropic materials α is a tensor quantity, but here for simplicity an isotropic solid is assumed, so that the polarisability reduces to a scalar. α is related to the dielectric function through the Clausius–Mossotti relation (Kittel, 2005):

$$\frac{\varepsilon(\omega) - 1}{\varepsilon(\omega) + 2} = \frac{n\alpha(\omega)}{3\varepsilon_o} \qquad (6.60)$$

---

[9]Other techniques, such as multipole expansion (García de Abajo, 1999; Kiewidt *et al.*, 2013) and T-matrix method (Matyssek *et al.*, 2012), have also been proposed, but there are comparatively fewer applications in the EELS literature.

where n is the number of dipoles per unit volume. If there are more than one type of dipoles the numerator on the right-hand side must be replaced by $\Sigma n_k \alpha_k$, where n_k is the number density of kth-type dipoles with polarisability α_k.

The local electric field in Eq. (6.59) is the sum of the electric fields due to the incident electron *in vacuum* ($\mathbf{E}_{vac}$) as well as neighbouring dipoles ($\mathbf{E}_{dipole}$):

$$\mathbf{p}_i(\omega) = \alpha(\omega) \left[\mathbf{E}_{vac}(\mathbf{r}_i, \omega) + \sum_{j \neq i} \mathbf{E}^j_{dipole}(\mathbf{r}_i, \omega) \right] \qquad (6.61)$$

$\mathbf{E}_{vac}(\mathbf{r}_i, \omega)$ can be derived from the electric scalar potential $\tilde{\phi}(\mathbf{q}, \omega)$ in Eq. (6.10), with $\varepsilon(\mathbf{q}, \omega) = 1$ due to the medium being vacuum, by inverse Fourier transforming with respect to $\mathbf{q}$. The resulting expressions can be found in Jackson (1998). $\mathbf{E}_{dipole}$ is given by (Jackson, 1998)

$$\mathbf{E}^j_{dipole}(\mathbf{r}_i, \omega) = \frac{3(\mathbf{p}_j(\omega) \cdot \mathbf{r})\mathbf{r} - r^2 \mathbf{p}_j(\omega)}{4\pi\varepsilon_o r^5} \qquad (6.62)$$

where $\mathbf{r} = \mathbf{r}_i - \mathbf{r}_j$, with $\mathbf{r}_j$ being the position vector of the jth dipole (r is the magnitude of the vector $\mathbf{r}$). The time consuming part of the calculation is to determine the polarisation of each dipole in a self-consistent manner, since the local electric field depends on the neighbouring dipoles. For N dipoles, and three degrees of freedom for each polarisation vector, this requires solving $3N$ coupled linear equations of the form given by Eqs. (6.61) and (6.62) (Henrard and Lambin, 1996; Geuquet and Henrard, 2010). The overall electric field at the incident electron position can then be determined from the potential of individual dipoles[10] using linear superposition, and the energy loss calculated using Eq. (6.11). The technique has been applied to carbon nanostructures (Henrard and Lambin, 1996; Rivacoba and García de Abajo, 2003) as well as silver nano-prisms (Geuquet and Henrard, 2010).

6.3.2 Boundary Element Method (BEM)

The BEM method calculates the surface charge density induced by the incident electron on the boundaries of the specimen; from this, the overall electric field and energy loss can be determined (Ouyang and Isaacson, 1989; García de Abajo and Aizpurua, 1997). Unlike DDA, which is a volume method, the dimensionality of the problem is reduced by one by

---

[10]For a dipole with polarisation vector $\mathbf{p}$ centred at the origin, the potential at position $\mathbf{r}$ is given by $\mathbf{p}\cdot\mathbf{r}/(4\pi\varepsilon_o r^3)$; (Jackson, 1998).

considering only the boundary, thereby potentially reducing the computation time. The BEM method has been generalised to include retardation (García de Abajo and Howie, 2002), although here only the non-retarded case will be discussed. It will be assumed that the incident electron is outside the specimen (i.e. aloof incidence) and furthermore any specimen support (e.g. holey carbon grid) is not taken into account. The derivation is based on that given in García de Abajo and Aizpurua (1997).

Fourier transforming Eq. (6.1) with respect to time gives

$$\vec{\nabla} \cdot [\varepsilon_o \varepsilon(\mathbf{r}, \omega)\vec{\nabla}\phi(\mathbf{r}, \omega)]$$

$$= \varepsilon_o \vec{\nabla}\varepsilon(\mathbf{r}, \omega) \cdot \vec{\nabla}\phi(\mathbf{r}, \omega) + \varepsilon_o \varepsilon(\mathbf{r}, \omega)\nabla^2\phi(\mathbf{r}, \omega) = -\rho_f(\mathbf{r}, \omega) \qquad (6.63)$$

Here, the dependence of $\varepsilon(\mathbf{r}, \omega)$ on the position vector $\mathbf{r}$ is due to the finite size of the specimen and does not imply that the local approximation for the dielectric function is not valid (Section 6.1.1). In this particular case, $\varepsilon(\mathbf{r}, \omega)$ is equal to the dielectric function of the specimen (vacuum) for all $\mathbf{r}$ within (outside) the specimen; at the specimen boundary it is the average of the dielectric functions for the specimen and vacuum. The scalar potential ϕ can be expressed as the sum of two terms, ϕ^{bulk} and $\phi^{boundary}$, given by

$$\phi^{bulk}(\mathbf{r}, \omega) = \frac{1}{4\pi\varepsilon_o} \int \frac{\rho_f(\mathbf{r}', \omega)}{\varepsilon(\mathbf{r}', \omega)|\mathbf{r} - \mathbf{r}'|} d\mathbf{r}' \qquad (6.64)$$

$$\phi^{boundary}(\mathbf{r}, \omega) = \frac{1}{4\pi} \int \frac{\vec{\nabla}\phi(\mathbf{r}', \omega) \cdot \vec{\nabla}\varepsilon(\mathbf{r}', \omega)}{\varepsilon(\mathbf{r}', \omega)|\mathbf{r} - \mathbf{r}'|} d\mathbf{r}' \qquad (6.65)$$

This is easily verified by substituting $\phi = \phi^{bulk} + \phi^{boundary}$ and the above expressions in Eq. (6.63) and using the relationship $\nabla^2(1/|\mathbf{r} - \mathbf{r}'|) = -4\pi\delta(\mathbf{r} - \mathbf{r}')$, where δ is the Dirac delta function (Jackson, 1998). For a homogeneous specimen Eq. (6.65) can only be non-zero at the specimen boundaries, where $\varepsilon(\mathbf{r}, \omega)$ changes abruptly. Following the standard definition of the potential due to a surface charge density $\sigma(\mathbf{s}, \omega)$, Eq. (6.65) can also be written as a surface integral over the boundary coordinates $\mathbf{s}$ (Jackson, 1998), that is,

$$\phi^{boundary}(\mathbf{r}, \omega) = \int \frac{\sigma(\mathbf{s}, \omega)}{|\mathbf{r} - \mathbf{s}|} d\mathbf{s} \qquad (6.66)$$

where

$$\sigma(\mathbf{s}, \omega) = \frac{1}{4\pi} \frac{\vec{\nabla}\phi(\mathbf{s}, \omega) \cdot \vec{\nabla}\varepsilon(\mathbf{s}, \omega)}{\varepsilon(\mathbf{s}, \omega)} = \frac{1}{4\pi\varepsilon_o}\left[\mathbf{D}(\mathbf{r}, \omega) \cdot \vec{\nabla}\left(\frac{1}{\varepsilon(\mathbf{r}, \omega)}\right)\right]_{\mathbf{r}=\mathbf{s}} \qquad (6.67)$$

The gradient of $(1/\varepsilon)$ is only non-zero along the boundary normal. Integrating Eq. (6.67) along this direction and noting that both $\sigma(s, \omega)$ and $\vec{\nabla}(1/\varepsilon)$ are similar to Dirac delta functions results in

$$\sigma(\mathbf{s}, \omega) = \frac{1}{4\pi\varepsilon_o} \frac{\varepsilon_{\text{spec}}(\omega) - \varepsilon_{\text{vac}}(\omega)}{\varepsilon_{\text{spec}}(\omega)\varepsilon_{\text{vac}}(\omega)} \mathbf{n} \cdot \mathbf{D}(\mathbf{s}, \omega)$$

$$= \frac{1}{4\pi\varepsilon_o} \frac{\varepsilon_{\text{spec}}(\omega) - 1}{\varepsilon_{\text{spec}}(\omega)} \mathbf{n} \cdot \mathbf{D}(\mathbf{s}, \omega) \qquad (6.68)$$

$\varepsilon_{\text{spec}}$, $\varepsilon_{\text{vac}} = 1$ are the dielectric functions for specimen and vacuum respectively and $\mathbf{n}$ is the unit vector along the boundary normal. $\mathbf{D}(\mathbf{s}, \omega)$ can be determined from $\mathbf{D}(\mathbf{r}, \omega) = -\varepsilon_o\varepsilon(\mathbf{r}, \omega)\vec{\nabla}\phi(\mathbf{r}, \omega)$ for either the vacuum or specimen and passing onto the limit $\mathbf{r} \to \mathbf{s}$. This, however, requires evaluating ϕ at the boundary. ϕ^{bulk} is well defined at the boundary provided that the incident electron does not cross the boundary along its trajectory, that is, ρ_f is always zero at the boundary (Eq. (6.64)). On the other hand, ϕ^{boundary} is not well defined at the boundary due to the surface charge density (Eq. (6.66)). It can be shown that (for a derivation, see Appendix A in García de Abajo and Howie, 2002):

$$\lim_{t \to 0} \mathbf{n} \cdot \vec{\nabla}_s \left(\frac{1}{|\mathbf{s} \pm t\mathbf{n} - \mathbf{s}'|} \right) = -\frac{\mathbf{n} \cdot (\mathbf{s} - \mathbf{s}')}{|\mathbf{s} - \mathbf{s}'|^3} \mp 2\pi\delta(\mathbf{s} - \mathbf{s}') \qquad (6.69)$$

where $\mathbf{s}'$ is another position vector at the boundary and $\vec{\nabla}_s$ is the gradient operator with respect to $\mathbf{s}$. The upper and lower signs in Eq. (6.69) correspond to approaching the boundary from the vacuum side and specimen side respectively (recall that $\mathbf{n}$ points towards the vacuum). It therefore follows that the displacement field at the boundary is given by

$$\mathbf{n} \cdot \mathbf{D}(\mathbf{s}, \omega) = -\varepsilon_o\varepsilon_{\text{spec}}(\omega)\mathbf{n} \cdot [\vec{\nabla}(\phi^{\text{bulk}} + \phi^{\text{boundary}})]_{\mathbf{r}=\mathbf{s}}$$

$$= -\varepsilon_o\varepsilon_{\text{spec}}(\omega) \left[\mathbf{n} \cdot \vec{\nabla}\phi^{\text{bulk}}(\mathbf{s}, \omega) - \int \frac{\mathbf{n} \cdot (\mathbf{s} - \mathbf{s}')}{|\mathbf{s} - \mathbf{s}'|^3} \sigma(\mathbf{s}', \omega) \, d\mathbf{s}' \right.$$

$$\left. + 2\pi\sigma(\mathbf{s}, \omega) \right] \qquad (6.70)$$

where Eq. (6.69) has been used to evaluate the gradient of ϕ^{boundary} when approaching from the specimen side. Substituting Eq. (6.70) in (6.68) and rearranging gives

$$\Lambda(\omega)\sigma(\mathbf{s}, \omega) = \mathbf{n} \cdot \vec{\nabla} \phi^{\text{bulk}}(\mathbf{s}, \omega) - \int \frac{\mathbf{n} \cdot (\mathbf{s} - \mathbf{s}')}{|\mathbf{s} - \mathbf{s}'|^3} \sigma(\mathbf{s}', \omega) d\mathbf{s}'$$

$$\Lambda(\omega) = 2\pi \left(\frac{1 + \varepsilon_{\text{spec}}}{1 - \varepsilon_{\text{spec}}} \right) \tag{6.71}$$

The above is the fundamental equation for calculating the surface charge density. The problem is simplified by considering the surface modes in the absence of an external field (i.e. $\vec{\nabla} \phi^{\text{bulk}} = 0$):

$$-\int \frac{\mathbf{n} \cdot (\mathbf{s} - \mathbf{s}')}{|\mathbf{s} - \mathbf{s}'|^3} \sigma_i(\mathbf{s}') d\mathbf{s}' = 2\pi \lambda_i \sigma_i(\mathbf{s}) \tag{6.72}$$

This is now in the form of a characteristic equation with λ_i being the eigenvalue for the ith eigenmode σ_i. Ouyang and Isaacson (1989) have shown that λ_i are real and that the eigenmodes are orthonormal, that is,

$$\iint \frac{\sigma_i(\mathbf{s})\sigma_j(\mathbf{s}') *}{|\mathbf{s} - \mathbf{s}'|} d\mathbf{s} \, d\mathbf{s}' = \delta_{ij} \tag{6.73}$$

where δ_{ij} is the Kronecker delta. The eigenmodes are a convenient basis set to express the electric field due to ϕ^{bulk}:

$$\mathbf{n} \cdot \vec{\nabla} \phi^{\text{bulk}}(\mathbf{s}, \omega) = \sum_i c_i(\omega)\sigma_i(\mathbf{s}) \tag{6.74}$$

The linear coefficient c_i can be determined by multiplying both sides of the above equation by $\sigma_i(\mathbf{s}') * / |\mathbf{s} - \mathbf{s}'|$ and integrating over $\mathbf{s}, \mathbf{s}'$ to give

$$c_i(\omega) = \iint \mathbf{n} \cdot \vec{\nabla} \phi^{\text{bulk}}(\mathbf{s}, \omega) \frac{\sigma_i(\mathbf{s}') *}{|\mathbf{s} - \mathbf{s}'|} d\mathbf{s} \, d\mathbf{s}'$$

$$= -\iint \mathbf{n} \cdot \left[\int \frac{\rho_f(\mathbf{r}', \omega)}{4\pi \varepsilon_o \varepsilon(\mathbf{r}', \omega)} \left(\frac{\mathbf{s} - \mathbf{r}'}{|\mathbf{s} - \mathbf{r}'|^3} \right) d\mathbf{r}' \right] \frac{\sigma_i(\mathbf{s}') *}{|\mathbf{s} - \mathbf{s}'|} d\mathbf{s} \, d\mathbf{s}' \tag{6.75}$$

where use has been made of the orthonormal property of the eigenmodes (Eq. (6.73)) and Eq. (6.64) has been substituted for ϕ^{bulk}. The surface charge density can also be represented as a linear combination of the eigenmodes:

$$\sigma(\mathbf{s}, \omega) = \sum_i f_i(\omega)\sigma_i(\mathbf{s}) \tag{6.76}$$

Substituting Eqs. (6.72), (6.74) and (6.76) in Eq. (6.71) gives finally

$$f_i(\omega) = \frac{c_i(\omega)}{\Lambda(\omega) - 2\pi\lambda_i} \tag{6.77}$$

Equation (6.72) together with Eqs. (6.75)–(6.77) fully determines the surface charge density. The total potential ϕ can therefore be calculated from Eqs. (6.64) and (6.66) and from this the energy loss (Eq. (6.11)) readily follows. The relevant expressions can be found in García de Abajo and Aizpurua (1997). For the particular case of aloof incidence only ϕ^{boundary} contributes to the energy loss. In order to achieve convergence a finer mesh must be used when sampling the boundary coordinate s in regions close to the electron beam path or in regions where boundaries lie close to one another (e.g. two neighbouring objects). For certain favourable geometries, such as translationally invariant boundaries, symmetry can be used to reduce the sampling and hence computation time (García de Abajo and Aizpurua, 1997). Figure 6.10 illustrates application of the BEM method to determine the spatial excitation of surface plasmon modes in a silver nano-prism (Nelayah $et~al.$, 2007). The simulated results are remarkably similar to experiment (Figure 6.4d), although the surface plasmon energies are slightly higher in the former. This is likely to be due to the mica support film, which was not taken into account in the simulations. Similar agreement has been obtained between energy filtered TEM (EFTEM) measurements of surface plasmon modes in gold nano-rods and BEM calculations (Schaffer $et~al.$, 2009).

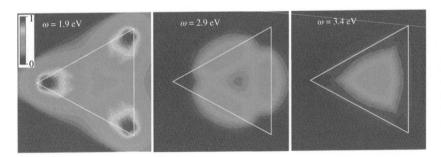

Figure 6.10 Excitation amplitudes for three surface plasmon modes at 1.9, 2.9 and 3.4 eV in a triangular silver nano-prism as simulated by the boundary element method. The specimen boundary is indicated in the figure. The results should be compared with the experimental measurements shown in Figure 6.4d (the nano-prism dimensions are identical for both simulation and experiment). From Nelayah $et~al.$ (2007). Reproduced with permission; copyright Nature Publishing Group.

6.4 SUMMARY

Electrodynamic theory, based on Maxwell's equations, is a versatile method for analysing low energy loss inelastic scattering events, which typically involve many electrons in the solid. The system response is characterised by the dielectric function. Interband transitions, bulk and surface (or interface) plasmons, phonons, delocalisation, etc. can be readily described within this framework. Furthermore, it is possible to extract the dielectric function of a material, and therefore its optical properties, from the measured energy loss spectrum through the Kramers–Kronig relations. An example where this might be useful is in band gap measurement at spatial resolutions significantly better than optical-based techniques. However, it is important to consider potential artefacts from losses due to Cerenkov radiation. Similarly, transition radiation is another artefact that could distort CL measurements. Fortunately, both Cerenkov and transition radiation can also be analysed using the same electrodynamic theory.

REFERENCES

Bolton, J.P.R. and Chen, M. (1995) *Ultramicroscopy* **60**, 247.

Brenny, B.J.M., Coenen, T. and Polman, A. (2014) *J. Appl. Phys.* **115**, 244307.

Couillard, M., Kociak, M., Stéphan, O., Botton, G.A. and Colliex, C. (2007) *Phys. Rev. B* **76**, 165131.

Couillard, M., Yurtsever, A. and Muller, D.A. (2008) *Phys. Rev. B* **77**, 085318.

Cundy, S.L., Metherell, A.J.F., Whelan, M.J., Unwin, P.N.T. and Nicholson, R.B. (1968) *Proc. R. Soc. Lond. A* **307**, 267.

Dass, T. and Sharma, S.K. (1998) Mathematical Methods in Classical and Quantum Physics, University Press (India) Ltd, Hyderabad, India.

Dwyer, C. (2014) *Phys. Rev. B* **89**, 054103.

Eberlain, T., Bangert, U., Nair, R.R., Jones, R., Gass, M., Bleloch, A.L., Novoselov, K.S., Geim, A. and Briddon, P.R. (2008) *Phys. Rev. B* **77**, 233406.

Egerton, R.F. (1996) *Electron Energy-Loss Spectroscopy in the Electron Microscope*, 2nd edition, Plenum Press, New York.

Egerton, R.F. (2014) *Microsc. Microanal.* **20**, 658.

Erni, R. and Browning, N.D. (2005) *Ultramicroscopy* **104**, 176.

Fermi, E. (1940) *Phys. Rev.* **57**, 485.

Ferrell, R.A. (1958) *Phys. Rev.* **111**, 1214.

Forbes, B.D. and Allen, L.J. (2016) *Phys. Rev. B* **94**, 014110.

Fox, M. (2008) *Optical Properties of Solids*, Oxford University Press, Oxford.

García de Abajo, F.J. (1999) *Phys. Rev. B* **59**, 3095.

García de Abajo, F.J. (2010) *Rev. Mod. Phys.* **82**, 209.

García de Abajo, F.J. and Aizpurua, J. (1997) *Phys. Rev. B* **56**, 15873.

García de Abajo, F.J. and Howie, A. (2002) *Phys. Rev. B* **65**, 115418.

García de Abajo, F.J., Rivacoba, A., Zabala, N. and Yamamoto, N. (2004) *Phys. Rev. B* **69**, 155420.

Gass, M.H., Papworth, A.J., Bullough, T.J. and Chalker, P.R. (2004) *Ultramicroscopy* **101**, 257.

Gass, M.H., Papworth, A.J., Beanland, R., Bullough, T.J. and Chalker, P.R. (2006) *Phys. Rev. B* **73**, 035312.

Geuquet, N. and Henrard, L. (2010) *Ultramicroscopy* **110**, 1075.

Gradshteyn, I.S. and Ryzhik, I.M. (1980) *Tables of Integrals, Series and Products*, Academic Press, New York.

Griffiths, D.J. (1999) *Introduction to Electrodynamics*, Prentice Hall, New Jersey.

Gu, L., Srot, V., Sigle, W., Koch, C., van Aken, P., Scholz, F., Thapa, S.B., KiGrifrchner, C., Jetter, M. and Rühle, M. (2007) *Phys. Rev. B* **75**, 195214.

Harris, D.C. and Bertolucci, M.D. (1989) *Symmetry and Spectroscopy: An Introduction to Vibrational and Electronic Spectroscopy*, Dover Publications, New York.

Hecht, E. (2002) *Optics*, 4th edition, Addison Wesley, California.

Henrard, L. and Lambin, Ph. (1996) *J. Phys. B: At. Mol. Opt. Phys.* **29**, 5127.

Howie, A. (1999) in *Topics in Electron Diffraction and Microscopy of Materials* (P.B. Hirsch Editor), Institute of Physics Publishing, Bristol.

Jackson, J.D. (1998) *Classical Electrodynamics*, John Wiley & Sons, New York.

Keast, V.J. (2013) *Micron* **44**, 93.

Kiewidt, L., Karamehmedović, M., Matyssek, C., Hergert, W., Mädler, L. and Wriedt, T. (2013) *Ultramicroscopy* **133**, 101.

Kittel, C. (2005) *Introduction to Solid State Physics*, 8th edition, John Wiley and Sons, USA.

Kociak, M., Henrard, L., Stéphan, O., Suenaga, K. and Colliex, C. (2000) *Phys. Rev. B* **61**, 13936.

Kohl, H. and Rose, H. (1985) *Adv. Electron. Electron Phys.* **65**, 173.

Krivanek, O.L., Ursin, J.P., Bacon, N.J., Corbin, G.J., Dellby, N., Hrncirik, P., Murfitt, M.F., Own, C.S. and Szilagyi, Z.S. (2009) *Phil. Trans. R. Soc. A*, **367**, 3683.

Krivanek, O.L., Lovejoy, T.C., Dellby, N., Aoki, T., Carpenter, R.W., Rez, P., Soignard, E., Zhu, J., Batson, P.E., Lagos, M.J., Egerton, R.F. and Crozier, P.A. (2014) *Nature* **514**, 209.

Kroger, E. (1968) *Z. Physik* **216**, 115.

Kroger, E. (1970) *Z. Physik* **235**, 403.

Lagos, M.J., Trügler, A., Hohenester, U. and Batson, P.E. (2017) *Nature* **543**, 529.

Losquin, A., Zagonel, L.F., Myroshnychenko, V., Rodríguez-González, B., Tencé, M., Scarabelli, L., Förstner, J., Liz-Marzán, L.M., García de Abajo, F.J., Stéphan, O. and Kociak, M. (2015) *Nano Lett.* **15**, 1229.

Maclean, E.D.W., Craven, A.J. and McComb, D.W. (2001), Proceedings of EMAG 2001 Conference, Dundee, p. 259.

Matyssek, C., Schmidt, V., Hergert, W. and Wriedt, T. (2012) *Ultramicroscopy* **117**, 46.

Mendis, B.G., Howkins, A., Stowe, D., Major, J.D. and Durose, K. (2016) *Ultramicroscopy* **167**, 31.

Mkhoyan, K.A., Babinec, T., Maccagnano, S.E., Kirkland, E.J. and Silcox, J. (2007) *Ultramicroscopy* **107**, 345.

Nelayah, J., Kociak, M., Stéphan, O., García de Abajo, F.J., Tencé, M., Henrard, L., Taverna, D., Pastoriza-Santos, I., Liz-Marzán, L.M. and Colliex, C. (2007) *Nature Phys.* 3, 348.

Ouyang, F. and Isaacson, M. (1989) *Phil. Mag. B* 60, 481.

Pitarke, J.M., Silkin, V.M., Chulkov, E.V. and Echenique, P.M. (2007) *Rep. Prog. Phys.* 70, 1.

Rafferty, B. and Brown, L.M. (1998) *Phys. Rev. B* 58, 10326.

Rez, P. (2014) *Microsc. Microanal.* 20, 671.

Ritchie, R.H. (1957) *Phys. Rev.* 106, 874.

Rivacoba, A. and García de Abajo, F.J. (2003) *Phys. Rev. B* 67, 085414.

Rocca, M. (1995) *Surf. Sci. Reports* 22, 1.

Sanchez, A.M., Beanland, R., Papworth, A.J., Goodhew, P.J. and Gass, M.H. (2006) *Appl. Phys. Lett.* 88, 051917.

Schaffer, B., Hohenester, U., Trügler, A. and Hofer, F. (2009) *Phys. Rev. B* 79, 041401.

Schattschneider, P. (1986) *Fundamentals of Inelastic Electron Scattering*, Springer-Verlag, New York.

Stöger-Pollach, M. (2008) *Micron* 39, 1092.

Stöger-Pollach, M. and Schattschneider, P. (2007) *Ultramicroscopy* 107, 1178.

Stöger-Pollach, M., Kachtík, L., Miesenberger, B. and Retzl, P. (2017) *Ultramicroscopy* 173, 31.

Vincent, R. and Silcox, J. (1973) *Phys. Rev. Lett.* 31, 1487.

Walls, M.G. and Howie, A. (1989) *Ultramicroscopy* 28, 40.

Wang, Z.L. (1996) *Micron* 27, 265.

Yamamoto, N., Sugiyama, H. and Toda, A. (1996a) *Proc. R. Soc. Lond. A* 452, 2279.

Yamamoto, N., Toda, A. and Araya, K. (1996b) *J. Electron Microsc.* 45, 64.

Zhou, W., Pennycook, S.J. and Idrobo, J.C. (2012) *Ultramicroscopy* 119, 51.

Appendix A

The First Born Approximation and Atom Scattering Factor

The atom scattering factor will be derived within the so-called *first Born approximation*, which is valid for high energy incident electrons and weak scattering potentials. The Schrödinger equation for the incident electron in the potential $V(\mathbf{r})$ for a single atom is given by

$$\nabla^2\psi(\mathbf{r}) + 4\pi^2 k^2 \psi(\mathbf{r}) = -\frac{8\pi^2 me}{h^2}V(\mathbf{r})\psi(\mathbf{r}) \qquad (A.1)$$

where h is the Planck's constant, m, e the relativistic mass and charge of an electron with wave number k and wavefunction $\psi(\mathbf{r})$ at the position vector $\mathbf{r}$. The wavefunction can be expressed in the integral form:

$$\psi(\mathbf{r}) = \exp\left(2\pi i k \mathbf{n}_0 \cdot \mathbf{r}\right) + \frac{2\pi me}{h^2}\int \frac{\exp\left(2\pi i k|\mathbf{r}-\mathbf{r}'|\right)}{|\mathbf{r}-\mathbf{r}'|}V(\mathbf{r}')\psi(\mathbf{r}')d\mathbf{r}'$$
$$(A.2)$$

This is easily verified by substituting this expression in Eq. (A.1) and making use of the result $\nabla^2(1/|\mathbf{r}-\mathbf{r}'|) = -4\pi\delta(\mathbf{r}-\mathbf{r}')$, where δ is the Dirac delta function (Jackson, 1998). The first term in Eq. (A.2) represents the unscattered wave, with incident wave vector $k\mathbf{n}_0$ (Figure A.1), while the second term is the (elastic) scattered wavefunction. At large r the latter has the asymptotic form of a (distorted) spherical wave $[\exp(2\pi i k r)/r]f(\theta)$, where $f(\theta)$ is the atom scattering factor. Assuming $|\mathbf{r}-\mathbf{r}'|$ is large compared to $|\mathbf{r}'|$, the $|\mathbf{r}-\mathbf{r}'|$ term in the numerator of the integrand in Eq. (A.2) is approximately $\mathbf{r} - (\mathbf{n}_1 \cdot \mathbf{r}')$, where $\mathbf{n}_1$ is the unit vector along the scattered direction (Figure A.1; Hirsch *et al.*, 1965). On

Electron Beam-Specimen Interactions and Simulation Methods in Microscopy,
First Edition. Budhika G. Mendis.
© 2018 John Wiley & Sons Ltd. Published 2018 by John Wiley & Sons Ltd.

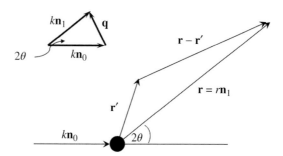

Figure A.1 Elastic scattering of an incident electron of wave number k by an atom (filled circle). $\mathbf{n}_0$ and $\mathbf{n}_1$ are unit vectors in the direction of incidence and scattering (i.e. r) respectively. The scattering angle is 2θ, and $\mathbf{q}$ is the scattering vector.

the other hand, r is substituted for the $|\mathbf{r} - \mathbf{r}'|$ term in the denominator. It therefore follows that

$$f(\theta) = \frac{2\pi me}{h^2} \int \exp(-2\pi i k \mathbf{n}_1 \cdot \mathbf{r}') V(\mathbf{r}') \psi(\mathbf{r}') d\mathbf{r}' \qquad (A.3)$$

The first Born approximation assumes that scattering is weak so that $\psi(\mathbf{r}')$ in Eq. (A.3) can be approximated by the unscattered wave. The atom scattering factor is then

$$f(\theta) \approx \frac{2\pi me}{h^2} \int \exp[2\pi i k (\mathbf{n}_0 - \mathbf{n}_1) \cdot \mathbf{r}'] V(\mathbf{r}') d\mathbf{r}'$$

$$= \frac{2\pi me}{h^2} \int \exp[-2\pi i \mathbf{q} \cdot \mathbf{r}'] V(\mathbf{r}') d\mathbf{r}' \qquad (A.4)$$

where $\mathbf{q} = k(\mathbf{n}_1 - \mathbf{n}_0)$ is the scattering vector of magnitude $2\sin\theta/\lambda$, with 2θ being the scattering angle and λ the wavelength of the incident electron (Figure A.1). The atom scattering factor can be calculated for any given potential using Eq. (A.4). Furthermore, the scattering cross-section is given by $d\sigma/d\Omega = |f(\theta)|^2$. The screened Rutherford cross-section (Eq. (2.10)) is derived in this manner (see McDaniel, 1989 for a derivation).

In practice, tabulated values of $f(\theta)$ are calculated by replacing the relativistic mass (m) of the electron in Eq. (A.4) by its rest mass (m_o). The atom scattering factor for any given electron beam energy is then obtained by simply multiplying the tabulated value by the relativistic correction term (m/m_o).

Finally, note that higher order Born approximations can be derived in a similar manner. For example, $\psi(\mathbf{r}')$ in the right-hand side of Eq. (A.2) can be approximated by the unscattered wavefunction, so that

$$
\psi(\mathbf{r}) \approx \exp(2\pi i k \mathbf{n}_0 \cdot \mathbf{r}) + \frac{2\pi m e}{h^2} \int \frac{\exp(2\pi i k |\mathbf{r} - \mathbf{r}'|)}{|\mathbf{r} - \mathbf{r}'|} V(\mathbf{r}')
$$
$$
\times \exp(2\pi i k \mathbf{n}_0 \cdot \mathbf{r}') d\mathbf{r}' \tag{A.5}
$$

This expression can be substituted in Eq. (A.3) to obtain a more accurate result for $f(\theta)$ and so on.

REFERENCES

Hirsch, P.B., Howie, A., Nicholson, R.B., Pashley, D.W. and Whelan, M.J. (1965) *Electron Microscopy of Thin Crystals*, Butterworths, Great Britain.

Jackson, J.D. (1998) *Classical Electrodynamics*, John Wiley & Sons, New York.

McDaniel, E.W. (1989) *Atomic Collisions: Electron and Photon Projectiles*, John Wiley & Sons, USA, p. 124.

Appendix B

Potential for an 'Infinite' Perfect Crystal

Here, the potential for an 'infinite' perfect crystal will be derived from that of a single atom at the origin, that is,

$$V(\mathbf{r}) = \frac{h^2}{2\pi m_o e} \int f(q) \exp(2\pi i \mathbf{q} \cdot \mathbf{r}) dq \tag{B.1}$$

where $V(\mathbf{r})$ is the atomic potential, h is the Planck's constant, m_o, e the rest mass and charge of an electron, $f(q)$ the atom scattering factor for scattering vector $\mathbf{q}$ and $\mathbf{r}$ is the position vector.

Consider the ith atom within the unit cell, which has position vector $\mathbf{r}_i$ with respect to the unit cell origin and atom scattering factor $f_i(q)$. The potential V_i due to all such atoms in the crystal is given by

$$V_i(\mathbf{r}) = \frac{h^2}{2\pi m_o e} \int f_i(q) \sum_{\mathbf{R}} \exp[2\pi i \mathbf{q} \cdot (\mathbf{r} - \mathbf{r}_i - \mathbf{R})] dq$$

$$= \frac{h^2}{2\pi m_o e} \int f_i(q) \left[\sum_{n_x, n_y, n_z} e^{-2\pi i q_x n_x a} e^{-2\pi i q_y n_y b} e^{-2\pi i q_z n_z c} \right]$$

$$\times \exp[2\pi i \mathbf{q} \cdot (\mathbf{r} - \mathbf{r}_i)] dq_x \, dq_y \, dq_z \tag{B.2}$$

where $\mathbf{R} = n_x \mathbf{a} + n_y \mathbf{b} + n_z \mathbf{c}$ is a lattice translation vector for integer values of n_x, n_y and n_z. $\mathbf{a}$, $\mathbf{b}$ and $\mathbf{c}$ are unit cell vectors along three orthogonal directions, while q_x, q_y and q_z are the corresponding components of the

Electron Beam-Specimen Interactions and Simulation Methods in Microscopy,
First Edition. Budhika G. Mendis.
© 2018 John Wiley & Sons Ltd. Published 2018 by John Wiley & Sons Ltd.

vector $\mathbf{q}$. Each summation in Eq. (B.2) is a geometric series, such that

$$\sum_{n_x=-N}^{N} e^{-2\pi i q_x n_x a} = \frac{\sin\left[\pi q_x a(2N+1)\right]}{\sin(\pi q_x a)} \tag{B.3}$$

Similar relationships hold for the other summations over n_y and n_z. For 'infinite' crystals (i.e. $N \to \infty$) the right-hand side of Eq. (B.3) is $\delta(q_x - M_x)$, where δ is the Dirac delta function and M_x is any integer. Non-zero values for Eq. (B.2) are therefore only obtained when $\mathbf{q}\cdot\mathbf{R}$ is an integer, or equivalently when $\mathbf{q}$ is a reciprocal lattice vector $\mathbf{g}$. Hence,

$$V_i(\mathbf{r}) = \frac{h^2}{2\pi m_o e(abc)} \int f_i(q)s(\mathbf{q})\exp[2\pi i\mathbf{q}\cdot(\mathbf{r}-\mathbf{r}_i)]d(q_x a)\ d(q_y b)\ d(q_z c)$$

$$= \frac{h^2}{2\pi m_o e\Omega} \sum_{\mathbf{g}} f_i(g)\exp[2\pi i\mathbf{g}\cdot(\mathbf{r}-\mathbf{r}_i)] \tag{B.4}$$

where Ω is the unit cell volume and $s(\mathbf{q}) = \Sigma\delta(q_x a - M_x)\delta(q_y b - M_y)\delta(q_z c - M_z)$, with the summation being over all integers M_x, M_y and M_z.

Next, summing over all atoms i in the unit cell finally gives the crystal potential V_{cryst} as

$$V_{cryst}(\mathbf{r}) = \frac{h^2}{2\pi m_o e\Omega} \sum_{\mathbf{g}} F_{\mathbf{g}}\exp(2\pi i\mathbf{g}\cdot\mathbf{r})$$

$$F_{\mathbf{g}} = \sum_i f_i(\mathbf{g})\exp(-2\pi i\mathbf{g}\cdot\mathbf{r}_i) \tag{B.5}$$

Appendix C

The Transition Matrix Element in the One Electron Approximation

In Chapter 5, it was stated that the transition matrix element (Eq. (5.69)) can be simplified using the one electron approximation to obtain Eq. (5.70). This result will be derived for the case where the system wavefunction can be expressed as a Slater determinant[1] (Ashcroft and Mermin, 1976). The initial (φ_i) and final (φ_f) state wavefunctions for a system consisting of N electrons is given by

$$\varphi_i(\mathbf{r}_1,\ldots,\mathbf{r}_N) = \frac{1}{\sqrt{N!}} \begin{vmatrix} \phi_i(\mathbf{r}_1) & \phi_i(\mathbf{r}_2) & \cdots & \phi_i(\mathbf{r}_N) \\ \phi_2(\mathbf{r}_1) & \phi_2(\mathbf{r}_2) & \cdots & \phi_2(\mathbf{r}_N) \\ \vdots & \vdots & \vdots & \vdots \\ \phi_N(\mathbf{r}_1) & \phi_N(\mathbf{r}_2) & \cdots & \phi_N(\mathbf{r}_N) \end{vmatrix} \quad (C.1)$$

$$\varphi_f(\mathbf{r}_1,\ldots,\mathbf{r}_N) = \frac{1}{\sqrt{N!}} \begin{vmatrix} \phi_f(\mathbf{r}_1) & \phi_f(\mathbf{r}_2) & \cdots & \phi_f(\mathbf{r}_N) \\ \phi_2(\mathbf{r}_1) & \phi_2(\mathbf{r}_2) & \cdots & \phi_2(\mathbf{r}_N) \\ \vdots & \vdots & \vdots & \vdots \\ \phi_N(\mathbf{r}_1) & \phi_N(\mathbf{r}_2) & \cdots & \phi_N(\mathbf{r}_N) \end{vmatrix} \quad (C.2)$$

The atomic electron undergoing the transition (arbitrarily) appears in the first row of the Slater determinant. A one electron wavefunction $\phi_\beta(\mathbf{r}_j)$ simply means the electron is in the quantum state denoted by β and has

---

[1] Strictly speaking, the system wavefunction can be represented by a Slater determinant only when the electrons are non-interacting, although for most calculations this is a reasonable approximation.

Electron Beam-Specimen Interactions and Simulation Methods in Microscopy, First Edition. Budhika G. Mendis.
© 2018 John Wiley & Sons Ltd. Published 2018 by John Wiley & Sons Ltd.

position vector coordinate $\mathbf{r}_j$ (there are N position coordinates for the N electrons). It is assumed that only the wavefunction for the electron undergoing the transition is altered between the initial and final states, and all other electrons remain unperturbed.

Substituting Eqs. (C.1) and (C.2) in the transition matrix element (Eq. (5.69)) and expanding, it follows that due to the orthonormal property of the one electron wavefunctions the only non-zero terms have the form:

$$\sum_\alpha \int \varphi_f^*(\mathbf{r}_1,\ldots,\mathbf{r}_N) \exp(-2\pi i \mathbf{q}\cdot\mathbf{r}_\alpha)\varphi_i\left(\mathbf{r}_1,\ldots,\mathbf{r}_N\right) d\mathbf{r}_1\ldots d\mathbf{r}_N$$

$$= \sum_\alpha \frac{1}{N!} \int \phi_f^*\left(\mathbf{r}_\alpha\right) \exp(-2\pi i \mathbf{q}\cdot\mathbf{r}_\alpha)\phi_i\left(\mathbf{r}_\alpha\right) d\mathbf{r}_\alpha$$

$$\times \begin{bmatrix} \int \phi_2^*\left(\mathbf{r}_m\right)\phi_2\left(\mathbf{r}_m\right) d\mathbf{r}_m \ldots \int \phi_N^*\left(\mathbf{r}_n\right)\phi_N\left(\mathbf{r}_n\right) d\mathbf{r}_n+ \\ \int \phi_N^*\left(\mathbf{r}_m\right)\phi_N\left(\mathbf{r}_m\right) d\mathbf{r}_m \ldots \int \phi_2^*\left(\mathbf{r}_n\right)\phi_2\left(\mathbf{r}_n\right) d\mathbf{r}_n + \cdots \end{bmatrix}$$

$$(m,\ldots,n) \neq \alpha \tag{C.3}$$

The general form of the non-zero term is a product of N integrals. The integral containing the $\exp(-2\pi i \mathbf{q}\cdot\mathbf{r}_\alpha)$ term links the initial and final state wavefunctions for the perturbed electron (this is the integral outside the square bracket in Eq. (C.3). The remaining $(N-1)$ integrals (i.e. the terms within the square bracket of Eq. (C.3) contain the wavefunction of the unperturbed electrons. For a given value of α there are $(N-1)!$ number of non-zero terms. These follow from the different permutations in position coordinate for the unperturbed electrons (see Eq. (C.3)). The integrals involving the unperturbed states are all unity. The final result is therefore

$$\sum_\alpha \int \varphi_f^*(\mathbf{r}_1,\ldots,\mathbf{r}_N) \exp(-2\pi i \mathbf{q}\cdot\mathbf{r}_\alpha)\varphi_i\left(\mathbf{r}_1,\ldots,\mathbf{r}_N\right) d\mathbf{r}_1\ldots d\mathbf{r}_N$$

$$= \sum_\alpha \frac{1}{N} \int \phi_f^*(\mathbf{r}_\alpha) \exp(-2\pi i \mathbf{q}\cdot\mathbf{r}_\alpha)\phi_i(\mathbf{r}_\alpha) d\mathbf{r}_\alpha$$

$$= \int \phi_f^*(\mathbf{r}) \exp(-2\pi i \mathbf{q}\cdot\mathbf{r})\phi_i(\mathbf{r}) d\mathbf{r} \tag{C.4}$$

This proves Eq. (5.70).

REFERENCE

Ashcroft, N.W. and Mermin, N.D. (1976) *Solid State Physics*, Holt-Saunders Japan Ltd, Tokyo.

Appendix D

Bulk Energy Loss in the Retarded Regime

Bulk energy loss is derived from classical electrodynamics in the retarded regime, where the finite speed of light is taken into account. In order to calculate the stopping power the electric field $\mathbf{E}$ of the incident electron must first be determined. The retarded electric field is expressed as $\mathbf{E} = -\vec{\nabla}\phi - \partial\mathbf{A}/\partial t$, where ϕ and $\mathbf{A}$ are the electric scalar and magnetic vector potentials respectively. The potentials can be derived from Maxwell's equations, which for a non-magnetic material have the form:

$$\vec{\nabla} \cdot \mathbf{D} = \rho_{\mathrm{f}} \tag{D.1}$$

$$\vec{\nabla} \cdot \mathbf{B} = 0 \tag{D.2}$$

$$\vec{\nabla} \times \mathbf{E} = -\partial\mathbf{B}/\partial t \tag{D.3}$$

$$\vec{\nabla} \times \mathbf{B} = \mu_{o}\mathbf{J} + \mu_{o}\partial\mathbf{D}/\partial t \tag{D.4}$$

Here, $\mathbf{D}$, $\mathbf{B}$ are the displacement and magnetic induction fields respectively, $\mathbf{J}$ is the electric current density, ρ_{f} is the free charge density, μ_{o} is the permeability of free space and t is time. Since $\mathbf{B} = \vec{\nabla} \times \mathbf{A}$ (Eq. (D.2)) the left-hand side of Eq. (D.4) is $\vec{\nabla} \times (\vec{\nabla} \times \mathbf{A}) = \vec{\nabla}(\vec{\nabla} \cdot \mathbf{A}) - \nabla^2\mathbf{A}$. Substituting into Eq. (D.4) and noting that $\mathbf{D} = \varepsilon_o\varepsilon(\omega)\mathbf{E}$ (local approximation) results in

$$\vec{\nabla}\left[\vec{\nabla} \cdot \mathbf{A} + \frac{\varepsilon(\omega)}{c^2}\frac{\partial\phi}{\partial t}\right] - \nabla^2\mathbf{A} + \frac{\varepsilon(\omega)}{c^2}\frac{\partial^2\mathbf{A}}{\partial t^2} = \mu_{o}\mathbf{J} \tag{D.5}$$

Electron Beam-Specimen Interactions and Simulation Methods in Microscopy, First Edition. Budhika G. Mendis. © 2018 John Wiley & Sons Ltd. Published 2018 by John Wiley & Sons Ltd.

where $c = (\varepsilon_o \mu_o)^{-\frac{1}{2}}$ is the speed of light in vacuum. The gauge condition is defined such that the term within the square brackets is zero.[1] The wave equation for $\mathbf{A}$ is therefore

$$\nabla^2 \mathbf{A} - \frac{\varepsilon(\omega)}{c^2} \frac{\partial^2 \mathbf{A}}{\partial t^2} = -\mu_o \mathbf{J} \tag{D.6}$$

Substituting $\mathbf{E} = -\vec{\nabla}\phi - \partial \mathbf{A}/\partial t$ in Eq. (D.1) and making use of the gauge, the equivalent wave equation for ϕ is obtained:

$$\nabla^2 \phi - \frac{\varepsilon(\omega)}{c^2} \frac{\partial^2 \phi}{\partial t^2} = -\frac{\rho_f}{\varepsilon_o \varepsilon(\omega)} \tag{D.7}$$

It is convenient to work with the Fourier transforms of Eqs. (D.6) and (D.7):

$$\left[4\pi^2 q^2 - \frac{\omega^2 \varepsilon(\omega)}{c^2} \right] \tilde{\mathbf{A}}(\mathbf{q}, \omega) = \mu_o \tilde{\mathbf{J}}(\mathbf{q}, \omega) \tag{D.8}$$

$$\left[4\pi^2 q^2 - \frac{\omega^2 \varepsilon(\omega)}{c^2} \right] \tilde{\phi}(\mathbf{q}, \omega) = \frac{\tilde{\rho}_f(\mathbf{q}, \omega)}{\varepsilon_o \varepsilon(\omega)} \tag{D.9}$$

For the incident electron $\rho_f(\mathbf{r}, t) = -e\delta(x)\delta(y)\delta(z - vt)$ and $\mathbf{J}(\mathbf{r},t) = \mathbf{v}\rho_f(\mathbf{r}, t)$, where $\mathbf{v}$ is the electron velocity. Their Fourier transforms are given by (see Eq. (6.9))

$$\tilde{\rho}_f(\mathbf{q}, \omega) = -e\delta(2\pi q_z v - \omega); \quad \tilde{\mathbf{J}}(\mathbf{q}, \omega) = \mathbf{v}\tilde{\rho}_f(\mathbf{q}, \omega) \tag{D.10}$$

The electric field $\mathbf{E} = -\vec{\nabla}\phi - \partial \mathbf{A}/\partial t$ in Fourier space is $\tilde{\mathbf{E}} = -2\pi i \mathbf{q}\tilde{\phi} + i\omega\tilde{\mathbf{A}}$. From Eqs. (D.8)–(D.10) it therefore follows that

$$\tilde{\mathbf{E}}(\mathbf{q}, \omega) = i \left[-2\pi \mathbf{q} + \frac{\omega \varepsilon(\omega)}{c^2} \mathbf{v} \right] \tilde{\phi}(\mathbf{q}, \omega)$$

$$= i \left[2\pi \mathbf{q} - \frac{\omega \varepsilon(\omega)}{c^2} \mathbf{v} \right] \frac{e\delta(2\pi q_z v - \omega)}{\varepsilon_o \varepsilon(\omega) \left[4\pi^2 q^2 - (\omega^2 \varepsilon(\omega)/c^2) \right]} \tag{D.11}$$

The important parameter is the electric field component ($\tilde{E}_z$) along the electron velocity direction, which is parallel to the z-axis. Hence,

$$\tilde{E}_z(\mathbf{q}, \omega) = i \left[2\pi q_z - \frac{\omega \varepsilon(\omega)}{c^2} v \right] \frac{e\delta(2\pi q_z v - \omega)}{\varepsilon_o \varepsilon(\omega) \left[4\pi^2 q^2 - (\omega^2 \varepsilon(\omega)/c^2) \right]} \tag{D.12}$$

---

[1]Note that the gauge condition differs from the Lorentz gauge in vacuum by the factor $\varepsilon(\omega)$.

The stopping power and energy loss probability can then be calculated along the lines outlined in Section 6.1.1. The final result is

$$\frac{\partial^2 P}{\partial \mathbf{q}_\perp \partial \omega} = \frac{e^2}{4\pi^3 v^2 \hbar \varepsilon_o} \mathrm{Im} \left\{ -\frac{1 - \varepsilon(\omega)\beta^2}{\varepsilon(\omega) \left[q_\perp^2 + (\omega/2\pi v)^2 (1 - \varepsilon(\omega)\beta^2) \right]} \right\} \quad \text{(D.13)}$$

where $\beta = (v/c)$.

Index

Electron Beam-Specimen Interactions and Simulation Methods in Microscopy, First Edition. Budhika G. Mendis.
© 2018 John Wiley & Sons Ltd. Published 2018 by John Wiley & Sons Ltd.